The

PARROT

and the

IGLOO

ALSO BY DAVID LIPSKY

Although Of Course You End Up Becoming Yourself

Absolutely American

The Art Fair

Three Thousand Dollars

THE
PARROT
AND THE
IGLOO

Climate and the Science of Denial

David Lipsky

W. W. NORTON & COMPANY
Celebrating a Century of Independent Publishing

For information about permission to reproduce selections from this book, write to
Permissions, W. W. Norton & Company, Inc., 500 Fifth Avenue, New York, NY 10110

For information about special discounts for bulk purchases, please contact
W. W. Norton Special Sales at specialsales@wwnorton.com or 800-233-4830

Manufacturing by Lake Book Manufacturing
Book design by Beth Steidle
Production manager: Lauren Abbate

ISBN 978-0-393-86670-4

W. W. Norton & Company, Inc., 500 Fifth Avenue, New York, N.Y. 10110
www.wwnorton.com

W. W. Norton & Company Ltd., 15 Carlisle Street, London W1D 3BS

1 2 3 4 5 6 7 8 9 0

For Lisa Gerard and Bill Clegg

Well, you know, I like a hustler.
—THOMAS EDISON

What's the use of having developed a science
well enough to make predictions, if in the
end all we're willing to do is stand around
and wait for them to come true?
—DR. SHERWOOD ROWLAND

History is full of stories that aren't actually true.
—LORD CHRISTOPHER MONCKTON

CONTENTS

PREFACE

THIS STORY IS ABOUT INGENUITY AND FOLLY.

It has three sections and an epilogue—which takes us up to the present day. I wanted it to be like a Netflix release, where all the episodes drop together.

You can read these parts in any order you'd like. If you're eager to know the history of the hired scientists who lied us into the problem, start with section three, Deniers. (They will surprise you.)

If you want to follow the detective work that identified carbon dioxide as our culprit, that would be episode two, Scientists.

If you'd like to browse the suite of gifts—the "work of the future will be the pressing of electric buttons," Nikola Tesla announced, when he understood how much he was about to change everybody's life—that led to our situation, your pick is Inventors, the series premiere.

The story this book tells is about the people who made our world; then the people who realized there might be a problem; then the people who lied about that problem.

There was a lot to read. It's a great story about the last seventy years, and I wanted it to work like a novel—like *Great Expectations*, if the hero were an idea: that one modest waste product of our power cords and long sweeping glides down the highway could eventually crowd our weather, which is the basis for everything. So the plan was to read all the articles, replay all the newscasts and antique NPR, and chart what a reasonably well-informed person might have been expected to know about climate during those seven decades. This is that story: the travels of one disturbing thought—in the words of one historian, "a theory more peculiar and unattractive than most"—its friends, enemies, and adventures, on the road between then and now.

I kept thinking the book was almost done. ("This story put a hole through my life," was my first plan for this preface. "Now it's your turn.") I had an old dog; then a very old dog. And I wanted her to still be alive when the book was complete—to know me again without this project left to do. I'd look down from my chair, see her long black-and-white body curled against my ankles. She died two years before I finished. She's buried now on a kind of low mountain, an ambitious hill, near West Point, New York, one more thing on the Earth.

Part I

INVENTORS

Well, you know, I like a hustler.

—THOMAS EDISON, 1885

THE MESSAGE

TECHNOLOGIES ARE LIKE STARS: THEY HANG AROUND the lot waiting for the right story, the proper vehicle. Electricity became a star with the telegraph. The telegraph needed a crime.

On New Year's Day in 1845, a sixty-year-old Quaker named John Tawell caught the Paddington train for Slough. (Slough, twenty miles from London, is one of those in-between places people like to make fun of. It's where the Ricky Gervais version of *The Office* is set.) John Tawell had been a chemist, a forger, a deportee to Australia; in 1841, he'd romanced and wed the dream bride of every reformed criminal: a prosperous widow. He also maintained a former mistress, Sarah Hart. They'd arrived at an informal palimony scheme. Fifty-two pounds a year, so long as Hart didn't spill the beans and ruin everything. Tawell was now running short of cash. Sarah Hart had a solution: Tawell could murder his wife. Instead, he bought a bottle of prussic acid and boarded the train for Slough.

Tawell poured the poison into a mug of beer, watched in Hart's kitchen as she drank it. Then he left. A neighbor—recognizing Tawell by his Quaker coat but not by name—saw a man so agitated he couldn't unlatch the front gate. The neighbor found Hart on the kitchen floor, foam around her lips. A second neighbor trailed Tawell to Slough station. Tawell was in time to catch the seven-forty for London—where he would melt into a world of coats and commuters. The business settled, Tawell had treated himself to a first-class ticket.

Tawell's misfortune was to have selected the one British rail line with a working telegraph. In the 1840s, the telegraph had been used as a royal novelty: to announce the birth of Queen Victoria's son and to fetch the Duke of Wellington some clothes he'd forgotten for a party. It had not caught on—made much imaginative headway—among nonroyals. By 1843, its British introducers were broke. They'd fallen to their "lowest point of depression."

The station manager sat down at the telegraph, wired a description to London. Tawell was shadowed from the station, then apprehended, then charged. He was tried—a tabloid bonanza, a dress rehearsal for our own podcasts and *Datelines*—convicted and executed. Here was a detective story with technology as the hero. Poles, wires, operators, and offices rose over the landscape. People would point and explain, "Them's the cords that hung John Tawell."

It took one train ride to demonstrate all the virtues scientists had been compiling for twenty-four centuries. Electricity was powerful, it was portable. It was life-altering at a dash.

From the beginning, electricity dawned as a sort of foggy new world—a coastline to be explored, mapped, then settled. As with lots of other things, the Greeks landed first. A sixth century BCE philosopher named Thales made the initial approach. The focus of his experiments was amber. Amber is fossilized tree resin, ancient sap. Thales discovered that when you rubbed a block of amber with cat fur—it's fun to imagine early experiments: different fur, fancier grips—the amber would attract light objects like feathers and straw. They'd slither across a table and stick.

What Thales was doing, of course, was generating static electricity. William Gilbert, Queen Elizabeth's physician—in portraits, he wears one of those *Jurassic Park* ruffs and a Brooklyn brunch stubble—reproduced Thales' experiments for the English court of 1600: the same time that Shakespeare was plotting sad ends for Hamlet and Othello. Gilbert roughed out the basic principles. He also gave the property a name. He took the Greek for "amber"—*elektron*. That's what electric means, basically: having the properties of amber.

Exploration stalled. Electricity became a stunt, a prank, a hobby for tinkering men of science and leisure worldwide. Hosts arranged dinner parties with sparking forks; men surprised women with electrified kisses. The

1730 exhibition The Electrified Boy required an underweight child to suit up with insulated clothes and get winched above the floor stomach-down, like a dolphin in the canvas between tanks. Static was applied to the boy's feet: his hair would rise, his fingertips made metal shavings hover; a crowd-pleasing spark could be elicited from his nose.

Benjamin Franklin left one Boston performance in a state. He immediately tried to put his hands on anything electrical. A primitive electric storage system—wire in water; this was called a Leyden Jar—had been introduced. Franklin showed off Leyden jars at his printing shop: throwing sparks, drawing crowds. Franklin created a hopping metal spider, cooked turkeys ("birds killed in this manner eat uncommonly tender"), shocked friends. "If there is no other use discovered of electricity," Franklin wrote, "this however is considerable."

Which is what sent Franklin into the Philadelphia rain with a kite, a key, and silk, an insulator. As failure insurance, Franklin brought along his son: If the experiment didn't come off, Franklin could always say the kite was the boy's. Two practical aims: He meant to demonstrate that the electricity generated by static, and the thunderous forks that sometimes singed trees and rooftops, were cousins; that here was a potentially awesome force. As churches, the tallest buildings, were often struck by lightning, Franklin's second aim was to popularize the lightning rod. Franklin became America's first celebrity scientist. Yale and Harvard offered honorary degrees. A Russian named Georg Richmann tried to repeat his work. The lightning burst Richmann's left shoe and pressed a small red circle into his forehead. Richmann became a casualty and a novelty: the first person electrocuted by love of science.

The technology proceeded by fits and starts. Also accidents and grudges, progress charted with entries in the dictionary. In 1781, an Italian professor named Luigi Galvani discovered that frogs' legs—relieved from their donors—would still kick under application of a current. Galvani's belief was that even the dead continued to generate some form of animal electricity. Alessandro Volta, his rival, thought this was just ridiculous. To demonstrate the ridiculousness ("What is left," he wrote, "of the animal Electricity claimed by Galvani?"), Volta invented a surer way to generate and store electricity: The voltaic pile, the battery. But there they are, preserved together in the dictionary—the conventioneers' hall where everyone shakes hands after death. A current that moves is "galvanic." The current is the volt.

Michael Faraday steps into the narrative like a hero out of Dickens: rough childhood, smooth features, virtuous character, shy achievement. (In the old ABC drama *Lost*, which featured a scary electromagnetic island, the sole character to have any idea what was going on is named Faraday.) Faraday, a London blacksmith's son, left school at twelve. He put in seven years as a bookbinder's apprentice. And it's funny that electricity, which would take a planet of readers and redirect them toward YouTube and touchscreens, was perfected by somebody from the trade. In 1812, a customer gave Faraday his tickets to a series of chemistry lectures. Faraday attended and was smitten. He composed a kind of love letter, using the tools of his shop—took perfect notes, bound them into a volume, presented the book to the lecturer, who happened to be head of the Royal Institute of Science.

Faraday became his apprentice—then began to outshine everybody. In 1822, Faraday wrote four words in his notebook: "convert magnetism into electricity." It took nine years. His solution was to rotate a metal disk inside the arms of a horseshoe magnet. Since the process was dynamic, his machine was called a Faraday dynamo, which meant another new word.

By 1831, electricity could be stored, it could be generated. It still didn't have a use.

SAMUEL MORSE WAS BORN—oddly—a mile from Ben Franklin's birthplace. He arrived in 1791, one year after the big Philadelphian departed, as if the tech were passing a sort of baton. Morse liked to draw— as a kid, he scratched a portrait into a school cabinet, got yelled at. Once he was famous, that chest became a relic.

Morse attended Andover, started Yale at age fourteen. He liked physics and chemistry. He played with Volta's battery, sat in on lectures about electricity: it was then understood as one of the world's secretions—"electrical effluvia," the "universal fluid."

But Morse was mostly an artist; he helped defray tuition by selling friends portraits of themselves. Then he headed for England, to study painting, enter competitions, grouse about money, return home broke after three years. London had been a series of wardrobe and beverage disappointments. "My drink is water," Morse complained to his parents. "I have had no new clothes for nearly a year."

In the States, Morse began an incredibly successful career as a failure. He became a kind of restlessly unsuccessful painter, changing lures and streams each time he couldn't catch fish, failing at a variety of locations and styles. He failed with mythological subjects in Boston. With patriotic images in Washington and mini-portraits in New Hampshire. (In 1827, he began a strange campaign in New York City against the ballet. "The plain fact of the matter is this," Morse explained in an editorial. "The exhibition in question is to all intents and purposes *the public exposure of a naked female.*" This failed.) For an uncustomary, four-year respite, Morse succeeded in Charleston, South Carolina, as a society painter. Then recession crept in, and he failed there, too.

So he headed back to Europe. To Italy and France. He had an idea for a kind of commercial anti-travel. Since most Americans knew about the Louvre, but were unlikely to ever get across, he would execute a group portrait of the museum's collection, which could then be exhibited for money throughout the United States. This failed. In 1832, Morse sailed home aboard the *Sully*—another Sully, Thomas Sully, was America's foremost painter. If Morse had been a luckless actor, this would have been as gloomy as slinking home aboard a boat called the *De Niro*, the *DiCaprio*, or the *Pitt*. It must have put him in a rotten mood.

One night at dinner he fell into a conversation about electricity. Another Franklin experiment had demonstrated how quickly current flashed over the wire. Even across rivers and miles: It happened in no time at all. Electricity, it later turned out, moves at nearly the speed of light. (As Martin Amis would write, "It ain't slow.") Franklin, a showboater, used this principle to set off gunpowder from remote locations. He liked a boom.

At the table, Morse posed his question: If the wire could communicate a spark—the idea that Franklin had just now decided to set off the gunpowder—why couldn't it also communicate information? If the information were simply on and off, couldn't you convert this into an alphabet? (Morse code, a sort of primitive binary, was the first spoken instance of what's now the most common language on Earth. Email, texts, Instagram; what gets exchanged more often over the course of a day than binary?)

Morse's father had been a minister, and the evening has the sound of one of those encounters from the Bible: the men fell to talking, and did not stop till morning. Morse wandered the deck, continuing the conversation in his head, and did not even notice when it was dawn. His invention would make use of everything then understood about electricity: battery, wires, speed.

He roughed out a sample message on deck—"War. Holland. Belgium. Alliance. France. England. Against. Russia. Prussia. Austria." Which reads like a stray bit of Nostradamus, a news broadcast from one century ahead. Debarking in New York City, Morse turned to the boat's chief officer. "Well, Captain, should you hear of the telegraph one of these days," he said, "remember the discovery was made on board the good ship *Sully*."

Then Morse began a twelve-year career of failing at the telegraph. (It's no accident his best biography is subtitled "The Accursed Life of Samuel Morse.") He mocked up devices that failed. He failed to patent his work. He couch-surfed with his brothers in Manhattan. He wrote tracts denouncing Catholics and immigrants—"The insolence of foreigners," he warned, "will no longer be endured," and "We shall soon have more papists in the North than they have slaves in the South." He took a position as professor of painting at New York University; unpaid. In his offices, Morse set up magnets, wires, batteries. To save money, he ate and slept by their side.

Public demonstrations of the telegraph failed. "His situation," explains an otherwise sober nineteenth-century biography, "was forlorn in the extreme." He ran for mayor of New York, on a strict anti-immigration, pro-nativist platform, and failed spectacularly. (Morse received just under a thousand close-the-pier votes.) Morse wrote to Washington, asking to demonstrate his apparatus. In February 1838, the inventor set up wires and receivers for senators and congressmen in the firelit chambers of the Commerce Committee. And a strange thing happened. It must have felt unnatural, like a spring breeze over snow.

Morse succeeded. The response was passionate, immediate. A kind of stunned babbling. The congressmen: "What would Jefferson think, should he rise up and witness what we have just seen?" "When will improvements and discoveries stop?" "Time and space are now annihilated." Another said simply, "The world is coming to an end." The committee voted funds, a $30,000 grant, for Morse to string forty miles of telegraph line connecting Washington to Baltimore.

And then, Morse being Morse, and politicians being politicians, they forgot. The bill needed full votes from the House and Senate; the grant was abandoned, and Morse resumed failing. He traveled for nine months through Europe on a capital hunt. Because you have to spend money to make money, he returned home flat broke. In New York, he failed at sending a telegraph message across the river. He stood again for mayor—another bitterly anti-immigrant platform—and face-planted once more. In 1841, one of his private art students asked if it wouldn't be okay to pay his fees a week late. "I shall be dead by that time," Morse said. He added, "By starvation." The student asked whether ten dollars might help. "Ten dollars would save my life," Morse said. "That's all it would do." They sat together at a restaurant, Morse explaining this was his first meal for a day. "Don't be an artist," he advised. "It means beggary. A house-dog lives better."

He wrote a friend, quoting his father's Bible. Proverbs: "'Hope deferred maketh the heart sick.' It is true, and I have known the full meaning of it."

In late 1842, Morse found himself back in Washington. A sort of last-chance commute. His savings were so meager—thirty-seven and a half cents—they had become entirely portable. They were riding in his pocket. This time, the appropriation came to a vote. Morse got his money. It took six months to wire the thirty-eight miles from a train station outside Baltimore to the chambers of the Supreme Court. The first intercity communication was sent successfully on May 24, 1844. A few days later, the machine made a bigger splash. Its debut in politics, in up-to-the-minute news-gathering. The Democratic Convention was being held in Baltimore.

The nominee for vice president, congratulated via telegraph, quickly wired back to refuse the honor.

After hours, Morse and his Baltimore operator had the first kick-around, interstate, late-night, nothing-much conversations in history. About food. FRIEND: Have you had your dinner? MORSE: Yes have you? FRIEND: Yes what had you? MORSE: Mutton chop and strawberries. You can feel the pastness—that menu—and the future in this conversation. A future of post-delivery texting, remote control under the thumb, sites and news channels scanning by. Word of the telegraph threw the country into a mix of excitement and anxiety. Electricity had found its first big use.

New York was soon wired to Washington. Americans—feeling that Morse had somehow trained lightning—called these wires "lightning lines." Newspapers wrote, "We stand wonder-stricken and confused." The telegraph, they explained, had abolished the idea of "elsewhere . . . it is all here." The machine was "the climax of all human might."

Morse's invention proved not just emotional and welcome, but retroactively *necessary*. As the country reached adolescence and kept on expanding—new states, territories, homesteaders—there were worries about keeping a nation as big and disparate as America united. "Doubt has been entertained by many patriotic minds," a government report noted, about how communication, "so necessary to a people living under a common representative republic, could be expected to take place throughout such immense bounds. That doubt can no longer exist."

By 1846, Boston, New York, and Washington had all been strung on the same wire. In 1846, the first national media service formed up, the Associated Press. (It's why these outfits are still called the wire services.) In 1857, the interstate telegraph company, Western Union, trailed the stagecoaches to California. People thought the poles would probably get chopped down by Indians or knocked over by buffalo. But the buffalo wandered away, and Native Americans weren't much interested. By 1861, San Francisco was wired to Manhattan: one long wire stretched above the country. By 1866—cable laid along the seabed—London was connected to America. There was now no place where a man like John Tawell could hide. By 1867, Western

Union had strung eighty-five thousand miles of wire; 5,800,000 messages were sent that year, at a cost of about a dollar each.

A few weeks after successfully demonstrating his telegraph, Morse failed to keep his footing, took a spill, and was laid up for six weeks. The first message he'd sent between Baltimore and DC was four words, a biblical quotation: "What hath God wrought?"

THE HUSTLER

HOW ELECTRICITY SLIPPED OFF THE WIRE AND INSIDE people's homes—hauling its big freight of power plants, light, and television, of pop culture and heavy machinery—is really the story of three men. One is very famous. Another put his name on washing machines. The third suffered from the combination of brilliance and misfortune that makes him famous among people who like to identify a dark unfairness at the heart of the world. The three men are Thomas Edison, George Westinghouse, and Nikola Tesla. It's Edison who threaded the wire under the streets and behind the wallboards. In a lot of ways, being alive now—this era's inheritance of problems and gifts—is like living inside a math question first posed in 1878. No matter how we spend our days and hours we're all pitching in on the implications: what would the world be like if Thomas Edison got his way?

Edison's whole life was about the telegraph. It's the answer sheet he cribbed from, the notes he snuck into every test. The telegraph made him rich—he nicknamed his first children "Dot" and "Dash"—and later in life, when he stopped being rich for a while, he said he could always go back "and earn my living again as a telegraph operator." Edison got famous in an innocent way nobody gets famous now. People called him "the Greatest Single Benefactor of the Race."

Starting out with the telegraph made Edison so famous that, when he was sixty-three, drawing his face on an envelope and dropping it in the mail would see your letter delivered to his hands.

He was born in the sort of incidental, beside-the-point town found-ers try to buff up with a grand European name: Milan, Ohio. When Edi-son was two, his uncle Snow Edison—a style of name that would have to wait for a century and movie stars to return to fashion—joined the wagon trains for the 1849 California Gold Rush, but died at the edge of the trip. Edison's life forms the bridge between gold-rush America and tech-rush America: Between value you gouge out of the earth and ideas you nudge into the air. His childhood got squeezed by processes he helped accelerate: technology shucking off industries, advances smothering towns, ecology taking the hit.

Milan was a canal on Lake Erie—a thriving grain port. Then railroads stole the business, and the Edisons relocated to Port Huron, Michigan, a tim-ber boomtown. Lumberjacks overcut the forests, and business dried up.

Edison's father tried real estate, farming, shingle making; he went bust with a grocery. A 1857 credit report notes that he'd been "indicted for sell-ing real estate not his own," and concludes with a BEWARE sign: "Should not like to trust him largely." Family life became a battle against middle-class slippage. Tight smiles in town, anxiety around the kitchen. At seven, his parents yanked Edison from school. Because a teacher had called him "addled"—also for a more homely reason: they were out of money. He ended up homeschooled by his mom. (Decades later, Edison in the national press, the former headmaster wrote him. He wanted to see about back tui-tion. "I am now almost seventy-seven, and am on the retired clergy list. As you have now a large income, I thought perhaps you would be glad to *ren-der me a little aid*." Being lively—as Edison always was—includes the talent for keeping grudges alive, too. Edison mailed back twenty-five dollars.)

As a boy, he read science, built a telegraph, fiddled with chemicals, blew the corner off a building. At twelve, Edison found a job with the railroads. Boarding the traincars at dawn, trawling the aisles, selling fruit, sandwiches and newspapers, stepping back into the kitchen at midnight, where he'd leave a dollar for his parents.

If Edison's life were a comic book, this would be the origin story. He was selling papers just as reporters, via the telegraph from the Civil War, were revolutionizing the news business: producing the first real-time battlefront coverage in history. Edison's train pulled into Detroit the morning of the Battle of Shiloh.

A mangling three days—reports fixed the dead and wounded at sixty thousand. Edison watched anxious families mill around the train station bulletin board, desperate for news.

Everything came together. Telegraph, media, money. Edison paid the station operator to wire the battle news all the way down the line. Instead of his usual one hundred papers, he took a thousand, on credit. Pulling into the next station, Edison found a panic he'd helped generate: worried spouses and parents rushing the bulletin board. Already, Edison had an inventor's pitilessness, the bartender spirit: whip up the thirst, sell the relief. "After one look at the crowd," Edison later recalled, "I raised the price to ten cents." At his next stop, Edison sold each paper for fifteen cents, then a quarter. By the final stop, no fixed price, Edison was accepting bids. "It struck me that the telegraph was just about the best thing going," he said. "I determined at once to become a telegrapher."

At sixteen, Edison left home entirely. As a *plug*, a junior telegraph operator. His fellow operators weren't the skinny men with visor caps and shirt garters you see in the old Westerns. ("Well, we already wired Kansas City about the new Marshall...") They were the adolescents who step off the elevator from IT to unstick your computer. What's important to remember about the telegraph is that it was the most advanced technology anyone could think of, the hottest thing going. It frightened adults. For them it was a hunk of machinery dropped into the center of their lives. But for their children it was a toy, part of the furniture they grew up with; they climbed all over it, took it apart, put it back together again better like a stockcar engine to see what it could do. Edison was a techie. His fellow plugs were techies—young men who, sitting with their hands on the telegraph key, could feel the whole world flash beneath their fingers. (Information theorist Marshall McLuhan called the telegraph the first mass media. A sea change. For the first time, McLuhan wrote, "messages could travel faster than the messenger." This change meant thunderheads. By "eliminating time and space," Morse's invention would eventually "dim privacy," "usher in the Age of Anxiety and Pervasive Dread." What McLuhan meant was that, by subtracting distance, the telegraph had turned Chicago and New York into rooms on the same hall, made every house the same house; what happened there now also happened here.)

Operators were helping pave the bumps and gulfs of America into a smooth, uniform country. Between press updates and financial reports, plugs passed the time wiring each other every interesting thing they heard. Funny stories, shock items from the local paper; gossip, riddles, news of the weird. Plugs would pin the best messages to the office corkboard. "Any joke originating anywhere," Edison later said, "was known the next day all over."

The teenaged Edison traveled to Fort Wayne, Cincinnati (known then as "Porkopolis"—the national hog-butcher). To Memphis, Louisville, Indianapolis ("Railroad City"). Midnight shifts, dingy bureaus in the business district, offices with holes in the ceiling. The Cincinnati office had formerly been a restaurant, and was overrun by nostalgic rats. One of Edison's first inventions was a rodent zapper, a kind of rat electric chair. His salary went to experiments. Edison filled boardinghouse rooms with tools and batteries, walked around with wire in his pocket, requested private time with the machine like an intern begging hours on the mainframe. Edison left the Nashville office at three in the morning carrying a sack of electrical magazines, and was shot at by a policeman, who believed he was pursuing a burglar.

After five years Edison washed ashore in Boston. He was wearing high-water jeans, a secondhand coat, a hat with a hole in it: One fellow plug described Edison as "the worst-looking specimen of humanity I ever saw." He was also something else: a new human type—somebody living a second, freer life inside the machine. Who could take the best of what he'd seen, then strip and recombine it in valuable and interesting ways. Edison placed an ad in a trade journal, announcing he would henceforth earn his living as an inventor. "I'm twenty-one," he told his Boston roommate, "I may live to be fifty. I have got so much to do and life is so short, I am going to hustle." It would take less than a decade to put himself in a position to perfect the light bulb, which would be the start of everything.

Edison opened a lab in Newark, New Jersey, where he argued with the gas company. (Gas companies held a monopoly on high-grade light. And were hated with something like the vehemence with which cellphone providers and wifi companies are today. Customers couldn't make sense of the bills but knew in their bones they were being overcharged.) He improved the telegraph printer, the stock-market telegraph—called a "ticker" because of the noise—the automated telegraph. Edison invented an electric pen that did not catch on but, years later, would become the basis for the first electric tattoo needle. Edison had a knack, what he considered a "logical mind's" ability to forge connections between unlike things. (Also how Marcel Proust defined the writing job: take two sets of phenomena, find their "common essence.") When Edison was twenty-four, his moneyman wrote him, "If you should tell me you could *make babies by machinery*, I shouldn't doubt it."

The wire at the time represented the broadcast spectrum. So the telegraph's big problem—just as it would become for the cell phone and internet—was bandwidth. How do you squeeze more information into the pipe? In 1874, Edison invented the first quadruplex—two messages at once, flashed in both directions. It made Edison's industrial reputation. Also his fortune: he sold the patent to Wall Street financier Jay Gould for thirty thousand dollars.

Edison cashed out of Newark, reinvested his winnings in a tiny town in the New Jersey woods called Menlo Park. Six houses, a train station so insignificant nobody bothered with a sign. Chug-over country. The isolation so unsettled his wife that Edison bought two big dogs, slept with a handgun under the pillow. He would eventually transform the town into one of the world's sites—a symbol. By changing the way research worked, Edison turned the spot holy. In 1929, Henry Ford would spend thousands transporting the remains of the Edison lab to Michigan; to collect all the magic, he'd even roll in trainloads of the *dirt*. Edison scholar Paul Israel ranks Menlo Park among Edison's innovations. The first "invention factory," a corporate lab. The Silicon Valley style—flying squads of researchers chasing down a single idea, joking and sharing takeout, pulling all-nighters that become days and weeks, stealing naps by the equipment—was Edison's.

Edison's plan, announced as he boxed up the Newark lab, reads as confident, innocent, and American as one of Jay Gatsby's self-improvement schedules. He would produce "a minor invention every ten days, and a big thing every six months or so."

HIS FOUR-WAY TELEGRAPH showed there was still room in the industry; like the best innovations, it found a door where others had felt only a wall. Researchers were struggling toward a new method for increased bandwidth. Musical tones, "acoustic telegraphy." It took a former speech therapist—Alexander Graham Bell's family had trained the deaf for generations—to pose the next question. If you could send messages by musical tone, why not skip the translation and send a human voice? "If I can make a deaf-mute talk," Bell said, "I can make metal talk."

In 1876, Bell offered his invention to its logical customer, Western Union. What he was really offering was a market. A monopoly at the price of a markdown—$100,000. The company rejected him flat.

Western Union's president couldn't see smart phones, cell towers, people yammering as they shopped and did laundry, whole days and outings turned into one floating, random, dreamy conversation. (Everybody becoming Samuel Morse, curious to know what had you for dinner.) The Western Union man asked, "What can we do with such an electrical toy?"

The slogan of the era had been coined by Andrew Carnegie, who was narrow-eyed and flinty and knew something about squeezing a dollar: "Pioneering don't pay." This is what gives teenagers their edge with fresh technology; their horizons aren't fenced in yet by experience.

Adult journalists met the telephone with stale gags: placed in the hands of "a distant and irate mother-in-law," it would end up a torture device. Even fellow inventor Elisha Gray, beaten to the phone patent by a heart-breaking two hours, saw a sterile masterpiece. Gray wrote his lawyer: The "talking telegraph" was "a beautiful thing from a scientific point of view," he said. "But if you look at it in a business light it is of no importance." In the 1950s, the cash-strapped outfit that would become Xerox offered to share its patent with IBM. Confronted by opportunity, IBM instead commissioned a study. Which determined the market for office photocopiers represented a few thousand units at most. IBM passed.

A few years later, Western Union's president fretted that he would gladly have paid twenty-five million for the speaking telegraph. In the interim, he hired Edison to improve it.

Bell's prototype had a single port. You yelled, listened, and repeated through the same hole. Every conversation became the worst call you've ever had, placed from a stairwell in the skyscraper basement one wall away from the subway tracks. Western Union's request, Edison recalled, was "to make it commercial."

Edison improved the telephone three ways. The most important, which resulted from a misunderstanding, would lead to light and the power industry. It would also fork unpredictably into entertainment—a path that extended to record players, Spotify, the movie business, Disney Plus, to staying at home because why bother when TV brings the performers to your couch for free? All because Edison had, in a fundamental way, misunderstood the device.

The first step was postnatal surgery: snipping the speaker and receiver. This took a year.

The mouthpiece Edison perfected made the phone a consumer-ready product, and would remain inside every handset sold until the end of the twentieth century. Edison brought to this work a decisive physical advantage. Bell had only tutored the deaf. Edison was himself partway deaf, and so uniquely sensitive to the benefits of crisp sound. His phone became the world standard. One of Edison's London advance men was the future playwright George Bernard Shaw—who in the 1870s suddenly had to work beside a raftload of *Americans*. Shaw grasped that these jumpy, electrified Yankees were themselves a new invention. "They were free-souled creatures," he wrote, "excellent company." Their moods darkened and lightened—glad-handing to touchy—like information flashing down the wire. Americans were "sensitive, cheerful, and profane; liars, braggarts and hustlers," Shaw wrote. "With an air of making slow old England hum which never left them."

Edison had, as he put it, "cured" the telephone. The therapy earned him a quarter million dollars.

In his second act, Edison faced down the drama of the incoming call. He went browsing and selected a word. Bell's plan had called for a sort of perpetually open line: Subscribers would be culled from the businesses that honeycomb every city's financial district. An office drone at J. P. Morgan would be desked beside the receiver. When the suits over at Jay Gould— Edison called these businessmen "sharps"—needed a conference, they'd shout "Ahoy-ahoy," and negotiations would commence.

A Philadelphia telegraph company's idea was to install bells. Edison, with his partial deafness, had an instinct for what might make the most clearly recognizable greeting. He did some experimenting, then proposed two syllables—in the river trade, raftmen used them for signaling boat to boat.

He wrote in August 1877, "I do not think we shall need a call bell as Hello! can be heard ten to twenty feet away. What do you think? Edison." "Hello" had made a handful of book appearances, but had not been picked up. It remained a kind of verbal litter, a Coke can left at twilight on an empty beach. ("The superior noise of Sam, Tom, and Charles"— Jane Austen, *Mansfield Park*, 1814—"tumbling about and hallooing.") By the 1880s, "Hello" arrived in the dictionaries. Before spreading to offices, lobbies, introductions, name tags; the first switchboard operators were called "Hello-girls." Edison had established our phone language. (The first two syllables spoken by Steve Jobs' first Macintosh were "Hello.") In England and America, every time a call gets answered, a front door opened, a friend greeted, the user is for that moment speaking Edison.

The third step, which would change Edison's life, then everybody's, resulted from *stuckness*. At thirty, Edison was as fenced in by experience as Andrew Carnegie or the Western Union president. When Edison looked at the phone, he didn't see a whole new communications device. He saw the same old thing: Exactly what everybody else was calling it—a "speaking telegraph." Only *less* dependable. Because with the telegraph, you always had both feet on the ground. There was always a written record to refer to. Telephone conversations—like every other sound in history—were sprinkled on the air. This would turn business into a highwire proposition, dealmakers working without a net.

So Edison set out to make the phone recordable too. He had a machine for recording telegraphs. A waxpaper disk, the telegraph key inscribing dents as it revolved on a platter. Run at high speed, Edison noted a weird, tinkling sound. A noise, he said, "resembling human talk heard indistinctly." Edison took his telephone microphone, affixed a small, blunt needle; he positioned the wax paper, then recorded the first word in history.

He shouted his testing word. And it's appropriate this first recorded sound should be a greeting. Edison and an engineer rigged a speaker. Then ran the needle back along the grooves.

You can see them: narrowed eyes, ears pitched. "We heard a distinct sound," Edison said, "which a strong imagination might have translated into the original 'Hello.'" Additional experiments, lab astonishment, Edison puzzled, mistrustful, proud. "I was always afraid," he said, "of things that worked the first time."

Word got out, and scientists smelled hoax. One exasperated professor wrote, "The idea of a talking machine is ridiculous." He urged Edison to quash rumors for the sake of his "good reputation." Any device with a dash of "telegraph" spliced in got an extra, commercial charge. For the same reason marketers later deployed the letter "I" for "internet": iMac, iPhone, iPad. Edison cobbled together the two words, and called his device the phonograph.

He offered. Again, Western Union refused. They saw "no conceivable use" for the product. Which is how Western Union ended up a company turned to for emergencies and novelty, a surprise for a sweetheart, a way to wire the kid money in the dorms. Business school boilerplate: When you see only the present, you're guaranteed to become part of the past. Your own fortifications lock you in. Edison brought his device to the offices of *Scientific American*. He unwrapped it, cranked the lever, and the machine spoke. Work stopped, desks were abandoned. So many bodies crowded around Edison, editors feared the floor would cave in.

Decades later, Edison phonograph ads would blaze a trail to the living room. With home entertainment, Edison had invented the majesty of programming and the binge-watch, the convenience of Netflix and Chill. "When the King of England wants to see a show," the Edison ad ran, "they bring the show to the castle. If *you* are a king..." Other ads worried the marriage angle, eye fixed on the wives: "When a man leaves home in the evening it is because he seeks amusement. The best way to keep him home..."

Within a decade, Edison would shoot for "an instrument that does for the eye what the phonograph does for the ear." And he'd step blinking from the lab with the motion picture, which he gave the clunky name the "kinetophonograph." The first commercially successful movies were sporting events—prize fights—then romances and Westerns. Faces, kisses, punches: performances could be stored and replayed at will, time kept fresh in a jar. The same way the phone would make whole days into conversations, the phonograph would turn the working day into an intermission between entertainments.

WHAT STRUCK THE JOURNALISTS at *Scientific American* was something spiritual. What wrung them out—left them, as Edison biographer Randall Stross writes, "so emotionally moved"—was an aspect Edison never considered. Hearing the dead speak after they'd gone. Parents, lovers, a president, a child. It went out in the headlines—"ECHOES FROM DEAD VOICES"—and made it seem as though Edison could command time. This gave Edison his nickname, "the Wizard of Menlo Park." And it is a startling experience, to head over to Wikipedia, click the link, and hear Edison's voice. An older man's cough and sandpaper, reciting "Mary Had a Little Lamb." It's like opening the door to a tiny, courteous, temporary ghost.

You can always recognize the big-league inventions. They're sized up as threats, bullies sharking through class to push around the shrimpy one. The *Times* worried that Edison's device would polish off book reading. Newspapers proposed replacing the Statue of Liberty's torch with a phonograph, to boom out welcomes to the immigrants in New York Harbor.

Edison was summoned to Washington, in the sort of long unfurl of triumph Samuel Morse once daydreamed about. He demonstrated his machine for the National Academy of Sciences and both houses of Congress. Then, at midnight, a command performance in front of President Rutherford Hayes. Finally, doggedly and sleepily, for the First Lady and some friends. The fever dream of every techie, rubbing their neck before stepping back to the problems on the worktable. Edison had climbed aboard the wire and ridden it all the way in—to the center from the outlands, from Milan, Ohio, to the Capitol.

Edison was about to define the way we'd power our lives, ushering in a century of benefits and problems. By inventing the music and film industries, he was fixing the way we'd receive performances, cutting a landing strip for television. (And for celebrity culture, since a celebrity only becomes a pure celebrity when they can happen in every room at once.) The movies would be populated by Shaw's "liars, hustlers and braggarts"—spreading the image of "sensitive, cheerful, and profane" Americans everywhere, plowing under local styles of what modern people ought to be.

And Edison himself was like that. How Americans imagine themselves: scrappy, resourceful, ingenious.

One thing successful Americans have in common is an ability to transform stories about themselves into stories about America: admiring them becomes a national endorsement. In the lab, Edison could live for weeks on apple pie. Like many American men, he remained a boy in his mind long after his body began to flash out the contradictory information. (Finishing up a lecture at a girls' school when he was thirty-one, he raced a friend down the campus hill in the rain.) At nineteen, Edison had been about to board a ship, to make his fortune via the telegraph in Brazil. An old man on the docks talked him out of it.

The man "had sailed the sea for fifty years," Edison recalled. He told Edison "there was no country like the U.S., that if there was anything in a man the U.S. would bring it out and that any man that left this country to better his condition was an ignorant damned fool." At twenty-two, Edison had arrived in Manhattan dressed like a rube from the cornfields: no bed, flat wallet, sleeping in the battery room of a Wall Street telegraph firm. First week, the tickers crashed. Panic on the sales floor, Edison edging forward like a kid from the bleachers; sign that kid up. This dream moment became Edison's first high-paid assignment. When reporters visited Menlo Park— and after the phonograph, they became daily members of the Edison cast— he showed the same formality as a kid in the *SNL* writers room. Slept-in pants, lunch going crusty on a desk; no-style as the true American style. In photos Edison looks modern—a member of our time, uncomfortable in the dusty setting.

In 1888, explorer Henry Stanley visited the lab. The two men shook hands. Stanley asked to examine the phonograph. They speculated about dead voices from history. Say the phonograph had existed forever: Whose voice would Edison most like to hear?

"Napoleon," Edison replied instantly.

Stanley was a churchgoer; this answer bugged him. "No, no," Stanley said. What he had meant was, "I should like to hear the voice of our Savior."

Edison wasn't a fan of the inherited title. Like him, Napoleon had started small, found his own way to make it big. "Well, you know," Edison said, "I like a hustler."

THE PROMISE

EDISON DIDN'T FACE THE LIGHT BULB ALONE: HE ASSEM-
bled a squad. This would be an assault—a rough scramble up a hill that
had already claimed many good men—and he was a captain recruiting commandos.

This one because he was skilled in the ways of metal; that one because
he could bend glass. Like a hip-hop crew, Edison gave them nicknames.
The bookish Princeton grad became "Culture"; the chemist who could
track the essentials he called "Basic"; the grizzled PhD veteran was "Doc."
Other experts—doughboys, whose skills the unit would rely on—Edison
called "sperts."

As Paul Israel wrote, "The invention of the phonograph was the most
important of Edison's career. It made his reputation." Fame operated like
money gravity, attracting venture capital to Menlo Park. And for the first
time in seven years, Edison took a vacation. After completing inventions,
he went out collecting experiences, one summer's long outdoor breath.

Edison traveled West. Stared at Indian country with the Wyo-
ming cavalry. Survived a rambling, late-night misunderstanding with a
marksman called Texas Jack, who only wanted to prove he could shoot a
weathervane through Edison's hotel window. (He could.) To observe the
country purely, Edison rode outside, above the train's cowcatcher, from
Omaha to the Sacramento Valley. It was the kind of trip that clears the
head, each day pounding out a little more of the old you. And after seeing the long stretches that the telegraph had made one country, pressing

as far as he could into the outdoors, Edison decided to perfect indoor culture with light. It was the sort of idea you bring back from vacation, like healthy eating or growing a beard. He returned to the lab wearing a black sombrero.

The story of the light is about electricity finally stepping inside the house. It's also about control. The telegraph snipped distance. The phonograph gave you a handle on time. The bulb offered programmable daylight, and electricity would cover the rest.

"Cities given," Scottish author Robert Louis Stevenson wrote in 1878, "the problem was to light them." This dogged the first settlers. Sixteen-nineties New York law required every seventh house to post an outdoor lantern. (Exemptions were granted for moonlit nights.) Lamps often ran on whale blubber. That's what *Moby Dick* is finally about: the deep-sea hunt for a light source. In 1781, the Earl of Dundonald hoped to reverse his family's declining fortunes by heating coal into tar. One of his giant kilns exploded. You can picture it like a cartoon. Stars and tornadoes orbiting a forehead, smoky hand massaging singed cheek, as the earl noted how the explosion's vapors gave off a sad bright waste light. He'd invented coal gas. In 1807, entrepreneurs circulated a prospectus through murky London. Sir Walter Scott—author of *Rob Roy* and *Ivanhoe*—sent a scandalized letter to a friend: "There's a madman proposing to light London with—what do you think? Why, with smoke!"

By the 1820s, there were streetlights on corners and three hundred miles of pipe snaking beneath London. It was like finding a new set of hours on the clock. This was the start of night school and the graveyard shift. Clubs flourished. Men and women took constitutionals in the nighttime streets. Gas made the classic downstairs technological journey. A status symbol for the rich; then a stretch for people hoping to appear rich; then a necessity for everyone. In 1821, a magazine editor returned home from a blazing dinner party, got out envelope and pen. "Dear lady," the editor wrote a friend, "spend all your fortune in a gas apparatus! Better to eat dry bread by the splendor of gas, then to dine on wild beef with wax candles."

By mid-century, electric lights arrived. Only they were the wrong kind. They spit, couldn't be turned off independently, had to be monitored like a big dog or small child. Arc lights. Carbon rods, separated by a small gulf. Powered up, electricity arched the gap and glowed.

How bright was like the sugar in a Slurpee—first a rush, then a problem. The French test-fired an arc system at nine in the evening. Women opened umbrellas, birds in the trees woke and sang. Arcs were so bright they became a weapon, as unsettling as stealth aircraft. British troops in Sudan built a ring of arc lights: enemy invaded, they flipped the switch, ranks broke, and the battle became "one of the strangest routs imaginable." The Eiffel Tower was originally conceived as the thousand-foot base for a giant arc light that would cast Paris in perpetual day.

That's where the technology stood, when Edison stepped onto the field. Gas was parlor light; wicks needed to be trimmed, glass cleaned, air circulated—a hobby, like tending a household garden. (The lamps made a soft gurgling sound. Dylan Thomas, *A Child's Christmas in Wales*: "The gaslight bubbled like a diver.") With two mortal aspects: first, open flame; second, if you killed the flame but left the gas, a possibility of never waking up.

Arcs presented a different problem set. They made a fizzy, spark-throwing racket. And their light was disturbing. (Robert Louis Stevenson called them "a lamp for a nightmare," which should be shone "only on murders and public crime." Police did experiment with arcs in the morgue, under the belief that seeing victims by their light would prompt murderers to confession.) And the light was too intense for home use—like reading magazines by the Bat Signal. Plus, there was no outlet to power them.

In the fall, Edison took a field trip to a Connecticut factory. Colleagues dropping in on colleagues. Two inventors were making a run at an indoor arc light. A reporter tagged along—"Edison" often meant "Edison plus reporter." He had become a subject for editorial page humor: "Edison has not invented anything since breakfast. A doctor has been called."

The reporter described Edison testing wires, sprawled on tables, "fairly gloating" in the workshop atmosphere of problems and solutions. Edison was an engineering athlete; the star who, stepping on court, can immediately read the state of play. He tallied up coal and transmission costs. Then he walked back to his fellow inventors. He announced, simply and honestly, "I believe I can beat you making the electric light. I do not think you are working in the right direction."

This was September 8, 1878. In a movie, you'd push close on Edison's face. Then cut to a montage of power plants, skyscrapers, a commuter closing his newspaper as the train arrives (breeze rattling the pages), a midnight city stretched bright and flat beneath an airplane window, a journalist writing about climate change on a computer screen. Edison didn't see only the bulb. He saw past it to a market. His model would be gaslight—the companies he'd fenced with in Newark, their understreet world of pipes and billing. Profit would come from the bulb; but it would really pour in by selling the current. It was one of those moments when the Earth's revolution slows, light bends, and you can see all the way around the curve and into a future.

Back at Menlo Park, Edison announced his plans. It set off a minicrash. It was like Jeff Bezos or Disney shouldering into a market. Light, as Edison envisioned things, would be the killer app for electricity—the foot in the door, the must-have that would finally run a line between the dynamo and the bedroom. Edison promised reporters the wire would power everything: current for elevators, warmth for homes, power for sewing machines, sizzle for dinners. Edison had personal reasons for this belief. He was turning the world into a reflection of his own life—which had been powered by electricity for fourteen years.

This generated an international round of scoffing. One engineer read the papers, warned, "More is promised than can possibly be delivered." A British physics professor said any system based on the incandescent light was doomed. Edison promised reporters he could get a prototype banged together in six weeks.

IT TOOK FOUR YEARS. Like an endless finals week, Edison had to keep cramming for the next real-life exam just as he was wrapping up the present one. He tutored himself in vacuums (the pop when a bulb breaks), metallurgy, power generation, digging. One of his commandos—the Princeton graduate, "Culture"—believed this was tactical. The overpromise as spur; a cement block wedged against the gas pedal. "I have often thought," he said, "that Edison got himself into trouble purposely. . . . So he would have a full incentive to get himself *out* of trouble." In an incandescent bulb, filament is heated until it produces a mild, steady glow. This approach beat the arc light. It had one disadvantage: temperatures nuked the filaments, which would buckle, melt, and burst into flame. Edison needed his bulb to burn hundreds of hours. The prototype lasted eight minutes.

He solved it like a man staring down an unpromising rack of Scrabble tiles. He knew his word was in there somewhere; you just had to plug away, keep rearranging the letters. Probably Edison's most famous remark is about the lightness of headwork versus the grind of the body's follow-through. Decades after the bulb's completion, he explained, "I never quit until I get what I'm after. That's the only difference between me, that's supposed to be lucky, and the fellows that think they are unlucky." Edison told a lab assistant, "Remember that nothing that's any good works by itself, just to please you. You got to *make* the damn thing work. . . . Genius is one percent inspiration, ninety-nine percent perspiration." Read backwards, that's the story of the light bulb. The sudden goggle—the expansion of possibility—becomes an assignment, then ends up a sentence, time served. You pay in years for the single, illuminating flash.

By Culture's estimate, the lab performed 2,774 lamp experiments. The filaments Edison auditioned sound like the Amazon history of a shopper going insane: platinum, nickel, iron, piano wire; wood, cork, hemp, coconut shells, coconut *hair*, grass; cotton, twine, fishing line, horsehair; tissue paper, flax, flour paste. Edison saw his first good results, he later claimed, from Limburger cheese. ("Reasonable things never work," he said. "You'll have to begin thinking up *un*reasonable things to try.") He tried spider webs, old carpet, leather, ginger, macaroni. Inventing isn't all science: when bamboo looked promising, Edison turned employees into prospectors, sent them to scour groves in Japan, Cuba, Brazil. The Cuba man died on the trip—yellow fever. Edison, hard-hearted, put the blame on alcoholism.

When Edison auditioned human hair, there was a lab competition to establish whose beard trimmings could withstand electricity the longest. And here is another Edison secret: His lab was fun. So long as the experiments got done, employees could make their own hours, come and go as they pleased. Edison took them fishing, sudsed them with beer, sang songs (bawdy ones and weepers like "My Heart Is Sad from Dreaming"), ordered in takeout. Weeks stretched to sixty hours, then eighty. One crew member later said he would gladly have paid for his spot in the lab.

In October 1879—after thirteen months—Edison saw a spool of thread on the worktable. He snipped a piece, pinned it inside the lamp. It glowed. Then he packed it with clay and fired it; this carbonized thread lasted thirteen and a half hours. Then, after a "death watch"—men would stare a few hours, hand off the work, sleep under a table, return—forty hours. Edison went even simpler: tightly rolled and charred paper. (Edison, who was not a believer, said, "God invented paper for the light bulb.") This bulb burned 170 hours. If it went that long, Edison knew he could make it last a thousand. In late December, he told the *New York Times*, "My light is at last a perfect one."

The story broke. New Year's Eve, the lab hosted tourists, a party scene from *The Great Gatsby*. The railroad had to schedule extra trains, swells in furs and leather gloves prowled the grounds, lab light spilling across the snow. New Year's Day, the rush became unmanageable (a peacetime version of Shiloh) crowds pushing and milling through the lab aisles. One man tried to short out four lights and had to be ejected. He was revealed as an employee of the gas companies, trying to catch and strangle the baby in its crib.

Edison retraced gaslight's social steps. He wired up investors, making electricity a status symbol. J. P. Morgan's Madison Avenue mansion received the trial version—it became an emblem for how the wealthy live different and better, like a collection of foreign motorcycles or an infinity pool. "Each room," a journalist marveled, "is supplied with it, and in order to illuminate a room you have simply to turn a knob."

Morgan asked Edison to wire his church, then his church gym. It took another two years to mock up and construct the generating plant; to dig the trenches, seal and plant the cable, Edison and his crews shoveling at night, beside the newspaper sellers and the dogcarts. For Edison, this must have seemed as far as a life could travel. He'd given his childhood to the railroads and Western Union, the first corporations. Now the crews and equipment belonged to him. He'd become a trademark: his company was called Edison Electric Light.

He wired up the brokerage houses and newspapers. Same strategy. Hook the opinion-makers on the stuff first. The *New York Times*, in the long years between dry idea and sweaty execution, had settled into the tone and pattern of a marriage. Compliments and optimism ("If these promises are fulfilled, its inventor will be a benefactor of the human race"). Then peevishness, skepticism ("Mr. Edison" is "an inventor as prolific in rosy promises as [he is] in ingenious devices that prove more curious than useful"). Finally resignation and acceptance ("Let us wait with minds open to convincing").

Just before three o'clock on September 4, 1882—afternoon light clear, vibrant, and basically unchallenged outside the windows—Edison and his team gathered at separate locations, their watches synchronized. Edison wore a white tie and derby for the occasion. He paced the J. P. Morgan offices. Another squad assembled by the switch at his generator plant on Pearl Street. It had taken four years. Edison had run through half a million dollars. Money that would not have been available without the phonograph—no phonograph equals no Edison. Coal, boiling water into steam, would generate the power: Edison would still, after all, be lighting a city with smoke. No supplier or catalog to order parts from. Every element of every system, Edison said, had to be "home-devised and home-made." At the last minute, one of the sperts bet Edison a hundred dollars the system wouldn't run. "Taken," Edison replied calmly. At three, the switches were thrown. The message didn't come in letters, like Samuel Morse's. It arrived as a force, a dress rehearsal for the future. Power, generated at an invisible, off-stage *there*, could be directed to perform work *here*.

The filaments glowed. Morgan's employees cheered. And, a few blocks away, fifty-two bulbs came to life at the *New York Times*.

The paper ran it the next morning as a news item. "Yesterday for the first time THE TIMES building was illuminated by electricity." Edison's invention was "simplicity itself," a great clean convenience. In the manner of conveniences, it was reckoned in negative terms, by what was no longer present. There were "no matches needed." Also "no nauseous smell, no flicker and no glare." The piece has a restaurant review's dazed, cozy sound of the full belly. Electricity had turned the offices "as bright as day." The lamp was "a hundred times steadier" than gas. It was also "soft, mellow, and grateful to the eye."

Inventors, like artists, live on reviews. At the end came the sentence you long for—the official release from duty. "The Edison electric light has proved in every way satisfactory."

Edison was thirty-five. Almost exactly four years earlier, he had removed his sombrero and stepped into the lab with brown hair; he'd come out gray, also famous and rich. Within a few years, people would be using the electrical terms *negative* and *positive* as classifications for people; *plugged in*, *charged up*, *wires crossed*, *dim bulb*, *short-circuit*, and *electrifying* would flash out of the outlets to recharge the conversations. By the end of the year, Edison factories would be turning out one hundred thousand bulbs. Within two decades, orders would pick up to forty-five million. Later that day, a journalist from the *New York Sun* caught up with Edison in front of the Pearl Street plant. The machines behind him were chugging away; lower Manhattan was being lit by his generators. Edison said, "I have accomplished all that I promised."

THE ELECTRICIAN

NIKOLA TESLA COULD BE SEEN AS A REPRESENTATIVE from a different future: the West Coast future of dietary restrictions, green politics, hotel turndown service, and complicated infirmities. Tesla liked to claim he'd been born—Serbia, 1856—during a midnight electrical storm. As a boy, he built a small motor powered by sixteen Junebugs, glued and stitched into harness. (This invention ran until a schoolmate—an officer's son Tesla described as "strange"—ate all the power sources.) As an adult, he tried to organize his life around the number three: waiters were instructed to approach no nearer than three feet; if he circled the block once, he needed to orbit it twice more; if he swam one lap in a pool, he had to splash out another twenty-six to make twenty-seven (three times three times three). He was incapacitated by the sight of earrings on women; also by peaches. He refused to shake hands (germs) and could be induced to touch another person's hair, he explained, only "at the point of a revolver." (Tesla never had a girlfriend or a boyfriend.) He thought he could live to be a hundred by restricting his food intake to the barest daily minimum. He was over six feet tall and 140 pounds. He referred to people generally as "meat machines." He sent the first song ever by radio wave: four miles in 1896, around the Colorado mountain Pike's Peak. It was called "Ben Bolt," and the first lines are about a brown-haired girl. When traveling, he signed his letters not "Nikola Tesla" but "Nikola Faraway." He disliked Jewish people. He was a vegetarian, a part-time Buddhist and—after he became friends with the influential hiking enthusiast John Muir—a passionate environmentalist. He believed electricity offered the cure for "melancholia," also for constipation. He claimed to have made his eyes bluer by power of concentration. He is the only world-famous inventor whose biographers are moved to stipulate things like, "I also examine such questions as whether Tesla received impulses from outer space." (Tesla was convinced he had.) In 1895, newspaper editor Charles Dana wrote, "It is no exaggeration to say that the men living at this time who are more important to the human race than Tesla can be counted on the fingers of one hand; perhaps on the thumb of one hand."

At twelve, he'd read about Niagara Falls. (Tesla's memories arrive wearing opera clothes. In his autobiography, he begins the story this way: "How extraordinary was my life an incident may illustrate." Later on the same page he details his ability to "digest cobblestones.") He imagined focusing the power of the falls: using water and gravity to spin a giant turbine. He promised his uncle he would go to America and turn this notion real. His uncle was not a believer; but "thirty years later I saw my ideas carried out."

He sailed to America—"the Land of Golden Promise"—in 1884, for a post at Edison Electric Light. Edison took in Tesla's clothes and airs and nicknamed him "our Parisian." (Once Tesla got famous, Edison came to dismiss him, with a commercial sneer, as a "poet of science.") Meeting Edison, the first thing Tesla could think to say was that his shoes were dirty; he told the inventor he meant to get a shine. Edison said, "Tesla, you will shine the shoes yourself, and like it."

They went at things differently. Edison's ideas came to him like a scatter of clues, a Ziploc bag of evidence from which he'd patiently reconstruct the crime scene. Tesla's visions arrived complete, like a police psychic's: he'd see weapon, blood, the killer's face. Tesla pitched the difference to an interviewer in terms of waste, comic spillage. "If Edison had to find a needle in a haystack," Tesla said, "he would proceed like a diligent bee to examine straw after straw. I was a sorry witness of such doings, knowing a little calculation and theory would have saved him ninety percent of the labor."

Tesla also understood the key problem with Edison's power and light system: its reliance on what's called direct current. All generated electricity is alternating current, the generator flipping a magnet's positive and negative fields. (Small electronics run on DC. Your MacBook power supply's heat means alternating current is getting converted into good old DC.) But DC operates at low voltage. This meant Edison could transmit power only a mile or two at a time. A city lit by Edison would need power plants spaced every twenty blocks like subway stations.

Engineers were scrambling to work out the AC kinks—lug generators out of the cities, into the fresh air, enlist the waterfalls.

Before leaving Europe—twenty-four, and living in Budapest—Tesla suffered a breakdown. Under ordinary conditions, he liked to demonstrate that his hearing was "thirteen times" sharper than any non-Tesla's. During breakdown, this became an excruciating superpower. Carriage wheels miles away "shook my whole body." Flies crossing a tabletop wore lead shoes; sunlight became a hangover, his heartrate hit 260 beats per minute, he developed a bat's creepy ability to perceive objects without sight or touch. (Even his sufferings did Tesla credit: a doctor pronounced this illness "incurable" but "unique.") Tesla rallied. He took brisk walks through the park with a friend. One winter afternoon, reciting poetry, Tesla was struck by the solution. In a flash, an entire working AC system; he was like a sudden human camera, receiving the complete print. He took a stick and drew a copy of what was in his head in the dirt. The plan never varied. He stared, sighed, told his friend he was now prepared to die happy. "But I must live," Tesla said. "So I can give it to the world. No more will men be slaves to hard tasks. My motor will set them free, it will do the work of the world." Tesla ran on 99 percent inspiration.

He presented his idea to Edison, who laughed. With a corporation behind him, Edison had become Western Union, shaking off the telephone. Edison warned him AC was a pipe dream; the technology had "no future."

Tesla left the company. For months, he supported himself by shovel, digging up Manhattan for two dollars a day. In 1888, he lectured on his AC ideas at Columbia College. The engineers understood they'd just received a weather report from the next century. Two months later, he sold the patents to George Westinghouse for four hundred thousand dollars.

The system he'd blueprinted in the dirt would power devices and buildings, push current not just under neighborhoods but across whole landscapes. Tesla joined society. He took up hotel living—the Waldorf, the Astor, never another pillow without roomkeys and a concierge desk—and became a fixture on celebrity to-meet lists. Mark Twain wrote in his journal of seeing plans for "an electrical machine lately patented by a Mr. Tesla . . . which will revolutionize the whole electric business of the world. It is the most valuable patent since the telephone." One man blocked the path with scare pamphlets, newspaper interviews, sympathetic legislators; the obstacle had become Edison.

THE WESTINGHOUSES CAME from Russia. They were originally called von Wistinghousen, which probably would have looked lousy on a washing machine. George Westinghouse had the look of a riverboat gambler—quick eyes above full cheeks, walrus mustache shielding his expression like a poker hand. He traveled the country by private railcar. Before the automobile, this was a CEO's luxury vehicle, the corporate jet. His electrical company competed with Edison for the big contracts. Unlike Edison, he was making a run at alternating current.

Their fight escalated from black eyes to court filings, a series of standoffs and skirmishes the papers called the War of the Currents. In cities and fields, shadow employees chopped electrical poles, slashed cable. Edison pinned his animosity on a dark invention: the electric chair. It would be powered by one of Westinghouse's generators. Edison wanted to shape the dictionary into a club—he hoped the slang for electrocution would be *getting Westinghoused*. Instead, the *New York Times* called the first 1890 electrocution "a disgrace to civilization." The initial jolt didn't take: the condemned man still twitched, steamed, breathed. A reporter from the Associated Press shouted, "For God's sake, kill him and have it over," then fainted. Next day, Edison offered mildly that he'd "glanced over" some press accounts. Westinghouse said, "They could have done better with an axe."

Tesla had the imagination of a press agent. When he strolled a hill, it became a mountain. Shaking hands with a tall man, he'd met a giant. To Tesla, Westinghouse appeared "like a lion in a forest." The "only man on this globe who could take my alternating-current system" and win "the battle against prejudice and money power." His Westinghouse stamped over fences wearing celestial boots. "Had he been transferred to another planet," Tesla wrote, "with everything against him, he would have worked out his salvation."

During the 1890 financial crisis—London's Barings Bank splashed into bankruptcy, sending ripples across America and Europe—Westinghouse Electric nearly went under. In early 1891, Westinghouse rode his private railcar to Manhattan.

The company's most valuable asset was the Tesla patent. It also represented their greatest liability. Tesla had negotiated a fee per watt; if the current landed, Westinghouse would be in the hole to him for millions. Suddenly there was Westinghouse, frowning down the aisles of Tesla's lab, making excuses. This scene comes from the inventor's memory; it's the Oscar clip of a film written, directed by, and starring Tesla.

"Your decision," the magnate announced, "determines the fate of the Westinghouse Company."

Tesla either asked or says he did, "And if I give up the contract you will save your company . . . so you can proceed with your plan to give my polyphase system to the world?"

Westinghouse nodded. If he made the next remark, he had read the room. "I believe your polyphase system is the greatest discovery in the field of electricity." But no matter what Tesla decided, Westinghouse said he would find some way to put America on an alternating current basis. It was a storybook test, the correct answer given at the bridge in a fairy tale.

Tesla says he drew himself up to his full skinny height. "Mr. Westinghouse, you have been my friend," he said. "The benefits that will come to civilization . . . mean more to me than the money involved. Here is your contract and here is my contract—I will tear both of them to pieces." And Tesla—or a Tesla in Tesla's mind—ripped up his fortune.

EDISON AND TESLA had different understandings of the planet. Edison saw endless abundance. One of his failed ventures, post-electricity, would be cement houses, slabbed with all-concrete furniture. America was chunks of raw material, populated by a race of tough-bottomed giants. Tesla had sailed from Europe, a world of cramp and limits. Lecturing to engineers,

Tesla fixed the questions of the next century. "For our existence and comfort we require heat, light, and mechanical power," he said. "What will man do, when the forests disappear, when the coal deposits are exhausted? Only one thing... men will go to the waterfalls, to the tides."

The country's most powerful falls—a drop of 167 feet—was at Niagara. By mid-century, it had hit its stride as a tourist trap. Bad restaurants, museums with dusty artifacts, rickety hotels. The sharpest use anyone could find for Niagara was as the backdrop for tightrope walks: Acrobats crossed on stilts, pausing to sip champagne and fry omelets. In the late 1870s the British engineer Sir William Siemens toured the area, breathed in spray, made excited calculations. Niagara represented 6.8 million horsepower— the equivalent force of all the coal mined every day from every crevice of the world. Newspapers began to speak of "the enormous power which for centuries has been running to waste," as if nature had left a light on in the garage forever.

In 1889, the Cataract Construction Company incorporated—New York money. A giant poker bet on electricity, two and a half million dollars to capitalize the falls. Their plan called for a hundred thousand horsepower, as much generation as every Edison power station combined. Famous engineers warned against the "awful," the "gigantic" mistake of using AC. Buffalo, the largest nearby city, was twenty-six miles away. Nobody in America had ever sent electricity more than a couple of miles.

An operation in Colorado ended up as the test case. Failure opens the mind. The Gold King mine was on its last legs, which meant a willingness to try anything. (It's also a conservation parable; diggers ran on wood, and Gold King had chopped the nearest forest bald.) The mine hired Westinghouse. Westinghouse ran the Tesla system. There was a waterfall three miles away; during the summer of 1891, a Tesla generator was installed in a shack by the cascade. Seven hundred dollars of wire connected the power to the mine. The system ran through mountain wind, Colorado snow, and mining continued. A Tesla generator in Oregon sent power thirteen miles to light up Portland.

In May 1893, the Niagara Company officially settled on AC; the Colorado operation had convinced them. ("The company," the *Times* reported, "had finally determined that it was by means of Tesla's inventions alone that its operations could be carried out.") Since he controlled the Tesla patents, they hired Westinghouse: he'd been invited to the party in order to bring his cool friend. Tesla's system would now be crossing the tightrope at Niagara.

Tesla was in a state: full of plans, hopes, dreams. In 1894, he predicted his invention would "solve the labor problem. . . . The hard work of the future will be the pressing of electric buttons." Journalists called Tesla "Tesla," the way Edison was called "Edison" or Beyoncé Knowles is "Beyoncé." Basic construction was complete by summer of 1895; the *Times* called Niagara "the unrivaled engineering triumph of the nineteenth century. Perhaps the most romantic part of the story," the paper added, was Tesla, the "man above all men who made it possible." (On the other hand, the same article also calls him "extremely modest and retiring.") Edison's nickname was "the wizard"—reporters called Tesla, simply, "the electrician."

In July 1896 Tesla traveled to Niagara. He came with Westinghouse. They toured the works. Reporters asked whether America could find a use for one hundred thousand horsepower of electricity. "This talk is ridiculous," Westinghouse said. "All the power here can and will be used." (As of our century, that much juice would illuminate the city of New York for a little under half a minute.) The two men inspected the water tunnels, sluices, generators. There was a new name now, a new metaphor for the conversations. Edison's plants were called "electric light stations." The brass plaque above the Niagara door said POWER HOUSE.

Tesla had read about the falls as a boy. Twelve years earlier, he'd drawn his plans in the sand; eight years earlier, he'd sold his patents; four years ago, surrendered his royalties, all to arrive here. In a sense, he was going to make it possible for people everywhere to duplicate the experiences of his breakdown. To pull the shades, silence the world, work from home: the power would give everyone the chance to live like mini-Teslas. Tesla told the reporters, "It is fully all that was promised." Maybe it's what inventors say—the most eloquent thing. It's almost exactly what Edison said the day his generators illuminated the *New York Times*.

On Sunday, November 15, 1896, the giant system was ready. Edison at

Pearl Street had spent five hundred thousand dollars. The Cataract Company had invested eight million. Edison had sent electricity one mile; it would now run twenty-six, the length of a marathon. Edison had sixty customers; Niagara would power a city. One of the company executives, like Samuel Morse, had grown up in a minister's household. He did not want to begin transmission, show off the latest thing the Lord had wrought, on the Sabbath. Staff killed time, eyeing the clock. Just after midnight, the Niagara Company's president threw three switches. (Tesla would have liked that.) The effect was instantaneous—electricity moves at nearly the speed of light. Benjamin Franklin had understood lightning as a branch of electricity. The *Buffalo Enquirer* wrote, "Electrical experts say the time was incapable of computation. It was the journey of God's own lightning bound over to the employ of man."

THE JUBILEE

AND THEN ELECTRICITY SHUCKED OFF TESLA, WESTING-house, Edison. They were the stages a rocket uses to climb into orbit.

Tesla fades from the story after Niagara. History is the faces you remember from the graduation photo. For the first years of transmission, the electrician's face stayed in focus. AC tromped across the landscape, displacing Edison; Tesla used fame as a cash trap. He raised capital for a new and more efficient lighting, fluorescent. "My bulb," he said, "will last forever." But he couldn't stay interested. What compelled him was what he called "wireless telegraphy"—a combination of radio and telephone that would someday allow communication to and from anywhere.

It became a dream he couldn't shake, and in the end it sucked every awake thing after it. In 1897, he boldly told the *New York Times* it would soon "be quite easy to communicate from any point with a ship in mid-ocean or a traveler at the North Pole." Tesla could see it clearly in his head. But the mechanism for transferring it from inside to outside was jammed. He patented the basic components for radio but never whipped them together; Guglielmo Marconi did, Tesla described him as "a virus." (The Supreme Court later awarded Tesla posthumous invention of radio.) It was as if, in a dream, someone had shown him a vision of the iPhone. He constructed a giant, 157-foot tower amid the scrub and wilds of Long Island. There was a round directional antenna on top, but who could it broadcast to?

In the beginning, he tapped the Westinghouse Company. He'd rescued them, enriched them, the least they could now do was provide lab equipment free of charge. "I believe," he wrote an officer, "there are gentlemen in that company who still believe in a hereafter." Never the sharpest bargaining strategy to take with a corporation. By the 1920s, the company was writing Tesla out of histories of electricity and Niagara.

He grew into a sort of nonspecific celebrity—feeding interviewers strange opinions. On the connection between feminism and entomology: "Sex equality will end up in a new sex order, with the females superior." This would "dissipate female sensibilities," then "choke the maternal instinct," and "human civilization will draw closer and closer to the perfect civilization of the bee." Just like his work: Sensible, interesting—then crazy.

He went broke. In 1916, the Waldorf Astoria sued Tesla for $20,000 in back rent. They evicted him, seized his property, demolished his Long Island antenna, selling off the scrap for lumber. The worst blow to an inventor—your invention has become less valuable than the sum of its parts. The St. Regis Hotel sued him for $3,000 in 1923. In 1930, Tesla was escorted onto Seventh Avenue from the Hotel Pennsylvania, $2,000 in the hole. In 1931, Tesla's seventy-fifth birthday, as electricity continued its march across the globe, Albert Einstein wrote him: "I congratulate you on the magnificent success of your life's work."

In 1934, he reached a kind of moral accommodation with the company he'd earned fortunes for. Westinghouse would pay Tesla $125 a month, plus rent at the Hotel New Yorker.

A famous life drifts off the front pages, settling farther and farther back in the paper, deeper into trivia. He became the kind of strange-news item Edison and the other plugs had once posted on the telegraph office walls. The old inventor with strange ideas: a death ray, an earthquake machine, a skinny man with antique manners, feeding pigeons in the New York City parks. (Tesla fell in love with one pigeon "as a man loves a woman," he told his biographer, "and she loved me.") He died in 1943, mathematically at peace. In room 27, in bed, on the 33rd floor. He was 86; dietary restrictions and all, he'd come within 14 years of his goal. A few years before Tesla's death, total capital investment in electricity was estimated at eleven billion dollars. His obituary noted, as one of its subheads, "Not Practical in Business."

THE SAME FATE awaited the man who'd made Tesla's plans real. In March 1907, the Dow Jones dropped a quarter of its value. The landscape opened, taking billions and industries. New York bankers cruised the shallows, settling scores, swallowing companies. George Westinghouse had been a roughly fair man: offering employees pensions, disability insurance, mortgages. The men who reorganized Westinghouse Electric could not make this square with a balance sheet. He fought for years, then watched his corporation lumber off without him, this strange thing he had built that carried away his name. Westinghouse never recovered. When he rode the train past his factory, he turned away from the signs, preferring the sad scrub hills on the opposite side of the railroad. He became subject to the illnesses of disappointment. Restlessness, anxiety. In 1914, Westinghouse traveled East, seeking treatment for whatever ailed him. He died at the age of sixty-four—like Tesla, in a New York City hotel.

EDISON LOST HIS COMPANY too, Edison General Electric. The early years of corporate evolution sound like the dinosaur times—lava and comets. In the Panic of 1892 and 1893, J. P. Morgan became the nation's one-person rescue and bailout team, lending and acquiring. Morgan had been Edison's key moneyman: the investor whose home and gym he wired first. It turned out Edison had nursed a crocodile. Morgan bundled up Edison Electric with a competitor. Then he amputated the stationery. Edison had started the industry; now even his name was no longer required. The new company was called, simply, General Electric.

Edison, who once hoped to make *getting Westinghoused* the slang for electrocution, had been decapitated by J. P. Morgan—in the nineteenth-century term, *Morganized*. He was left to watch his former company slouch away without him.

It turned him against electricity. The day Edison lost the company, his secretary found him looking pale and demoralized. "I've come to the conclusion," Edison said a few months later, "that I never did know anything about it." Edison promised to do something "so much bigger than anything I've ever done before that people will forget that my name was ever connected with anything electrical."

He never found it. Nothing in Edison's later life matched the triumphs of the first part. It was all wrap-up—the second half was the long sad drive from the ballpark, when the game has been called on account of rain. He couldn't find a steady dependable project to settle down with. Edison tried mining, cement suburbs, synthetic rubber, electric cars. He lost millions in the record business. He thought anybody with taste would hate loud music. He couldn't understand why listeners wanted to hear stars. Edison's doubts belong on music industry walls, beneath block letters saying FAIL. "Why so much for a vocal cord?" he asked. "What is it but a bit of meat?"

He found himself more and more beloved as he did less and less. Tesla and Westinghouse had perfected the electricity Edison found the use for. But the rest of the country still felt their lives had been constructed from the gears and switches on Edison's workbench; they believed they were residents of Planet Edison. All inventions blurred to his features. A 1930 *New Yorker* profile described him, without fuss, as "America's greatest man." The magazine added, "The airplane is almost the only major modern invention on which Edison has left no mark."

Edison had teased Tesla as a scientific poet—a man who never mastered the fundamental invention, which is the one that turns notions into money. But Edison wrote Henry Ford, the big friend of the last part of his life, that his own relations with Wall Street deserved a gloom soundtrack: "My experience over there is as sad as Chopin's funeral march." (Ford told reporters he considered Edison "the greatest man in the world"; also "the world's worst businessman.") Always a quick study, Edison could see where the corporate idea was headed. He read Igor Stravinsky's remark, "The tempo of America is greater than the rest of the world. It moves at a wonderfully swift pace." And he scribbled in the margin, "Yes, with a metronome of money."

But Edison remained an American story: Optimism rewarded. A young man from the provinces arrives in the city—its narrows and shadows—with light for everybody. In 1921, a Kansas housewife sent Edison a global thank-you:

> *I hear my wash machine chugging along, down in the laundry, as I write this it does seem as though I am entirely depen-*

> *dent on the fertile brain of one thousands of miles away for*
> *every pleasure and labor-saving device I have. The house is*
> *lighted by electricity. I cook on a Westinghouse electric range,*
> *wash dishes in an electric dishwasher. An electric fan even*
> *helps to distribute the heat over part of the house. . . . I wash*
> *clothes in an electric machine and I clean house with electric*
> *cleaners. I rest, take an electric massage and curl my hair on*
> *an electric iron. Dress in a gown sewed on a machine run by*
> *a motor. . . . Please accept the thanks Mr. Edison of one most*
> *truly appreciative woman.*

You wonder if Edison remembered. Here was every use—light, heating, cooking, power—he'd once predicted for electricity. But that's all there is: the letter in the Edison files, a note handwritten above it, "Thank her very much etc."

IN 1930, the *New York Times* cast a how-did-we-get-here glance over the power industry. There were now four thousand generating plants, sending current to twenty-five million customers. "The pace throughout has probably been faster than in any previous application of a new economic factor in the experience of the world." The change had taken place across fifty years—the span of a single life. "Power to most people today means but one thing," the paper explained. "Electricity."

The Kansas Edison fan's appliances weren't a consequence; they were a cause—how electricity made the leap. By the second decade of the century, there was the non-Tesla and de-Westinghoused Westinghouse Company, and the Edison-free General Electric. They had a product, electricity. The gadget, the irresistible convenience (the iron, the toaster, vacuums) turned out to be the way to move it.

How'd it work? They used Edison's bulb. The bulb was the knock on the door, the salesman's card slipped under the frame. Years later, a General Electric executive would explain, "Every branch of the electrical industry owes much to electric lighting." They had an idea to express this: the bulb was "the entering wedge," clearing a path for "many labor-saving devices."

Turn-of-the-century power ads explained, "A Home Without ELECTRIC LIGHT is like a coat without a lining." Other spots worked the romance angle, husband as hopeful, bashful provider: "Make Your Wife Happy With This Gift . . . Electric Light in the Home Means Much to Women."

But once people got the wire into the house, it remained a high-tech candle, on an extra-long wick. In Chicago, home use was a two-hour show. Six to eight p.m.: lights came up, tables were set, dinners eaten, plates cleared, lights down, billing stopped. A General Electric handbook tallied up the problem: "The public is buying light, and has no feeling for energy."

The next wedge came from one homely appliance. In 1908, the head of the Chicago utility mounted his electric iron offensive. Salesmen packed a single truck with ten thousand irons—a chunky, clunking mountain—and steered it through the nicer parts of town. A sign promised one iron free with installation. Customer base tripled. Salesmen put the effect matter-of-factly; as you might describe the corporate marketing of a substance like cocaine. Encourage any installation "on the assumption that the use of electricity even in the most trivial manner soon becomes a habit, and inevitably leads to a more extended use."

This became the pattern, how electricity grew. You had the capacity; so drive up the demand. In Indiana, the General Electric affiliate sent salesmen door-to-door, pushing hot plates, toasters, percolators, each one another draw on the socket. They sold at a loss but made the money back in billings.

The magic of corporate branding took it from there. In 1911, General Electric was charged with monopoly. So the company had withdrawn, becoming faceless, fingerprint-less. For a decade, it approached publicity with the enthusiasm of a man throwing a raincoat over his face as he's put in the squadcar. The result: "The General Electric name," complained one executive at a 1922 crisis meeting, "means very little to the average house-wife." The corporation needed its face recognized: General Electric turned to an ad man named Bruce Fairchild Barton.

Barton is one of the great, weird figures of the American corporate adolescence. The casual lawgiver who strolls down from the mountaintop with the lenient stone tablets. In a teasing mood, Edison might have called Barton a saint of business—except that Barton made a ton of money.

Bruce Barton grew up in the church backstage. His father was a circuit rider—a saddlebag minister, riding horseback between flocks. (He ended at the parish where Ernest Hemingway was raised, and soured the writer on religion forever.) In the same way that Edison laid a bridge between gold-rush America and tech-rush America, Barton connects two different types of faith: religion and business. "All work is worship," Barton said. He understood the place the corporation might win in consumer hearts. Advertising, he explained—already a torrent, hard to focus on any particular splash or twig—made it impossible for buyers to reach each decision independently. Shopping would also become belief. "We will pin our faith and our buying habits," consumers would decide, "to a few big names we know and can trust."

Barton was a man with the ability to rebrand an entire religion. He's most famous now as the author of a 1924 bestseller, *The Man Nobody Knows*. Religion, to him, was at a crisis. And when you're locked into a formula, the only course is change the packaging.

If, say, businessmen were ignoring Jesus, why not turn Jesus into a glad-hander, dress him in a suit and tie? Barton explains in the book how he grew up hating the Sunday School Jesus—the "sissified" man with "flabby forearms." This Jesus made no sense because "only strong magnetic men inspire great enthusiasm and build great organizations." Barton's Jesus was somebody you'd want at the board meeting and company picnic. "A physical weakling? Where did they get that idea? He was a successful carpenter." Also a talent-spotter and an excellent mixer. "A killjoy? He was the most popular dinner-guest in Jerusalem! . . . A failure? He picked up twelve men from the bottom ranks of business and forged them into an organization that conquered the world!" Barton explains he wrote the book for "every businessman," because Jesus was "the founder of modern business." His book remained on bestseller lists for years. Outsold Hemingway, outsold *Gatsby*.

These were the miracles and wonders Bruce Barton was summoned by General Electric Company to perform. Barton suggested they do away with that chilly third word: *Company*. Because corporations were big, and faceless, and might not have your best interests at heart. In fact, who could you really count on? And what did you do, when your friend's name had too many syllables?

That's how the campaign rolled out. "G-E: The Initials of a Friend." General Electric had a disunited product line. Barton advised them to stamp the GE logo on every item: "Reputation," he explained, "is repetition."

He advised ads front-loaded with people and faces; never another spot with just the product. As Barton explained, one media baron had a rule: "No photograph shall ever be printed in his newspapers unless it contains human beings." This was because "you and I are interested most of all in ourselves," Barton wrote. "Next to that we are interested in other people." After all, what GE sold was cool and remote: an appliance, a spark flying down a wire. People would turn this local and warm. (It's a ratio you can still chart, in ads for giants like Microsoft and Chevron. The more faces, the warmer the glow, the slower the motion—the more children running with flags down a cobblestoned street—the more impersonal the product.) Edison and Tesla had given the industry its product; Barton settled on its features, its image-making style.

It's interesting to imagine the Edison response to another Barton suggestion: that his former company, which three decades earlier had squeezed Edison out, should insist on the connection, and chain itself to the old wizard. As corporate historian Roland Marchand asks, "Why should a manufacturer settle for hawking light bulbs . . . when it could claim credit for light itself?"

Since Edison was the wizard, GE became "the House of Magic," the "Trustees of Light." Within a decade, rivals understood just how effective this new kind of image-making could be. Marchand quotes a J. Walter Thompson executive, griping that buyers were conditioned to respond positively whenever "G.E. puts out anything."

On October 22, 1929, GE sponsored light's Golden Jubilee—the fiftieth anniversary of the bulb Edison had jimmied together out of electricity and cotton thread, the first that burned through the night. Edison was eighty-two. White-haired (*he* was Snow Edison now), fighting his health, totally deaf. His wife had to communicate by the language of his teen years; she tapped incoming messages on his hand in Morse code.

It must have been a surreal week for Edison. National time travel—
millions of strangers on a guided tour of his past. He was responsible, the
New York Times wrote, for fifteen billion dollars of economic activity, nearly
"the value of all the gold dug from the mines of the earth since America was
discovered." One-and-a-half million people were employed by industries
that had seen their start on the "Edison brain-value table." ("The adding
machine rebels," wrote the paper.)

Henry Ford hosted Edison at his Dearborn, Michigan, recreation of
the old Menlo Park lab. Edison strolled the grounds, murmured "It's amazing." And "I wouldn't have believed it." He recognized the dirt Ford had
shipped in, turned it over with the toe of his shoe. Then he smiled: "The
same damn old New Jersey clay." As he walked toward the resurrected buildings, his wife tried to reshape and rebutton his clothes: Edison brushed her
away. "I'm just as young," he said, "as I was when I worked there in that
old laboratory."

Next evening, Edison, Ford, Harvey Firestone, and President Herbert
Hoover boarded a special vintage train to chug from Detroit to the banquet. Edison didn't sit. He picked up a box of snacks and called, as if he
were twelve years old again, "Candy and fruits for sale!" Hoover waved him
over—the most surreal moment of all, a president co-starring in your memories. "I'll take a peach," the president said. In the lab, for a radio audience
of millions, the bulb's discovery—globe, thread, power—was reenacted. A
breathless announcer turned the proceedings biblical: "And Edison said,
'Let there be light!' " And in cities across the country, special gold lights
were timed to spark at that moment.

President Hoover was to speak; GE's chairman would serve as toast-
master. Edison sat down at the door of the banquet hall and wept.

"I won't go in," he said. His wife fetched him a glass of warm milk.
Orville Wright was inside, along with Madame Curie, the heads of America's
corporations, all people somehow implicated by those fifteen billion dollars.
The GE chairman thanked Edison for giving birth to an industry, President Hoover said every American owed Edison a debt, for "repelling darkness." By special radio hookup, from a secret Berlin location—the physicist
was pursuing lab work for Germany—Albert Einstein addressed the hall.

He sounded the night's lone note of caution. "The great creators of technics," Einstein said, "among whom you are one of the most successful, have put mankind into a perfectly new situation, to which it has as yet not at all adapted itself." Next day, the Edison Jubilee Handicap was won by a horse named Old Kickapoo. A week later the Great Depression began.

When Edison died, in 1931, Hoover arranged for every city to dim its lights—a reminder of what the world had been like before, and one of the first occasions that the country had been midnight for decades.

By the turn of the next century, the value of the electrical grid was about one trillion dollars: The power industry had become America's largest single earner. Westinghouse and GE had spread. In 1960, the first commercial nuclear reactor, built by General Electric, came online outside Chicago; the second, in New England, was constructed by Westinghouse. In the 1960s, a critical turn; what utility providers call *peak demand* used to arrive in winter. Christmas Eve: heat, holiday lights, electric train sets, gifts unwrapped. Now and after, it came in summer, a relief from heat: air-conditioning. Westinghouse is CBS TV now; Tesla's great insight, and Westinghouse's tenacity, are the foundations for an antenna that broadcasts *Survivor, Sixty Minutes*, and *Young Sheldon*. GE pushed its way forward, became America's second largest corporation, bought and sold NBC and Universal studios. TV, radio, movies—the pop industries Tesla and Edison helped create—had become the cultural weather system, the human climate. As the Department of Energy put things on its website, "Electric power arrived barely a hundred years ago, but it has radically transformed and expanded our energy use. To a large extent, electricity defines modern technological civilization." This is what Edison and Tesla had foreseen, in 1878 and 1885; standing beside their generators, it's the thing both men once promised.

Part II

SCIENTISTS

What's the use of having developed a
science well enough to make predictions, if
in the end all we're willing to do is stand
around and wait for them to come true?
—DR. SHERWOOD ROWLAND, 1986

THE WAYWARD WIND

AND THEN YOU HAVE ROGER REVELLE, FAMOUS OCEANOG-rapher. And 1956—a hinge year in a swing decade, old and new, storm fronts colliding. Repetition (it was going to be Dwight Eisenhower again, mincemeating Senator Adlai Stevenson for the presidency) and fresh problems (America taking over the French lease on Vietnam). Elvis Presley's "Heartbreak Hotel" had been replaced as the number-one song by a sort of rearguard musical action: Gogi Grant's "The Wayward Wind." You could still see where the pop culture was heading. Presley's songwriters had worked from a *Miami Herald* newspaper story about a five-word suicide note: "I walk a lonely street." (Elvis' producer called the song "a morbid mess.") "The Wayward Wind" was a breeze. "Oh the Wayward Wind, is a restless wind."

According to *Time* magazine's issue of May 28, 1956, the worldwide argument—communism versus capitalism—had strayed onto treacherous, boggy ground. The Russians no longer wanted to loom. They intended to welcome. "The Soviet policy of smiles," the magazine explained, was "picking up mileage and momentum." In the era's great, overcommitted, news-on-patrol language, this offensive was "relaxing freedom's watchfulness."

In that May 28th issue, a milestone got crossed. America no longer contained enough audience to support two circuses. In a story called "End of the Trail," the Clyde Beatty Circus ("the number-two big-topper in the U.S., after Ringling Brothers") had filed for bankruptcy. It was the box—just as the Edison ads had asked, why not let the show come to you? "Suckers may still be born every minute," a circusman told *Time*. "But TV gets 'em first."

Roger Revelle was in the Science section, with a discovery he'd remade. The title was "One Big Greenhouse." The article totted up the residue of all Henry Ford's cars, Tesla's generators, Edison's current. "Since the start of industrial revolution," *Time* explained, "mankind has been burning fossil fuel," adding its "carbon to the atmosphere as carbon dioxide."

The first paragraph continued, "In fifty years or so this process, says Director Roger Revelle of the Scripps Institution of Oceanography, may have a violent effect on the earth's climate."

The piece didn't have a breaking news, the-barn's-on-fire sound: by the mid-fifties, the science was already well understood. Once the carbon exceeded the absorption and dissolution capacities of the oceans and forests, it would crowd the atmosphere, trap sunlight, raise temperatures.

Even in 1956, it was clear you wouldn't need an extreme rise to get extreme effects. "In the future," *Time* explained, "if the blanket of CO_2 produces a temperature rise of only one or two degrees, a chain of secondary effects may come into play." (This process would later receive an official name: "feedback." Another, slightly more bullying term is "forcing.") Once the air warmed, "sea water will get warmer too, and CO_2 dissolved in it will return to the atmosphere," the magazine continued. "This will increase the greenhouse effect of the CO_2. Each effect will reinforce the other, possibly raising the temperature enough to melt the icecaps of Antarctica and Greenland, which would flood the earth's coastal lands."

So Revelle and fellow scientists would be withdrawing to the atmospheric stock room, taking a carbon dioxide inventory. As the piece concluded, "When all their data have been studied, they may be able to predict whether man's factory chimneys and auto exhausts will eventually cause salt water to flow in the streets of New York and London."

Global warming predictions and news reports have gotten a lot more detailed, without much altering the basic science *Time* and Roger Revelle laid out in 1956. Lots of changes have come and gone since. America fought in Vietnam; and Grenada, Panama, Afghanistan, twice in Iraq, many subtler and quieter times with drones. Communism, that campaign of smiles, switched back to frowns, then back to smiles, then to a look of farewell surprise, then to nothing. In the United States, the questions of African American and gay civil rights were partway answered; abortion became a legal right, then an exhausting conversation. TV changed its address to cable, to Netflix, then your phone. There was room for *Star Wars*, two sequels, five prequels, three additional sequels, and much merchandising. And no major legislative step had been taken in America to reduce carbon dioxide, though the basic science had been public for sixty years.

WHEN ROGER REVELLE SAT down with "the carbon dioxide problem," he was meeting what was already, science-wise, a prosperous senior. The greenhouse theory had rounded 130 years old.

Jean-Baptiste Joseph Fourier turned twenty-one at the start of the French Revolution. In portraits, he looks like the mild man at the next restaurant table: someone who might ask you to keep it down but isn't going to press the point. The young mathematician watched idols and mentors— a generation of people he wanted to become—put to the guillotine. In 1798, he opened a letter from the interior minister. He was instructed to pack up, make his goodbyes, report to the docks.

In unsettled times, the safest course is do as you're told. Fourier had never been aboard ship in his life. When he stepped from the dock, there was Napoleon, along with his officer corps, plus a kind of academic superstaff— France's greatest scientists. In a fleet of 180 boats, escorted by fifty thousand troops and mariners, they set sail for the sun and deserts of Egypt.

It must have been a gritty transition. Egypt was all flatness, brightness. France, as Gustave Flaubert complained, was a "country where you see the sun in the sky about as often as a diamond in a pig's asshole." For three years, Fourier worked as an administrative moonlighter: governor of Lower Egypt and secretary of the French scientific mission. The team dug up, among other items, the Rosetta Stone, which is how hieroglyphics eventually got solved. (In an inventory, the find is modestly described as "a stone of black granite . . . found near Rosetta.")

And he learned about heat. Another visitor might have sampled regional cuisine and scaled pyramids. For Fourier, Egypt was a crash course in burning. Learning just what the sun, uncut by trees and latitude, was capable of. Half of Napoleon's force succumbed to weather. Back in France, Fourier wore heavy coats, kept his rooms stuffy—part home, part still in the desert. And he studied the science of heat. In 1824, Fourier approached the biggest question: What kept the Earth warm at all? When sunlight sidestepped clouds, bounced off the ground, what kept it from bouncing all the way back out again? The solution Fourier came up with was the first greenhouse theory. The planet's atmosphere must act like the "glass cover" of a vessel, designed to receive and conserve heat.

WARMING MOVES IN A kind of weird processional lockstep with the inventors. The Irish physicist John Tyndall took over from Michael Faraday as head of the Royal Institute—as lead brain, England's scientist laureate. (When Edison's phonograph was demonstrated for a "large and fashionable" London crowd, John Tyndall was on stage working the apparatus.) Tyndall had the odd Victorian look: sensible features above, crazy sling of a beard below. In a lot of other ways, Tyndall is recognizable as a modern man. He was wild about mountaineering—became the first climber up the Weisshorn, fourth tallest of the Swiss Alps. And he suffered from the anxious modern ailments, indigestion and insomnia, for which he swallowed two daily medications.

Tyndall was tough-minded, ironic. He lectured, to a roomful of Victorian optimists, that you could expect any new invention to bring fresh problems. "Science, like other things, is subject to the operation of this polar law," Tyndall said. "What is good for it under one aspect being bad for it under another." He drilled home this face-things point later in the speech: "It is as fatal as it is cowardly, to blink facts because they are not to our taste."

In late 1858, Tyndall got curious about which gases might be trapping solar energy. Tyndall began his research during a fairly soft winter. By May 1859, he was crowing in his journal—as if he'd hauled and yanked himself to the top of a research mountain. "Experimented all day; the subject is completely in my hands!"

Tyndall knew something atmospheric was trapping heat; he'd run through a number of "perfectly colorless and invisible gases," hunting the trappers. Tyndall was able to clear oxygen and nitrogen. For sunlight, they were an open window; easy in, easy out. Tyndall was on the verge of dropping the experiment when he thought to test coal gas—a modern invention, Walter Scott's "madman" smoke, lighting streetcorners and homes. When it came to infrared radiation, coal gas turned out to be as transparent as cement. Sunlight could just click through. But once it warmed the Earth, the heat had to stay put.

Tyndall tested carbon dioxide; same result. The atmosphere *did* trap the sun's heat. He knew which gases: water vapor and carbon dioxide. Tyndall experienced the research scientist's thrill—first up the hill, first to understand a basic why. Together, the gases were the planet's cold-weather gear, he wrote, a "blanket more necessary to the vegetable life of England than clothing is to a man." Strip that blanket, even for just "a single summer evening," and the "warmth of our fields and gardens would pour itself unrequited into space, and the sun would rise upon an island held fast in the iron grip of frost."

Tyndall, at fifty-five, had married a much younger woman, Louisa. As a conjugal hobby, Louisa took over Tyndall's medications. A slug of magnesia in the morning, to ease the scientist's digestion; chloral hydrate at night, to tilt the scientist into sleep. One winter morning in 1893, Tyndall lived out his polar law of science—the inventions that sustain us can also do damage. In an awful Laurel and Hardy mix-up, Louisa got the bottles confused.

Tyndall took a deep swig of chloral hydrate. Both understood immediately—and gallantly, Tyndall's sympathies leapt to her. "Yes, my poor darling," he told his wife, ten hours before he died. "You have killed your John." The next year, a Swedish scientist began the calculations that are at the base of our own anxious world.

GLOBAL WARMING HISTORY is a long afternoon spent in the Hall of Ironies. The ground-floor calculations would end up the cultural property of the anti-splurgers, the enemies of high living; but they were worked out by a scientist who loved high living. Svante Arrhenius looked like a guest attending the costume party as Winston Churchill—droop cheeks, mushy chin. He could make even work a party; colleagues said he "spread joy in the lab." At conferences, he toasted fellow researchers with their favorite cocktails; he'd stop a dinner to hustle everybody outside to admire the aurora borealis. His official Nobel Prize biography takes the trouble to describe him as not just "contented" but "happy."

So the first greenhouse numbers were run by a cold-weather Swede, who liked to imagine the artificially enhanced summers of the future. And there's another hazards-of-marriage story—global warming was the product of a local domestic cooling.

Svante Arrhenius was just who classmates would've picked to be the first Swedish recipient of the Nobel, which he collected in 1903. He grew up a math prodigy—taught himself reading at age three, graduated youngest and brightest in his high school class. In July 1894, Arrhenius married for beauty, intelligence, and proximity. Sofia Rudbeck was his sharpest university student. The marriage seems to have soured quickly. Arrhenius picked Christmas Eve—a young couple's social time, invitations and fireside drinks in the cool piney basement of the year—to sit down with greenhouse theory.

He basically didn't get up for twelve months. Fourier had tapped the atmosphere as the lid for Earth's warmth; John Tyndall had tagged water vapor and carbon dioxide as the warmers. It was simple grinding math no one had ever thought to do before. If you raised and lowered carbon dioxide, what effect would that have on heat?

Arrhenius was putting in fourteen-hour days, just as Tesla and Westinghouse's Niagara Falls generator was being smoothed and adjusted for its 1896 debut. That summer, Sofia filled her suitcases, left him. She was five months pregnant, and spent the rest of the year sending Arrhenius letters, cataloging every benefit of a Svante-free life. In December, Arrhenius finally stood up. He had the blinking quality of someone who has lost their weekend to Netflix. He told friends he hadn't crammed as hard for anything since college; they'd been "the most tedious" of his life's calculations. Arrhenius looked back in grudging wonder—it was "unbelievable so trifling a matter has cost me a full year." He published in a London journal the following spring.

Arrhenius' first question is so familiar it's like sitting down at the opera, then realizing you already know the score from a Bugs Bunny cartoon. "Is the mean temperature on the ground in any way influenced by heat-absorbing gases in the air?"

It would have surprised Arrhenius to know that this sentence would become one of the next century's most dependable argument-starters—that his basic question would light up books, editorials, magazines, documentaries, research panels, protests, political campaigns. That it would power bumper stickers, fuel international spats and cable-news alley fights—a jumpy scientist at one end, a suave lobbyist cracking his knuckles at the other.

It was just science. Halve carbon dioxide, and temperatures would needle down toward the ice age. Double it, and they'd rise between five and eight degrees. Arrhenius relied on the old schoolroom tackle, pencil and paper—compared with computers, this is like heading into LA traffic at the wheel of a homemade car. But his doubling estimate held. It's about the same as a United Nations scientific panel came up with in 2013.

Not that Arrhenius was troubled. He was an optimist, and it was a big atmosphere. A carbon dioxide rise would come only from coal burning—from "the industrial development of our time." If "modern industry" stuck with its 1896 rate, doubling wouldn't arrive for thousands of years.

And Arrhenius was a Northerner. On the whole, a balmier Sweden sounded good to him.

Arrhenius pocketed his Nobel, remarried. Gingerly—he chose the sister of a friend. In 1908, he put his carbon dioxide research into a book, *Worlds in the Making*.

Another milestone: the first uptick in global warming numbers. (Arrhenius' book is probably the source of our phrase *the greenhouse effect*. The term Arrhenius actually used was *hot-house*—which brings to mind Tennessee Williams, sensibilities, and fainting spells. It wasn't his preferred metaphor. Tyndall, an outdoorsman, pictured the atmosphere as rugged and necessary, a blanket. Arrhenius was an urbane Swede. He saw carbon dioxide as wardrobe: he called it an overcoat.) The 1890 world had burned through about 500 million annual tons of coal. That number was "rapidly increasing"—in 1904, to 890 million.

So a carbon dioxide rise to "a noticeable degree," Arrhenius calculated, would now come "within a few centuries." Still, not a problem: the oceans would take care of it. By Arrhenius' reckoning, seawater removed about five-sixths of all carbon dioxide, which would powder the surface, settle, roll its way to the depths. (The scientific term for oceans is a "sink," a giant disposal. Forests, which filter air on land, are sinks, too.)

The 1908 world already had the look of a fuel hog. And this is how Arrhenius ended the chapter. People were complaining that "the coal stored up in the earth is wasted by the present generation without any thought of the future." In a sort of reverse-Tyndall, Arrhenius located the silver lining. "We may find a kind of consolation in the consideration that here, as in every other case, there is good mixed with the evil," he wrote. Through "the increasing percentage of carbon dioxide in the atmosphere, we may hope to enjoy ages with more equable and better climates . . . for the benefit of rapidly propagating mankind."

Nobody paid much attention. Part of the blame rests with Arrhenius. He concluded *Worlds in the Making* with another theory dear to his heart: that interplanetary spores, tipped with DNA, had wandered the galaxies, crashed our atmosphere, seeded the planet. (This would become the basis for another modern anxiety: *Invasion of the Body Snatchers*, with its many reflections and remakes. You never know what your work will fertilize.) He called this idea "panspermia." Reviewers ignored the climate material. But *Scientific American* called the spore notion "ultra-modern in every respect"; it would excite "the most interest." A *New York Times* editorial called the idea "dazzling."

Arrhenius became director of the Nobel Institute. He began turning up in odd venues for a scientist. On the "Literary Gossip" page of the *Los Angeles Times*. As a symbol of the life of the mind in the Sinclair Lewis novel *Arrowsmith*. By the 1920s, the *New York Times* could describe Arrhenius as "the best known and probably the most distinguished chemist alive."

The First World War seems to have lowered Arrhenius' emotional thermostat. He stared at fossil fuels and couldn't find the silver lining. There'd been only about two hundred thousand cars in the America of 1908. Two decades later, that number had climbed a bar graph straight up, to twenty-two million. And every year, more factories, power plants, coal.

So Arrhenius sat down to another round of tedious calculation. The world's gasoline supply was likely to dry up around the twenty-first century; England's coal would go a century after that.

Arrhenius wrote in his last book, 1925's *Chemistry in Modern Life*, that he meant to scare readers. Dwindling energy resources would be the cause of the future's wars. With the "States which lack" throwing "lustful glances at their neighbors who happen to have more than they use." Nations grabbing oil from "lands on the other side of the seas." (That's how Arrhenius understood the First World War: as a mining expedition in uniform.) One solution would be to dilute the power of individual companies and nations. "It is clear that some day we must come to forbid national egotism [and] the profit-seeking industries from seeking a solution to the problem of the proper use of raw materials," Arrhenius wrote.

The other answer was conservation—the first lab commandment, he pointed out, was "Thou shalt not waste!" But Arrhenius didn't expect much. Conservation could only be organized by politicians; and "statesmen are only in exceptional cases interested in nature and science."

It'd taken Arrhenius just three decades to chart the century's entire energy course, from green optimism to veteran gloom. He died two years later. He's famous now only because of the greenhouse theory. (Though there's also a big lunar crater named after him: planetary graffiti.) His 1927 *Times* obit doesn't even mention the hothouse—though in 1938, it was still calling his panspermia idea "ingenious." In the early fifties, the proto-ecologist Loren Eiseley wrote of gazing "into the star-sprinkled evenings." Where he would remember "that almost forgotten theory of Arrhenius, that the spores of life came originally from outer space."

IF LOREN EISELEY HAD brought a telescope, and gone on peering all the way to Venus and Mars, he'd have spotted the other theory. Decades later, after landings and readings, the two planets would turn out to be proof of the calculations Arrhenius had handwritten in 1895.

Mars has only about one-one-hundredth of our atmosphere. Not enough to generate a good greenhouse; it's freezing. The placid red-desert look is a misleading travel brochure, the palms and veranda on the hotel website. Temperatures are subzero, eighty below.

And Venus, with an atmosphere sixty-six times as dense as ours, is about 97 percent carbon dioxide; it's 850 degrees. Lead melts on Venus, oceans puddle. When the Soviets sent heavy probes—bruisers that pushed through our atmosphere, then long-hauled the 180 million miles—they survived Venus for about an hour. Venus is worst case, all feedback, what climate scientists call a runaway greenhouse.

The planet can have a great, lip-unbuttoning effect on headline writers. *Washington Post*, 1967: "Venus Called a Boiling Hell-Hole with Weird Psychedelic Imagery." (Though it's hard to beat Associated Press, 1961: "Planet Venus Called Stifling, Choking Hades.") Every night they rise up, reminders of what Arrhenius, Tyndall, and Fourier estimated. Not enough greenhouse, iron frost; too much, fire. Astronomers compare the three planets with a nursery-room wink: Venus, Mars, Earth—hot, cold, just right. They call it the Goldilocks Problem. Up they go, every night, anti-role models: the kids who went wrong in the class, boiling and shivering.

THE SODA MACHINE

WE TEND TO THINK OF CLIMATE CHANGE AS MODERN, AN entry in our sophisticated ledger of worries and regret. We tend to think heating started in the eighties. That's when the greenhouse effect had its IPO on the market of anxieties, splashing itself across *Time* covers and the nightly news. That's when we rolled over in bed and thought: oh no, another problem—let's hope the fools got this one wrong.

The carbon dioxide accumulation actually began earlier: around the time Arrhenius first decided to celebrate Christmas at his desk. (For an idea of how many generations and fashions have come and gone since: F. Scott Fitzgerald was born five months after Arrhenius published. He had time for a glamorous career at Princeton, to write, get famous, get drunk, be forgotten, die in Hollywood, and become famous again, all before anybody started having anxious conversations about Svante Arrhenius.) The carbon dioxide from the wood and coal people burned through the 1800s was bite-sized. It matched the carbon that could be socked away by oceans and trees. People began to notice a warming in the twenties; they just didn't have a name for it. (Decades later, the NASA scientist Jim Hansen—whose midnight nickname was "the Paul Revere of Global Warming"—would fix the heating's start at 1880. He received an immediate reward: the Department of Energy under Ronald Reagan cut his funding.) Nineteen twenty-one was history's then-warmest year for cities as climatically distant as Portland, Oregon, and Washington, DC; 1927 the hottest for Cape Town, South Africa.

In 1975, the *Times* would report that, worldwide, the decades between 1900 and 1950 had apparently been Earth's warmest in five thousand years. The first "unprecedented heat" articles ran in the twenties. By the thirties, it had become a head scratcher. At a 1932 meeting of the American Meteorological Society, a scientist with the Weather Bureau issued a novel warning: just by raising temperatures "a few degrees," polar ice would melt, cities flood. That melt wasn't on the schedule, though; it wouldn't be coming for "a few million years."

The next year, warmth made the front page of the *New York Times*. There it was, above the fold, flanked by an article about a German claim that bad behavior was behind them ("Nazis End Attacks on Jews in Reich"), and another about the drive to put speakeasies out of business. "America in Longest Warm Spell Since 1776; Temperature Line Records a 25-Year Rise." The Weather Bureau was said to be "closely" watching a "red line that has been rising since 1908."

Meteorologists wondered whether that line had "passed the peak." But 1934 was the worst dry spell in American history—61 percent of counties in drought—with 1936 close on its heels. (The *Washington Post* ran a long story on the birth of the air conditioner. It turns out the phrase "It's not the heat, it's the humidity" was already such a groaner that you could make a joke by leaving off the last three words.) That red line nosed upward. By 1939, the *Chicago Tribune* could run its own "Experts Puzzle" story. Twenty warmer years, worldwide. "The reason? No one seems to know." A team of MIT scientists had been on the hunt, but had come back with no theories to be proved or disproven. ("It's a mystery. It will be some time," a University of Chicago meteorologist predicted, "before anyone gets a workable theory on this.") At a three-day symposium in New York, J. C. Kincer of the Weather Bureau told the *Times* the trend was so "worldwide in scope as to suggest that the orthodox conception of the stability of climate" would need a revision.

Soviet meteorologists, along with California's Scripps Institution of Oceanography, were already reporting warming in the Arctic. As *Time* said, "The reason for such climate changes is obscure."

In these years, there were occasional articles about carbon dioxide: column fillers, novelty items, carbon dioxide as goof. The 1937 *Los Angeles*

Times reported that hog butchers had adopted the gas for killing swine, with noticeable improvements to lard and meat. The *Washington Post*, in a 1938 "Milestones of Science" squib, let readers in on a discovery. Since tail-pipes dispensed carbon dioxide, every car was also a " 'soda water' plant": "two of the exhaust products, water and carbon dioxide, if cooled and mixed, would make soda."

G. S. Callendar understood something else. Callendar is the climate science Tesla; underdog and overlooked, the man who pulls the load but gets excluded from the banquet. In photos, he looks both optimistic and defeated; he wears the half-smile of a man watching a party through the windows.

He walked into climate science as a weekender, in the British tradition of unlikely hobbies. But he also came at it from the trade: he was a steam engineer with the electric industry, a coal man; Callendar saw just how much fuel was going up the stacks, at the same time that he felt the country heating up around him. So he traveled to old county offices, slapped open dusty books, chased figures, made estimates.

Callendar compiled the first how-much list: how much smoke, how much temperature rise. In 1938, he got invited to play the big room. He delivered his paper "The Artificial Production of Carbon Dioxide and Its Influence on Climate" at the Royal Meteorological Society. By his reckoning, in the decades after 1890, mankind had spilled 150 billion tons of new carbon dioxide into the atmosphere. The rate was extraordinary. People, Callendar wrote, were "throwing some 9,000 tons of carbon dioxide into the air every minute." (That rate is 2.57 million pounds per second today.)

This wasn't a question of changing climate in some future. It was a question of changing the climate now. Few people, he allowed, "would be prepared to admit that the activities of man could have any influence on activities of so vast a scale." However, "such influence is not only possible, but is actually occurring at the present time."

Callendar probably read with a colony of butterflies in his stomach. Response on-site, according to historians, was "condescending." Callendar wasn't even a scientist; the credentialed saw the twin rises as coincidence. Another historian cooed over his failure: "Poor Callendar." Callendar's work—you can see from his photo he'd half-expected this—"excited widespread indifference." He'd been smiled out of court.

The reason was a kind of expert respect for the world. The planet was seen as a perfectly balanced machine—a car that could repair itself. Those who remembered Arrhenius agreed with his fail-safe. The oceans would take care of it. The American biologist Alfred Lotka, in an influential textbook, had pegged the amount of carbon dioxide filtered by the oceans at 95 percent. The theory cooled. As Spencer Weart, official historian at the American Institute of Physics, later wrote, "The idea that humans were influencing climate by emitting carbon dioxide sat on the shelf." Mothballed with "the other bric-a-brac, a theory more peculiar and unattractive than most."

Maybe stepping onto the field as a part-timer—the way Edison did with light—helped Callendar read the old plays in a new way. The following year, Callendar published again: Temperature's "marked increase commenced from about the time that carbon dioxide production became rapid." He wrote, "Man is now changing the atmosphere at a rate which must be very exceptional on the geological time scale." The years between 1934 and 1938 had been the warmest in nearly two centuries.

But by then there was a war on. Callendar was in the employ of the British government, testing armaments, aiding the effort. (When a 1942 book drew the connection between carbon dioxide and heat, a reviewer gruffly cautioned against taking on "an extra bogey. . . . Personally, I'll begin to worry about that after the Japanese and the Germans are out of the way.") Still, Callendar kept the science alive; in the long relay, when the theory was leaking stamina, he picked up and carried the baton. In the 1950s, Roger Revelle read the British hobbyist. In his honor, Revelle called the heating—one more name to be folded away with the blanket and the overcoat—the Callendar Effect. "Callendar," a British climate expert told the BBC five decades later, "was the first to discover that the planet had warmed."

IN SCIENCE CIRCLES, the term for evidence is one you probably already know: signal. The classic experimenter's ratio—signal to noise. You might suspect your brakes need an overhaul. That's what you're listening for. But as you turn out of the driveway, you also face a great deal of background info: the slither of tires on cement, the steering waggle, general stop-and-go. That nonspecific data is noise. The singed-iron smell, the persistent metal-on-metal scrape, the crunchy traffic stop—that's signal. ("Real data is messy," explains a mathematician in the Tom Stoppard play *Arcadia*. Like a piano in the next room: "It's playing your song, but unfortunately it's out of whack, some of the strings are missing, and the pianist is stone deaf and drunk—I mean, the *noise*!") Nearly all global warming arguments, after about 1980, turn on how much signal scientists are willing to agree we've received.

In the forties, signal seemed to be piling up. *Time* and the *New York Times* were running old-timer quotes: Decembers had become easier, summers sweatier. In the northern cities, subzero days were occurring at half their nineteenth-century rate. In the waters around the island of Spitsbergen—the setting for Hans Christian Andersen's "The Snow Queen" and Disney's *Frozen* movies—temperatures had warmed 18 degrees in four decades. Lake Victoria, the world's largest tropical body, had sunk seven inches. Other African lakes had melted away.

Warm-weather animals—cardinals, possum—that never felt much urge to venture above New Jersey were staking territories farther north. (In our own era, the *Birds of North America* guide would write about the declining novelty value of the cardinal. This "common bird is a winter fixture at snow-covered feeders throughout the northeast, but it only spread to New York and New England in the mid-twentieth century.") Birds of southern Europe were nesting in Scandinavia, Iceland, Greenland. The cod was a fish the Inuit had never much encountered before 1920. By midcentury, it had become a staple. Schools were swimming farther north than ever; one observer wrote it wasn't necessary to bait the hooks. Whales, seal, and walrus had deserted traditional Inuit hunting grounds for cooler waters. A McGill University zoologist had been studying Inuit response to warming temperature for fifteen years. It turned out Inuit children disliked cod. They missed whale blubber, and sulked when their fathers came home with more fish. The zoologist told the *Chicago Tribune*, "The Eskimo child's chunk of blubber corresponds to the bar of candy of the city youngster."

And there was the ice. Throughout climate-change history, ice has been the fingerprint—what scientists call a "sensitive indicator of climate." Ice is fragile. And it's also tremendously important. Which makes for a strange reversal of priorities. Antarctica and Greenland—the shy landmasses, the ones that didn't much feature in the lesson plan—have become the stars of the class.

It has a lot to do with feedback. Another way of looking at feedback is as self-reinforcing change, change that produces more of the same. If you suddenly decide to eat more, your belly gets larger. This leaves you more belly to fill, which makes you want to put more stuff in. If the weight gain makes you depressed, you'll hang around the house, fridge nearby, which . . . etc. Every element on Earth has a color temperature called an *albedo*. (It comes from the same Latin root as *albino*. It means whiteness.) Albedo is the degree to which an object will retain or reflect sunlight. A white T-shirt has an albedo around 90 percent. That's why you want to wear one in August; it bounces the light back off. A black T-shirt has an albedo around zero; bad summer idea, light and heat stay put. (I finally got this when I was outside in the sun petting my dog; her black fur was warm, the white parts cool. Magic. If no pets are available, place your hand on a dark and light car.) Snow and glaciers are white. Millions of acres of ice are one way the planet stays cool. But as ice melts, it slushes away to water and soil, which absorb more heat—which melts more ice, which . . . etc. Once the process really gets going, it's nearly impossible to stop.

Scientists have calculated the amount of sunlight that shines down on every square meter: 240 watts. (As the *New Yorker* climate writer Elizabeth Kolbert explains, "roughly the energy of four household light bulbs." Edison would have liked that.) That's all the heat we get—a figure that's at once comfortingly measurable and terrifyingly small. Ice and albedo are a big determinant for how much stays in the system.

In 1947, Dr. Hans Ahlmann, director of the Swedish Geographical Institute, reported on his own two decades of research to a conference at the University of California. Here's the *New York Times* lead: "A mysterious warming of the climate is slowly manifesting itself in the Arctic, engendering a 'serious international problem.'" (Writers pick up odd frequencies.

When Humbert Humbert joins a mid-century science expedition in *Lolita* to rid himself of disastrous cravings, it's to study Arctic climate change.)

Glaciers were shrinking; the waters around Russia, once navigable for only a season, could now be sailed eight months of the year. More Soviet ships trawling European waters! This was a geopolitical problem. Air temperatures in the Arctic had warmed ten degrees—from a scientific standpoint, Ahlmann explained, this change was "enormous." (In Greenland, melting ice was disgorging relics, kicking up whole ancient farmhouses; in Finland, farmers plowed fields that had been ice for centuries.) Ahlmann drew a connection for the *Times* that's become a familiar chill, one of our era's campfire stories: melting ice caps, rising sea levels, "catastrophic proportions."

For Ahlmann, a threshold had been crossed—where a question floods the lab, swells into politics. He told the *Times* the changes were "so serious that I hope an international agency can be formed to study conditions on a global basis. That is most urgent."

That word, "Urgent," would become a sort of slogan in government climate studies, and news reports about them, after the 1970s. If the reports came with gift bags, and the gift bags contained ball caps, those caps would have been stamped with the word "Urgent."

THE TIRE PRINTS
AND THE SMOKE MENACE

BY THE TWENTIETH CENTURY'S HALFWAY POINT, A NEW generation of scientists was circulating to the head of the classrooms and institutes. For them, the shifting weather had an explanation. It was nearly a hundred years old; you could read about it in Tyndall, Arrhenius, Callendar. The answer was in the smokestacks they passed on their morning commute, in the bright cities where they ate their dinners. If their area codes were New York or Chicago, it was in the smudgeprints on their windowsills.

To someone from our time, the fifties are like stepping into a scrambled house, where furniture makes you trip. Some hallways you recognize—the Russia problem, the Israel problem, the climate problem. Then you round the corner and where you expect a window there's a bricked-over wall. Newspaper stories like "The Negro in the Navy" and "Liquor Industry's Ban on Women May End." (You couldn't use female models to sell whiskey or vodka. The great era of the poolside ad still reclined in the future.) Another big difference was the air: it was incredibly grimy.

Our pollution experience runs closer to scientists before mid-century— one reason some people would find global warming hard to accept. Our cities are livable, breathable. In a Hollywood story from the 1928 *Saturday Evening Post,* F. Scott Fitzgerald could write about a movie star stepping into his car, driving into "the everlasting hazeless sunlight" of Los Angeles. Two decades later, that sentence would have been a dream or a punchline. (When Raymond Chandler does the city it's all smog, blurring and stinging.) One Los Angeles day in seven was a smog alert; a pavement walk felt like swallowing a gallon of salt water. In Chicago, there were days when smoke blocked out 40 percent of sunlight; St. Louis had to flip on the nine a.m. streetlamps to help drivers pick their way through the murk; in New York, the loss of sunlight to smoke could hit 42 percent. An August 1950 smog was so thick cars grew toupees of soot and wheels left tire prints in the grit on Madison Avenue.

This was the air the new scientists breathed, nine thousand quarts a day. (Sylvia Plath on the summer of 1953: "The dry cindery dust blew into my eyes and down my throat.") United Nations Headquarters opened in midtown Manhattan to an immediate problem: not security, stalemate, or the parking nightmare. Smoke from a coal-burning Con Edison power plant. It seeped through the air-conditioning, and outside was staining one side of the white tower brown—the present puffing smoke on the internationalist future. A 1951 study in *Harper's* put costs of "the smoke menace" at $1.5 billion. If you were a scientist who'd ever set foot in a city, you knew just what smoke was capable of. There was also Donora.

Donora is a town in the Pennsylvania steel belt; backstop, high school, factory, and you're on to the next town. Most residents worked in the wire and steel mills across the river. On Wednesday morning, October 27, 1948, people opened their doors to find long smeary ribbons of carbon hanging in the air. It was an eerie undersea sight that made getting to work difficult. Kids used flashlights. Whole corners had been erased. The factory stacks kept running, smoke limping up a few feet, then rolling down into the streets. The soot had a penetratingly sweet flavor, and could be tasted by air travelers a mile up.

Friday was the Halloween parade—costumes and floats, a big night for the kids. Saturday morning, Donora's high school football team suited up and played in the queer light before a packed stadium. Saturday night people started dying. Nearly a third of the population—four thousand people—became sick. Four hundred were desperately sick. Twenty died, before a Sunday rainstorm crumpled away the smoke. By then, doctors had run out of medicine, firemen were pitching oxygen tents from blankets, and anyone able had fled the town. No rain, a doctor with the medical board estimated, and another thousand would have been lost. "It was plain murder," the doctor said. "There's nothing else you could call it." (A few days later, US Steel declared itself "certain" there'd been no connection between its factories and deaths.) An inversion front—cold air pressing warm air down—had sealed in the smoke at ground level. International headlines. Donora was, to the time, America's single worst pollution event.

If smoke could do that, it was no longer such a stretch to imagine what feats it might be pulling off in the upper atmosphere. In 1950, for the first time, Callendar's name and idea appeared in American newspapers.

This wasn't an amateur reading a freak notion into the record, going wet-lipped before a panel of experts. It was the head of the US Weather Bureau's extended forecast section explaining to the Associated Press about "the hot-house theory" and how carbon dioxide could be making air warmer. (The East Coast was then in the grip of an "amazing" two-year heat spell that had brought the spring frogs out of hibernation in January.) Gilbert Plass looked just the way you imagine a scientist: wonky glasses, way too much forehead. He was a Harvard-and-Princeton-trained theoretical physicist at Johns Hopkins, where he had access to swifter equipment than Svante Arrhenius and John Tyndall could have dreamed of. There could now be no question that carbon dioxide held in heat.

In 1953, Plass was standing and delivering a paper before the American Geophysical Union. And there was that word, one of the first American times: *greenhouse*. There were also the numbers. Carbon dioxide was warming the world by 1.5 degrees a century; six billion new tons of CO_2 a year. If anything, previous research had underestimated the climate-changing effects.

Plass' remarks got picked up by newspapers around the country; there it was, in the 1953 *Washington Post*, a sentence now familiar to any news fan. "World industry, pouring its exhausts into the air, may be making the earth's climate warmer." It was in the *New York Times*. And it was right up top, one of your first sights in the *Christian Science Monito*r, a quote from Plass: "The carbon dioxide acts in the air in the same manner as the glass in a greenhouse." It had taken Arrhenius' theory sixty years to make the front page.

THE GEOPHYSICAL EXPERIMENT

IT HELPED THAT THE WEATHER HAD TURNED SO *STRANGE.*
Between 1950 and 1953, the *New York Times* ran more than half a dozen
articles about climate change. The headlines read as if the paper is conducting an argument with itself: mumbling, retracting, restating. "Is Climate Changing?" then "Is the Climate Changing?" Answered definitively
by "No, the Weather Isn't Changing," then "Our Changing Climate," then
"Oldtimers May Be Right when They Tell Us that the Climate Is Getting
Warmer," then "How Industry May Change Climate." And, at last, "The
Weather Is Really Changing."

Weather stories were everywhere. The weather had, unfortunately,
turned incredibly interesting.

In 1954 and 1955, three separate hurricanes (period names: Carol, Edna,
Diane) smashed into New England. Two hundred sixty-four deaths, $1.3
billion in damage, water ten feet deep on the streets of Providence, Rhode
Island. Atlantic hurricane frequency—warm air is hurricane food—was up
50 percent over two decades. Tornadoes are fed by warm surface air. In 1916,
the Weather Bureau reported around a hundred tornadoes. The *Times* was
calling 1953 "a year for tornadoes" with a record-breaking 266 by summer.
Four years later, there were 230 in May alone. (Sixty Mays later, the continent would manage three hundred tornadoes in a whirlwind two weeks.)
It wasn't just America. In Libya, the desert summer dripped with rain; a
record heat wave struck Czechoslovakia in February; there was snow on
the Portuguese fields in May.

In 1956, *Life* magazine—the absolute compass point of American middle opinion—ran its own climate piece. "Our New Weather: Scientists believe more hurricanes, more tornadoes, higher temperatures, and unseasonal storms are part of a long-term change in world climate."

The weather, *Life* explained, had "just plain gone nuts." The magazine weighed cynic opinion: nothing had changed, childhood snowbanks only seemed taller because you were small. "Who's right?" *Life* asked. "Oddly enough, not the wise cynics. . . . The weather is *not* the same."

All these articles included the greenhouse material, and the word had grown a tail: it was now the greenhouse *effect*. Finally, the first Washington testimony, in the middle of March 1956. Cherry blossoms, spring breeze, open windows, reporters' pencils. There was Dr. Roger Revelle, the head of Scripps Oceanographic—a tanned six-feet-four, foghorn voice—warning the ties and microphones of the House Subcommittee on Appropriations. From the beginning of its second life, climate-change theory was public: "I would estimate that by the year 2010 we will have added something like 70 percent of the present atmospheric carbon dioxide to the atmosphere. This is an enormous quantity," Revelle said. "It may, in fact, cause a remarkable change in climate."

And with that there was also an immediate grasp of the problem, the ruts and thorns on the path between a solution and here. Gilbert Plass had done his Hopkins research under auspices of the military. It gave his work an establishment résumé, the college sticker for the rear window. In a new paper, Plass yardsticked the problem. If by the century's end, "measurements show that the carbon dioxide of the atmosphere has risen appreciably, and at the same time the temperature has continued to rise throughout the world," you'd have firm proof, signal from noise.

That October, the *Times* did a follow-up on Plass. As of 1956, the issue was so thoroughly known the paper could write of "the familiar greenhouse analogy." The *Times* nodded at worst-cases—glaciers gone, "striking changes," an Arctic thawed to "tropical deserts and jungles, with tigers roaming around and gaudy parrots squawking in the trees." Picturing solutions was tougher. The paper gave the situation a worldly headshake. "The introduction of nuclear energy will not make much difference. Coal and oil are still plentiful and cheap in many parts of the world, and there is every reason to believe that both will be consumed by industry so long as it pays to do so."

BY THE EIGHTIES and nineties, climate change would suffer a sort of discredit-by-association: the new student introduced by uncool friends. It would seem an idea hatched by the environmentalists: a creature rising from the bog of slogans and vegans and a political grudge against technology.

Actually, the environmentalists took it from the scientists. It came from the establishment, the committees, budgets and uniforms. John Tyndall had been the preeminent British scientist of his time. Svante Arrhenius won a Nobel and directed the Nobel Institute. The scientists of the seventies and eighties would be funded by the government, which had every reason to wish the problem gone. (James Hansen, muffled by administrations from Reagan to the second Bush, spent nearly every work week drawing a government paycheck.) Gilbert Plass and Roger Revelle were funded by the navy. That's another of the ironies. Climate change isn't an eighties idea, or even a sixties idea; it came from the race against the Soviets. The Cold War rediscovered global warming.

And it wasn't a crisis. In 1956, scientists still had faith in the fallback, the big sink Arrhenius relied on. The ocean. So long as the oceans were a giant blue rug the carbon got swept under, the gas was just trivia, the remainder in an equation. Lots of scientists would've agreed with what conservative Congressman Dana Rohrabacher said half a century later: that to imagine something as big as climate being pushed around by something as small as people was "grandiose." (The congressman had his own theory for the source of warming: dinosaur farts.)

Roger Revelle, in middle age, looked like William Holden playing an oceangoing scientist—pouchy eyes, square jaw. Later in life, after his hair went snowy, he'd resemble Tony Bennett.

Revelle is one of those people who turn up at crossroads. In 1946, the navy tasked him with sampling the waters around Bikini Atoll, where the first H-bombs would be test-fired. (The name of the swimsuit comes from the island. A detonation on a beach towel.) Revelle was thus present at the birth of two modern anxieties: the nuclear age and climate change. In the sixties, he'd teach at Harvard, where his students would include the son of a senator, Albert Gore, who'd do a lot of thinking-over about stuff he heard in class. Revelle would serve in the Kennedy White House, chair study groups for Presidents Johnson and Carter. In 1990, George Bush awarded Revelle the National Medal of Science. When he died a year later, believers and disbelievers would struggle over his legacy, like two sects tugging at the remains of a saint.

The navy began to flirt with oceanography in the first days of World War II. Sonar didn't function properly in the tropics. Conditions gave it fits: it missed pings, heard ghosts. Someone thought to call in the boys from Woods Hole, who discovered that warmer water would conceal pings—so the oceanographers designed a special instrument, American subs learned to hide in thermal dips and valleys, saving sailors, torpedoing Germans.

After that, it was love. Oceanographers predicted the tides at Normandy. They designed a barnacle-busting paint, which saved the navy millions on fuel. During the war, they trained a generation of rush-job navy oceanographers, sending them out to spread the gospel. With the German and Japanese fleets dry-docked or sunk, the armed forces kept pouring millions into research. Could you check for Russian nuclear tests? How did clouds affect nose cones? Oceanography had become one of the world's fastest-growing sciences; and the military had become the whale, the big spender who wanders onto the casino floor, tossing money in every research direction.

Revelle had served as a commander during the war, directed oceanography for the Bureau of Ships, helped found the Office of Naval Research, then headed back to Scripps Oceanographic. So he was a natural, in the fifties, when the navy needed a scientist to check whether nuclear tests were poisoning fish. Suddenly, no diner in Japan wanted anything to do with a catch from the Pacific.

Revelle tested fallout. He checked whether the ocean bottom was a safe place to store nuclear waste. He'd also read his Plass and Callendar. (The names, in this second global warming phase, have a clunky, broad-shoulder symbolism you wouldn't accept in a book. *Plass* is Norwegian for public square; there's *Callendar*, to remind you of the days; *Revelle*, with an "I," becomes a morning bugle. Spooky.) With navy funding, he decided to check the fallback position: whether the waters really were the planet's giant air filter. Whether the oceans had room, along with fish, coral, and undersea bomb tests, for all that carbon dioxide.

It turned out they did. Revelle and a colleague wrote up their results for publication.

Then Revelle remembered what he'd learned wading the lagoons at Bikini. How touchy seawater was, how it greeted any change quickly and allergically. He had one of those moments where you reverse field, and everything turns simple. He'd been asking the wrong question. It wasn't, Would the oceans absorb carbon dioxide? The question was, Would they hold it?

It turned out they didn't. New carbon dioxide would settle on the surface—and the oceans would evaporate it right back out again. I imagine Revelle on a research ketch in the middle of the Pacific. It's night; water lapping, desk swaying. It turned out the oceans would hold only about 10 percent of what earlier scientists had anticipated. I imagine his hands going cold, his footsteps stamping back down the deck, fingers scrabbling for the paper. Revelle was so excited he wrote the new data on a separate sheet, then scotch-taped it into the manuscript: as hot-off-the-presses as science gets.

The rest of that carbon dioxide would have to be going somewhere. On a fresh sheet, Revelle wrote what would become the most famous sentence in warming science.

"Human beings are now carrying out a large-scale geophysical experiment of a kind that could not have happened in the past nor be reproduced in the future," Revelle said. He added, "Within a few centuries, we are returning to the atmosphere and oceans the concentrated organic carbon stored in sedimentary rocks over hundreds of millions of years." He published in 1957.

The science journalist Bill McKibben, who'd write the first big bestseller on global warming three decades later—with an eye-popper title, *The End of Nature*—has a nice gloss on this. It had taken the planet ages to bank all that fossil fuel. Now, "it was as if someone had scrimped and saved his entire life and then spent everything on one fantastic week's debauch."

McKibben heard something chilly in Revelle's words, a "morbid dispassion." But Revelle had found a gigantic new thing, and for those first moments he had the mountaineer's pleasure of being the one person at a new view, the only figure in a landscape.

PHYSICIST AND HISTORIAN Spencer Weart would write that, in fact, Revelle had made a "gross underestimate" of the situation. Because in 2001, Weart knew what Revelle didn't: just how quickly fossil fuel use would grow. "Between the start of the twentieth century and its end the world's population would quadruple, and the use of energy by an average person would quadruple." This meant, by 2000, "a sixteen-fold increase in the emission of CO_2."

The future always knows more. Because it's the future. There's every chance that any position now taken on climate will look provisional, crazily wrong-headed, from the high ground of the future. The obvious fact missed, the correct path stubbornly refused. In 1951, one of its first "Is the Climate Changing?" pieces, the *New York Times* editorial board had advised readers to discount the "scientific romancers"—the old H. G. Wells term for science-fiction writers—"who attribute climatic changes to vagaries of ocean circulation, to variations in the carbon dioxide content of the atmosphere, to changes in the eccentricity of the earth's orbit."

Each of these statements has since been proved more or less true. The Serbian mathematician Milutin Milankovitch worked out the orbital idea: the one-hundred-thousand-year swings—ice age to warm—are known in his honor as Milankovitch cycles. The paper writes about them all the time.

The Columbia University geophysicist Wallace Broecker demonstrated the profound effect ocean currents have on climate. Broecker has received just about every honor a scientist is eligible for—including the National Medal of Science and a long, admiring *Times* profile, which ran in 1998. (It included a Broecker remark that's become famous, and gets a lot of grim nods in weather circles: "The climate system is an angry beast, and we are poking it with sticks.")

In 2005, when an actual science-fiction romancer, Michael Crichton, wrote a novel suggesting climate change wasn't true, the *Times* called his argument "laughably rigged" and "ludicrous." *Time* criticized Crichton as "a science-fiction fabulist"—a complete revolution, a sort of cultural Milankovitch cycle, in just fifty years.

Another problem for the public-relations wing of global warming was amnesia. Each time climate reappeared as an issue, it seemed all the steps had to be taken again: a language forgotten in a car accident, then relearned from scratch. The whole process, from debatable headlines ("Is the Climate Changing?") to worried acceptance. In 1956, one of Revelle's colleagues told reporters we'd have the answer—how bad would warming be?—in thirty years; the picture would clear up by the mid-eighties. In 1979, the chairman of an Energy Department study on carbon dioxide—who'd go for broke and call emissions "potentially the most important environmental issue facing mankind"—would tell the *Times* that "some effects should be felt by 1990."

But when, arriving right on time at the station, a scientist named James Hansen announced to Congress, in a hot hearing room during the drought summer of 1988, that the warming had begun, the response was universal surprise. It wasn't, *Yes, of course, right on schedule.* It was, *So soon?*

Every political rule was stacked against warming. The problem took ages to accumulate. Small changes piling up: exhaust after exhaust, molecule by molecule, like counting flakes in a snowfall. There was always something more immediate to get fixed. And fixing would mean either brute overhaul—or the third-act appearance of a stunning gadget. It meant projecting beyond the next decade, planning the election after the next three; like betting on poker hands a hundred deals in the future. And people don't like thinking long-term, since the longest term has an expiring look to it. That's a future Roger Revelle also failed to see. As an economist at MIT could explain in 1995, "If you said, 'Let's design a problem that human institutions can't deal with,' you couldn't find one better than global warming."

THE OVERWHELMING DESIRE

SCIENCE CLIMBS LIKE A VIDEO GAME. CLEARING ONE rack doesn't mean rest; it cuts the path for a whole new set of problems.

Two big answers had been graded and posted. Did carbon dioxide trap heat (yes); could you rely on the oceans (no). Revelle understood the next and most important question needed to be asked and answered. His estimate was about half of all new CO_2 was pooling in the atmosphere. If he was right, this would be measurable, and growing every year.

Down the road from Scripps, there was a Cal Tech postgraduate named Charles David Keeling. This name makes him sound like a newspaper magnate; in the Midwestern style, he preferred to be called "Dave." He used the word *darn*. A problem came up, he'd do a head shake, say *darn*.

Keeling's big ambition was to work outdoors. He had oversize ears, a crewcut, friendly eyes—in a suit, he looked like he'd been sent to body jail. He'd taken a bet from his department chairman, to measure carbon dioxide, that became a job. And as long as he was measuring, he could avoid labs, indoor lamps, mimeographs. Instead, wind, the sun on his hands. He was a geologist—but he was more like whatever the landlocked version of an oceanographer would be. He measured with a Zen frenzy. He chased down a deer in Yosemite after it stole his carbon dioxide notebook. He was up on the roof, measuring carbon dioxide, while his wife was in a hospital delivery room. "Keeling's a peculiar guy," Revelle told an interviewer years later. "His outstanding characteristic is that he has an overwhelming desire to measure carbon dioxide." He ordered $10,000 measuring kits, tinkering and perfecting. He lived in the fractions of fractions: he wanted the number nailed down to the millionth part, then the five-hundredth of a million.

He was flown to Washington, DC, put in a chair, quizzed about carbon dioxide by the head of meteorological research at the Weather Bureau. (Then, because he was in the neighborhood, he detoured south to the islands of Virginia, and measured more carbon dioxide.) He was flown to Scripps, inspected by Roger Revelle, hired. In 1958, Revelle got his hands on more government money, and sent Keeling to the top of a Hawaii volcano, Mauna Loa, where the air was purest and sweetest. The Weather Service had pitched some cement huts and stood up the antenna for an observatory.

Keeling opened his measuring kit. He was eleven thousand feet up, the middle of the Pacific, trade winds blowing. He measured. It was like getting the first sip of soda before the bottle gets passed round the room.

Years later, the Pulitzer Prize-winning science writer Jonathan Weiner asked if there'd been a Eureka moment. Weiner explained what he meant. The word *eureka* is associated with the Greek mathematician Archimedes, who'd been trying to solve a tricky joint physics-and-government problem. A locked-box puzzle: he was supposed to measure a king's crown for gold content, without doing any damage to the king's crown. The solution hit him in his bathtub. Archimedes was so excited he ran naked through the streets. Shouting "Eureka!"—"I have found it!" In the 1870s, an amateur archeologist had the notion that the biblical flood must have been an actual, storms-and-records event. He was one of those people with a fixity: it *had* to be true. If he kept searching the British Museum's stone tablets, he was bound to find some mention—the tablets were the ancient world's journalism, and such a flood would have made the papers. One afternoon, somebody handed him a new stone. He set it on the table, read about a great flood. And in his excitement, he began tearing off his clothes in the British Museum.

It took Keeling four years, and the clothes stayed on. G. S. Callendar had pinned the CO_2 content of the preindustrial atmosphere at 0.029 percent—that's the tiny number, a cinder in the eye. The official way to describe this figure is as 290 parts per million, or 290 ppm. In 1953, Keeling measured 310 ppm by the waters of Washington State and in the Arizona mountains. In 1958 and Hawaii, Keeling's first reading was 313 ppm. Next year, it was 314; then 315. (When "it was higher the fourth year," Keeling explained, "we knew something was going on.") He began a graph.

If Revelle's words are the most famous sentence in global warming, Keeling's graph—it's called the Keeling Curve—is its most famous visual. It's the set of chalk marks on the wall, where the children's heights are measured. Fifty years later, the Weather Channel would call the day Keeling unstoppered his kit on top of Mauna Loa history's Biggest Weather Moment. (It beat out the creation of the Environmental Protection Agency and the D-Day invasion.) The BBC would name the Keeling Curve one of its icons of science—a world-rattling measurement you could shelve with $E = mc^2$ and the DNA double helix.

In 2002, Keeling received the National Medal of Science from the second President Bush. The emotional air currents, the dips and rises, would have been interesting for a specialist like Keeling to measure. A leader who didn't believe in climate change delivering an award to the man who helped verify it. When Keeling died in 2005, his son Ralph—good, no-fuss name—was carrying on the family business, measuring carbon dioxide for Scripps Oceanographic. Dave Keeling died after a Montana hike: an outdoors sort of death. He'd been measuring carbon dioxide the day before. If you can call living out your objectives happiness, Keeling probably died happy. The carbon dioxide number is 421 ppm today.

BY 1957, if you were a newspaper reader, you could find Revelle's language in the columns.

There it was in the *Times*: "Scientists are preparing to study the effects of an experiment in which the world itself is the laboratory. It is being carried out inadvertently by mankind through the ever-increasing use of combustion to drive the wheels of industry and transport." It led a long *Christian Science Monitor* piece—"Every time you start a car, light a fire, or turn on a furnace you are joining the greatest weather 'experiment' man has ever launched"—and it made the *Washington Post*, where a Revelle colleague gave an icecaps heads-up and glumly observed, "We're changing the earth's atmosphere."

For the first time, the greenhouse appeared in a *Times* front-page story, a spot it would make lots of visits to in the upcoming six decades: "U.S. Is Urged to Seek Methods to Control the World's Weather." (If you were an "Urgency" fan, your word was in here, too.) A committee reporting to President Eisenhower warned that weather control would soon become "vital" to the national future; research would one day outrank the development of the bomb. The *Times* expressed the hope that, with government money, someone would come up with "a gas" to reduce the greenhouse effect.

And if you were a TV fan, warming had already made it onto the box. The first network science program came out of Baltimore—"The Johns Hopkins Science Review," a hit. In 1955, the Gilbert Plass material went out over the air: he was local talent, under contract to the studio. In 1958, Frank Capra—director of *Mr. Smith Goes to Washington* and *It's a Wonderful Life*—made a climate program, a sort of *Mr. Deeds Meets the Weather*. He'd been phoned by the president of AT&T, asked to write and produce science broadcasts. (Capra had a chemical engineering degree from Cal Tech.) The program—*The Unchained Goddess*—aired in 1958 on NBC.

SCIENTIST: Even now, man may be unwittingly changing the world's climate through the waste products of his civilization. Due to our release through factories and automobiles every year—[Anxious music under montage: traffic, factory, smoky harbor]—of more than six *billion* tons of carbon dioxide, which helps absorb heat from the sun, our atmosphere seems to be getting warmer.

BAFFLED EVERYMAN: And this is *bad*?

SCIENTIST: Well, it's been calculated a few degrees rise in the Earth's temperature would melt the polar ice caps. [Nervous violin, crumbling glacier] And if this happens . . .

Floods. Inland seas. The scientist drew a future where "tourists in glass-bottomed boats" took snapshots of the watery remains of Miami, a cityscape in an aquarium. You can still find *Unchained Goddess* clips on YouTube.

G. S. Callendar had warned in 1939 that if the theory was correct, you'd catch "a gradual increase in the mean temperature of the colder regions of the earth." By 1958, the Weather Bureau was reporting "dramatic change" in the Arctic; in 1959, warming at both poles. (When a Rhode Island-size chunk of West Antarctic ice cracked into the sea in 1995, a climatologist told *Newsweek* the first thing he did was cry. "It turns out we had it wrong," explained a polar expert. "The ice sheets are shrinking, and they're doing it almost a hundred years ahead of schedule.") In a long oceanography roundup, Roger Revelle explained to *Time*, "Man is moving and shaking the great globe in spite of himself. We may be disastrously changing the climate." In 1960, the new president of the American Association for the Advancement of Science proposed one of the first warming solutions. A forest offensive, millions of trees. Engines and factories exhaled carbon dioxide; trees breathed the same gas in, perfect match. You could plant your way out, groves of filtration. He had a formula, like a coach circling a power forward with extra defense. Ten trees per car, one hundred per bus, one thousand for every airplane. Otherwise, the scientist said, "We just get hotter and hotter."

By the early sixties, the *Times* could write of carbon dioxide warming not as a minority viewpoint, not as a worrywart fad or a surprise, but as a mainstream preoccupation—"one of the chief concerns of climatologists in recent years." When it wrote about *the greenhouse effect*, the paper left off the apologetic word *theory*. The story had reached the stage where it climbs out of its own cubby, goes to visit other stories. When *Time* covered nuclear power, it listed one great plus: reactors could cheat the greenhouse. "They excrete no carbon dioxide."

It's interesting to imagine what might have come next: A Senate hearing? An international committee like the one proposed in 1948?

As a political issue, global warming has just about the worst luck. It's snakebit. At moments when the momentum builds—reporters up-to-date, science coming together—and the roll goes toward Washington, it's found itself pitched into the shadows by something taller. An energy crisis, a vol-

cano, a new Congress, West Wing marital trouble, terror attacks, recession, health care, a Twitter presidency, impeachment. In 1961, the Weather
Bureau announced the warming had stopped. So the media attention also
stopped. The world appeared to have solved the problem itself. What was
the answer? Pollution.

THE FINE NOSES

RUSSIA TACKLES PROBLEMS HEAD-ON. IT HAS A SOFT spot for the frontal assault, and the Russians never stopped believing in global warming. In the mid-seventies, the Soviet dean of climatology, M. I. Budyko, had the idea of dumping sulfur dioxide directly into the stratosphere. Sulfur dioxide is a coal by-product. Coal spins the turbines, and so powers everything. (Sir Walter Scott groused about this when gaslight first arrived in England. "There's a madman proposing to light London with—what do you think? Why, with smoke!" We're still lighting our cities with smoke.) Sulfur dioxide had its big moment in the eighties, when it became famous as the cause of acid rain. Poisoned fish, fizzled crops, chewed-away housepaint: that was sulfur dioxide.

But up in the atmosphere, it acts as anti-carbon dioxide. Carbon dioxide blocks heat from getting out. Sulfur dioxide dust is highly reflective—tiny mirrors, sending the sun back at itself. It stops the warmth from ever getting in. With a black asterisk: according to the World Health Organization, acid rain is responsible for about half a million premature deaths every year.

At Columbia University in the mid-eighties, scientist Wallace Broecker read over the Soviet paper. Interesting proposition, lousy side effects. It sounded, he thought, like "the sort of experiment that might easily blow up in your face and blacken your shirt."

But global warming gives old facts new shadows. A few years later, heat persisting, Broecker found himself acting as a shirt-blackening advocate in the *New York Times*.

The headline was "Scientists Dream Up Bold Remedies for Ailing Atmosphere." (The subhead—"Mimic Effect of Volcanoes"—was a fine rare X-Men note to hear in the *Times*.) When volcanoes pump sulfur dioxide into the air, reflectivity causes "a slight planetary cooling." So Broecker's plan was to field an air fleet, route the jets through the stratosphere, sprinkle 35 million tons of sulfur dust.

Travel agency costs would amount to $40 billion. "This is not a big expense, compared to totally changing our reliance on fossil fuels." Broecker gave things an actuarial spin: "A rational society needs some sort of insurance policy."

The paper called the idea "exotic" and "fantastic"—in the old and nongood, scientific romancer sense—and noted three drawbacks. First, flights would need to be repeated every year. Second, acid rain. Last, sulfur dioxide dust would turn the blue sky whitish.

In 1988, this seemed unwarranted, freakish, unlikely. (Another idea was to float the oceans with millions of Styrofoam chips—like a person doing a titanically messy job opening a package—which would create a bobbing layer of albedo. An additional plan was painting every rooftop white.)

But global warming is such a circular issue that eighteen years later, the *Times* could run more or less the exact same article. It was even written by the same reporter. The issue was like a bar, where you could step outside, head back to work, raise kids, send them off to college, step back into the bar two decades later, and it's still the same argument, starring the same debaters.

This time, the piece was called "How to Cool a Planet (Maybe)." But plans no longer sounded exotic and fantastic. By 2006, they had become "big, futuristic."

What had changed? Celebrity endorsements. In 1992, the National Academy of Sciences got behind the basic concept. (The science's name, "Geoengineering," sounds both bland and slightly coercive, like a good brisk nurse.) A physicist with the government's Lawrence Berkeley labs had floated a plan that cut out the airplanes. You could train navy howitzers to shoot sulfur dioxide shells straight up, an artillery strike on the sky. In 1997, Edward Teller, father of the H-bomb, bestowed what amounted to an approving, fatherly shrug: It "appears to be a promising approach. Why not do that?"

The other change was a stalled mood, a funk. No progress—except the carbon dioxide had increased, the warming continued, the melting and trickling had taken on the air of an emergency.

In 2006, the Nobel scientist Paul Crutzen signed on with a kind of high-minded, public-spirited disgust. Attempts at government emissions control, he wrote, had been "grossly unsuccessful." Because nothing had worked; because, essentially, the country's chief news source could run the same article twice in twenty years with no action in between. "I repeat," Crutzen said. "The very best would be if emissions of greenhouse gases could be reduced so much that the stratospheric sulfur release experiment would not need to take place. Currently, that looks like a pious wish."

And now Wallace Broecker, shirt fears outgrown, was all for it. "People used to say, 'Shut up, the world isn't ready for this.' Maybe the world has changed." So when the time comes for solutions, we'll probably see Dr. Broecker's air wing take to the skies.

TO THE *TIMES* the idea was, essentially, fume wrestling; send one pollutant to fight another. This is what climatologists believe had happened; by mid-century, the air had become so thick it was canceling the greenhouse. The concept is called atmospheric turbidity: a measure of just how much skyborne, solar-reflecting dust is silting the air above rivers, forests, cities.

Turbidity became important in 1816, when it killed a season. In the spring of 1815, the Indonesian volcano Mount Tambora went up: A four-month, smoke-and-red plume. Effects were immediate. They were also incredibly far-flung, like the same movie opening the same day everywhere. That winter, there was yellow snow in Italy, and brown snow—sometimes, eerily, flesh-colored snow—in Hungary.

The dust lingered for a year, June rolled around, and summer wouldn't start. Which meant no July and August, and also no harvest. Summer snows in New England, where people ate clovers; the Swiss ate moss, in France and England there were grain riots, two hundred thousand dead across Europe. The strange lighting effects also produced the sunset pictures of J. M. W. Turner, one path to Impressionism. Foul weather drove a group of English writers indoors in a dark mood. And when they emerged, they'd launched the horror genre: *Frankenstein*, and the first version of the Dracula tale, *The Vampyre*. Because there weren't a lot of oats for horses, the Year Without a Summer also led to the introduction of the bicycle, a ride you didn't have to feed.

A non-fantastic way of putting sulfur dioxide into the air is by burning coal. English use began in the thirteenth century. From the beginning, it was clear coal was going to be ambidextrous: A problem-solver and a problem-creator, the mobster you hire for protection who scares away customers by hanging around the bar, the cat who scatters mice, then needs to be run off by the dog. In 1257, Queen Eleanor fled her castle, to get away from coal smoke. She called the dust and stench "unendurable."

From there, a series of firsts: her son King Edward instituted London's first anti-pollution laws, in 1307. No more coal. The punishment schedule progressed backward, from sensible to medieval. First offense, a fine. Second offense, "demolishment of all his furniture." Third offense, death.

None of it worked; because coal was cheap, it was convenient, it was warm. The Queen Elizabeth of Shakespeare's time gave the coal ban another go. In 1644, one of the first anti-environmentalist parodies: Londoners had griped for years about coal smoke, then civil war cut off their supply. The focus was women, as if concern for air quality contained something fickle, suspect, unmanly.

> Some fine-nosed city dames used to tell their husbands, "Oh Husband! We shall never be well, we nor our children, whilst we live in the smell of this city's coal smoke; Pray a country house for our health, that we may get out of this stinking coal smell."
>
> But how many of these fine-nosed dames now cry, "Would God we had coal, O the want of fire undoes us! O the sweet coal fires we used to have, how we want them now."

In 1775, the first smoke-to-cancer links—an epidemic among London's teenaged chimney sweeps. (It was, awfully, cancer of the scrotum.) In 1873, the first deaths. In 1905, first use of the word *smog*. In 1952, largest death toll. In 1956, the country's first modern pollution bill, the Clean Air Act.

ANOTHER GOOD WAY to generate turbidity is by starting your car. Los Angeles had been America's City of Sunshine, where the California air kept you in a vacation mood year-round: a long fluffy towel unrolled across the seasons. Balmy days, clear nights. (An 1898 bestseller by Mark Twain's old writing partner Charles Dudley Warner had pinned Southern California to the national dream map. It was all mellow weather, placid moods, life as a seventies album cover. His book was called *Our Italy*.) In the forties, war came. Overnight, the city was in the defense-contracting business. Los Angeles swelled from 1,500 factories to around 8,500. Blast furnaces, refineries, smokestacks, diesel trucks hauling and parking seven days a week.

In 1943, for the first time, a strange brownish smoke rolled across the city. Smog. Noses ran, throats itched—plus a new phenomenon, smog tears. This was Los Angeles' war weather. When it rained in October 1946, a journalist noted that the sudden clearing "gave Southern Californians an unaccustomed view of the object known as the sun."

No one could remember smog before 1943. There were many theories. Some authorities blamed garbage fires; others fingered specific chimneys. One expert spent two years studying the problem, pinned the blame on rug beating and dust mopping, then left government. The city understood Hollywood economics: Los Angeles with bad air and no sun was like a great property without stars attached. It was the climate that made transplants decide all at once they'd had enough Minnesota and pack up their cars. In 1946, the *Los Angeles Times* declared a city emergency.

Readers wrote the paper: "It's said we don't take the atom bomb seriously enough.... After living through Los Angeles' smog, any kind of bomb that would instantly and painlessly put me out of this coughing, strangling, throat, eye, and nostril-burning misery at least would be a change!" Committees auditioned a pollution czar. They were looking for an air sheriff—technical know-how plus personal force. And in October 1947, they imported a man from Washington named Louis McCabe.

He was Elliot Ness, the sworn G-Man dispatched with orders to clean up a dirty town. He was tall, soft-spoken, with a broad forehead and bitter-looking mouth—a mouth used to giving orders. He'd trained as an engineer, during the war he'd become a colonel, running captured coal mines for the Allies.

It made a kind of sense: The war had created the problem, a veteran was going to solve it. At the time, McCabe was the government's highest-ranking anti-pollution man.

His first visit, supervisors brought McCabe to the top of City Hall tower. They led him onto the balcony, gestured at the vista, as at an enemy camped in siege positions outside the town gates: brown-gray smoke.

Dr. McCabe relocated to Los Angeles with his wife and four children. First day there was a smog alert—and on the cover of the *Times*, there was the new air marshal, glasses pushed up on his forehead, dabbing his eyes. Not looking like any kind of authority: just another beat-up commuter, suffering and miserable. Welcome to LA.

If Revelle and Keeling roughed out the basic science of global warming, McCabe did a walk-through of the basic, bruising politics. His smog cleanup was a test run for the exchanges of charge and counter-charge, the lab gladiator matches—regular scientist versus industry scientist—that would set the pattern for climate change in our era. It was junior prom: first corsage, first limo, first dance.

Pollution had started with industry. There were anxieties about whether McCabe had the stomach to "step on toes." Large and small oil refineries were operating throughout Los Angeles; they emitted about eight hundred thousand pounds of daily sulfur dioxide. The oil companies braced themselves. Sulfur, the classic urban pollutant, was where any cleanup would start. After a few months on the job, McCabe demanded they control their sulfur.

What would you do, if you were an oil company? When a scientist is saying you're at fault, there turn out to be three good options. First is to demand a second opinion. (And then a third, a fourth. Even when the next round confirms the initial diagnosis, you get a breather, more years to move your product; you've also put doubts and questions in the public mind.) Second is recruitment: hire a batch of scientists, with a finer nose for exculpatory evidence. The oil companies leafleted schools and communities: they were blameless!

Option three has to do with power—the hard eye and cool display. In April 1948, the oil companies invited Dr. McCabe to a sit-down at the California Club. Not the type of dark alley you want to walk down alone. McCabe arrived with his scientists; the oil companies were waiting with theirs. Narrowed eyes, labcoats, pollution fight at the OK Corral. Then tense agreement, to go forward in harmony. One thing every negotiator learns is control the location. The California Club was a lounge for the city elite. Waiters, silver. It was a gesture: tangle with us, and you're fighting marble.

Toward the end of the century, scientists would come up with a non-catchy phrase for the industry approach to climate change: Studying the Problem to Death. Historian Spencer Weart would express this as a sort of formula: More-Money-Should-Be-Spent-on-Research. You're not saying no problem—you're saying let's explore, let's be 100 percent.

At the California Club, the *Los Angeles Times* quoted a petroleum chairman with the industry position: the "belief that all the elements in the smog situation are not known scientifically, and that it will take time to ferret them out."

If you'd like to see how durable a strategy this is, flip ahead to our century—after five international decades and tens of billions in spending, an industry chairman calling for More Research. "There are too many complexities around climate science for anybody to fully understand all of the causes and effects and consequences," Exxon's CEO said in 2008. "We have to let scientists continue their investigative work."

McCabe's absolute background was science. But his immediate background was military: size up a problem, pick a solution, execute. The army is not an especially hospitable place when it comes to second opinions. Two weeks after the California Club, McCabe broke the truce.

The pollution czar sent a public letter—and what he wrote could serve as a mission statement for the second half of the century. McCabe's experience stepping on toes had taught him, in six months, what it would take everybody else decades to learn about More Research.

He pointed out that nearly twenty thousand experiments on Los Angeles air had already been conducted. McCabe wrote that research is "necessary." But you "should not pursue research for research's sake, and must studiedly avoid ephemeral research designed to divert or postpone enforcement activities." He said, "Research properly used is a valuable tool. It is not a panacea, and it is a dangerous sedative."

After that, range war. McCabe cleaned up the oil refineries. As one contemporary journalist put it, the pollution man was "obstructed and harassed; within a year he was being heckled, investigated, sniped at, and defied."

A hearing was called. McCabe promised the smog would improve. He put it on a timetable: eighteen months. ("I wouldn't advise anybody to sell their home," McCabe assured a wavering householder, "because we are going to lick smog.") By May 1949, he'd had enough. He scooped up his family, accepted the top Washington pollution job, and abandoned Los Angeles. Louis McCabe had been run out of town.

That fall, oil industry scientists released their own findings: they'd need another year's research, but their product bore no "specific responsibility" for smog.

In November, McCabe addressed the country's first National Air Pollution Symposium. He was reporting on his own unintentional research, as a man leading a pollution fight. Here was another roadmap to the next seventy years. Going up against industry on pollution, McCabe said, you could expect three things: "There were 'co-operative' programs with the dual objectives of delay and defeat," he said. "Engineers were assigned to write diverting papers on minutiae of the problem, and trade journals editorialized on the unreasonableness of 'do-gooders.'" If the aim was delay and defeat, here was your battle plan.

An end note: A few weeks after McCabe's speech, the president of the West Coast petroleum trade group announced, "the oil industry is fed up with being the whipping boy." But autumn 1950, as if sticking to McCabe's .

timetable, was virtually smog-free. "Where is the smog?" the *Los Angeles Times* asked that October, and called the smogless 1951 "astounding."

In the mid-fifties, scientists began a series of essential connections: gas stations, tailpipes, atmosphere. To us, it's amazing this wasn't evident, but it wasn't. People had after all been behind the wheel for decades. In 1954, scientists listed the clues: Smog got much lighter on Sundays; weekdays, it toplined between six and nine in the morning, then toplined again from four to seven at night. What else happened at those hours? The first big LA freeways had been cut across the city in 1952; traffic had doubled. In 1954, even the oil company scientists—now hinting they'd had their suspicions—were forced to admit the problem. The largest single generator in Los Angeles smog was the main consumer of their product. It was the gasoline combusted in two million engines; it was the car. It took twenty more skirmish years with the automobile makers—who after all had their own lobbyists and scientists on the payroll, and could invite lawmakers to their own intimidating versions of the California Club—to install the $130 paperback-size device that would thin emissions, the catalytic converter.

THE MOLES

HOW POLLUTED WAS THE AIR IN THE FIFTIES? THE FIRST weekend of December 1952, an epic smog blacked out London. Saturday morning, Oxford's cross-country team laced up to run against Cambridge—and the challenge became simply to complete the race. Spotters along the route shouted, "This way! This way!"

Buses collided; a suburban train ran down three workmen. Motorists cut their engines, abandoned their cars, and groped their way along the shoulder. Ambulances and firetrucks were escorted through the streets by men on foot, seeing-eye guides. Livestock suffocated in their stalls. (A breeder tried to save prize heifers by rigging a gas mask from grain sacks and whiskey.) The smog lasted four days. Since people are people, there was also a rash of muggings, home invasions, purses snatched right out of the mist. Gangs held up shops, even a post office—the smog made a perfect, climatic accomplice. And hospital beds started to fill.

People had begun dying on the first day. It was a classic London smog—smoke and fog, the smoke component being sulfur dioxide. London factories and power plants were belching out about eight hundred thousand pounds a day. The smoke was yellow and white and was described as lung-stinging. It made people choke; it also caused acute bronchitis and pneumonia.

London was then the world's largest city, with a weekly death average of around eight hundred people. An additional four thousand died that week—eight thousand more died from smoke-related illness in the three months after. Headlines carried a single word: MASSACRE. The death toll topped the worst months of the Blitz, Londoners killed by German bombs on purpose. A January 1956 smog took one thousand more people, and that year England passed its Clean Air Act. When a December 1962 smog killed only seven hundred, it was seen as a kind of progress.

A ten-day 1953 New York smog—in its prickly way, the *Times* thought the phenomenon ought to be renamed *Smaze*, for smoke and haze; it's not a name that's made much headway—killed 240 people. A set of surreal postcards, from a city long gone: bridges erased mid-span, skyscrapers without tops, ferries circling a downtown that had vanished into shadows and outline.

Every major city faced a smoke problem. The 1951 national cleanup figure of $1.5 billion exceeded what was then getting spent on medicine and private education. By 1954, that number had hit $5 billion; 1963, $12 billion, lost to crusted buildings, junked machines, bum harvests. The smoke problem was worldwide. Just by itself Germany's Ruhr Valley—the nation's industrial zone—was fuming out a staggering four million tons of daily sulfur dioxide.

In 1966, another smog visited New York for the four-day Thanksgiving holiday, this time taking 168 people. In his keynote address at the National Conference on Air Pollution, the Secretary of Health, Education and Welfare foresaw a gray, all-protection future: Gas masks, cities kept under sanitary domes, people living "like moles" indoors. "There is not a major metropolitan area in the United States without an air pollution problem today." Vice President Hubert Humphrey promised strong action. "Just as we need a [nuclear] nonproliferation treaty among nations," he said, "so we need nondiffusion treaties." At the end of the conference, the surgeon general nodded with familiar bitterness at all the years lost to debate—whether smog was really a problem, whether there could ever be any solution. Without knowing it, he was relearning Louis McCabe's California lesson: that research without action is a dangerous sedative.

Los Angeles, with its back-and-forth brawls between oil companies, pollution czars, and automakers, was now seen as a model. The endless scrapes had hardened California—made it the veteran who knows the feints and scars of the battlefield. You set your rules, took the heat from industry. Even now, in legislative conferences, if you want to see a lobbyist flinch, you only have to use the words "California Standard."

In 1967, the first links. Two studies noting the rise in turbidity, figures suggesting it would be a cooling factor. Dr. Reid Bryson at the University of Wisconsin's Climatic Research Center found that the "rapid and catastrophic increase in the amount of atmospheric dustiness around the world" had cooled the planet by about two-thirds of a degree. Enough dust, enough cooling, would retrigger the clock, call in another ice age. That same year, federal researchers with the National Center for Air Pollution Control published in *Science*. Atmospheric dirt was indeed counteracting the greenhouse. They'd sampled two cities—Cincinnati, Ohio, and Washington, DC. In just three years, turbidity had dirtied up the skies above each by about 60 percent. The dirt was collecting faster than carbon dioxide—ironic, the scientists noted, since both came from smokestacks and tailpipes. This effect would in the nineties receive a cool name of its own: global dimming. By then, pollution controls had kicked in, the dust cleared, warming resumed.

In the late sixties, most people had forgotten about global warming. What they knew was that there was pollution, and pollution was bad. As the *New Yorker* put things in 1968, "The average person, however, is not worrying about melting ice caps when he looks up at the murky sky but is simply wondering what the air is doing to *him*."

It wasn't just the air. It was the water (poisoned by chemicals), it was the food (gimmicked with preservatives). It was in the sense of the world on the brink of something irreversible and bad. There were overall scare articles. Like a long 1968 piece by a Cornell ecology professor in the *New York Times Magazine*: "Can the World Be Saved?" (The professor wrote there was evidence "that the answer . . . is in the negative.") In 1969, *Times* columnist Anthony Lewis sat down to dinner in Brussels with "a leading European biologist." Lewis asked how seriously he ought to take the new environmental consciousness. The biologist smiled. "I suppose," he said, "we have between thirty-five and a hundred years before the end of life on the earth." (So arrange all farewells and balloons no later than 2069.)

A headline in the *Washington Post* read, simply, "'Progress' Ruins Earth, Panel Warns." You have to appreciate that sober, workaday word *Panel*. It was, as it happens, the Congressional Science Subcommittee.

In 1969, *Time* could describe the waters of the capital's Potomac River as "stinking." In the West, as the Missouri River rounded Omaha's slaughterhouses, it was gunked with balls of animal fat the size of oranges. In the summer of 1969, Cleveland's Cuyahoga River—whose waters ran brown with oil and sewage—caught fire. It was like one of the prophetic and unsettling episodes from the Bible: water, something that shouldn't normally be burning, was burning. "What a terrible reflection," said the mayor of Cleveland, "on our city."

Apocalypto. After years of pollution, political crunches, energy crises, the national mood had turned end-times. There was going to be another ice age; we could look forward to a population explosion, killer bees, famine. In his 1971 bestseller *Rabbit Redux*, John Updike turned it global: "Nature is what we're running out of." (There was Charlton Heston's last guy on Earth movie trilogy. *Planet of the Apes*: nukes did it. *The Omega Man*: epidemic. *Soylent Green*: Heston's character remembers hard, then says, "Global warming.") In the 1975 thriller *Three Days of the Condor*, a CIA station chief coolly prophesies to Robert Redford not just the end of fuel but a menacing, overall lack. "It's simple economics. Today it's oil, right? In 10, 15 years, *food*. Plutonium. Ask them when there's no heat in their homes and they're cold. . . . You want to know something? They won't want us to ask them. They'll just want us to *get* it for them."

It was bad luck that global warming rejoined the national conversation in this moment: it was like a real demon showing up at a Halloween party. It was during this decade—the disaster movie era—that "the greenhouse effect" began its competition with the phrase "global warming." "The greenhouse effect" was a terrible name. It had the sound of a Hollywood project with B-list talent. *The Towering Inferno. The Poseidon Adventure. The Greenhouse Effect.* As a name, "global warming" got carbon dioxide off the movie pages—it was a phrase that could exit a senator's mouth with dignity.

It would prove too evocative. In a 2002 strategy guide, the influential

Republican pollster Frank Luntz would encourage his candidates to side-step the term, opt for something less punchy. As Luntz wrote,

> It's time for us to start talking about "climate change" instead of global warming. . . . "Climate change" is less frightening than "global warming." . . . Climate change sounds like you're going from Pittsburgh to Fort Lauderdale. While global warming has catastrophic connotations attached to it, climate change suggests a more controllable and less emotional challenge.

That's how fraught and political global warming would become. It could deputize conversations in car pools, office parks, supermarkets, student assemblies. Every time someone used the words "climate change" instead of "global warming" they were, without meaning to, spreading a talking point. They were helping make carbon dioxide sound like winter vacation.

THE BRAKES AND THE INDIAN

MAYBE IT WAS THE SOOT IN THE STREETS. MAYBE IT WAS
the reports of pesticide in human breast milk, or the unusual sight of a
river on fire. Or maybe President Nixon, stuck in Vietnam, just wanted to
change the subject. He started 1970 by establishing the Environmental Protection Agency, and ended it by signing the Clean Air Act.

You read his quotes, your brain supplies the man's woofy, basset-hound
voice. "It is particularly fitting that my first official act in this new decade
is to approve the National Environmental Protection Act," he said. "The
nineteen-seventies absolutely must be the years when America pays its
debt to the past by reclaiming the purity of its air, its waters, and our living
environment. It is literally now or never." Nixon called 1970 "the year of
the beginning."

The environment became what political reporters call a motherhood
issue—who's against motherhood? Years earlier, Ronald Reagan had said,
"Seen one redwood, you've seen them all"; now the California governor
called for all-out war "against the debauching of the environment."

The seventies closes out the military's part of the story. The classic
pattern. The armed forces clears the land—stands up barracks, maps the
territory—and then the settlers take over. From here out, the issue would
belong to federal employees, to civilians and scientists. And from the seventies on, science became the voice that steers you away from deluxe nachos
and into a nice green salad, that cheers you on as you reshelve the beer. The

voice of no fun. Jonathan Weiner once spent an evening with pioneer Harvard biologist Edward O. Wilson. A fascinating man who won two Pulitzer Prizes, recommended *Blade Runner* as an environmental primer, got named one of the twenty-five most influential people in the nation by *Time* magazine, and invented the field of evolutionary psychology after a career-long study of ants. (They have much to teach us.) Weiner asked why environmentalism rarely generates the enthusiasm of a slam-bang discipline like economics. The biologist considered it. "Because ecology," he said, "is seen as the brake on the wheel." From 1970 on, from the year of the beginning, environmentalism became the guilty second thought twitching your foot away from the gas.

Nixon embraced the environment in the manner of an older man buying his first Beatles album. Not to play the tracks, as an up-to-date signal to leave on the coffee table for guests. Two months after signing the EPA, the president told aides the environment was a fad: "I think interest in this will recede." He specified in memos, "Spell it out very clear that I expect [staff] to drop the substantive activity and move to the more important thing of what word we get out." It's a brutal pleasure to find people taking in private just the rough approach you imagine. For another aide, the president made the politics crystal: "In a flat choice between smoke and jobs, we're for jobs." On the first day of 1971, the *Times* called the new cleanup laws one bright spot in an otherwise "uniquely dismal" year. Nixon, meanwhile, was telling two Ford Motor executives that environmentalists wanted to "go back and live like a bunch of damned animals . . . What they're interested in is destroying the system."

Nixon felt so at sea in the natural world he wore dress shoes to the beach. Yet here was the man who helped scour the country. "No other chief executive," the journal *Environmental History* later marveled, "approved as much important environmental legislation."

When Nixon toured a Santa Barbara beach front after an oil spill, he traveled in mixed, symbolic company: the chief executive of the oil company, who told trailing news crews that there was no mess, the waters were clean—no harm, no foul. The president's second companion was geophysicist Dr. Gordon MacDonald. That's what it is to be president: a double-barrel escort, business at one elbow, science the other. MacDonald complained the walk was "choreographed." So on impulse, the scientist later said, "I kicked at the sand, sending an oily glob onto a highly strategic area of the President's trousers." Dirty slacks, the natural mess right there in front of the cameras. That glob is one symbol of the president-environmental scientist relationship. The presidents wanting the scientists there as models of concern, as set dressing. And the scientists staining them with evidence, with news.

The greenhouse was so well-known it had become a subject for Walter Cronkite, in a 1970 CBS News series called "Can the Planet Be Saved?" Now the anchorman was giving a quick frown as a scientist warned we had twenty-five to thirty years; now he was explaining to camera, "The scientists I spoke with disagree about the schedule of disaster, but we found not one who disagreed that some disaster portends."

A physicist in the *Los Angeles Times* could now call it "the famous greenhouse effect"—a science celebrity. Famous enough to spark an industry pushback: one of the first anti-global warming editorials, which ran in the *New York Times* on August 28, 1971. A kind of anniversary. Skeptics would later learn to question where the warming was, whether the warming was, when the warming was: after a certain point, you could no longer dispute the carbon dioxide. This was the bell to the first round. Where climatologists saw science—calculations aimed to match the swirl of the natural world—skeptics recognized politics, the never-ending campaign, your story versus ours. Here's one of the trial balloons, floated by the editor of *Engineering and Mining Journal*. (In this case, *mining* equals coal.)

> *Then there is the Greenhouse Effect Theory. The build-up of carbon dioxide in the atmosphere, so goes this particular idiocy, will cause a temperature increase throughout the planet.... [We won't] roast because at the present level of carbon dioxide in the atmosphere it would take about 957 years to triple the current level. Such speculations have no more validity than the prediction that my puppy dog, at his present growth rate, would be fifteen feet long and weigh 900 pounds at age five.*

Industry fought two environmental battles in the run-up to global warming. Skirmishes, evaluations of the debate hardware, field tests of the armaments. The second was the ozone layer—the sky up there as a potential menace. The first, of all things, was over recycling.

Environmentalists would eventually come up with a name for the new strategy: *Greenwash*. (The word arrived in dictionaries around 1991.) You don't change the old facts. Instead, you spruce them up with a fresh coat of opinion. It was prankish, sophisticated. And it was also like fighting gum. How do you argue, when your opponent is marching right there beside you, shoulder to shoulder?

There's one ad most people alive and at their sets in the 1970s remember. Open on a river, at sunset, with thumping drums. An Indian in buckskins paddles his canoe downstream. He's concerned, ancient, dignified—someone who knew the land before. His canoe skims newspapers and nudges paper cups. He beaches on a junkyard shore.

Then, superhighway: the something gray and hopeless of car after car. Somebody chucks a fast-food bag out their window. Napkins and french fries splash across moccasins. The Indian turns directly to us, camera pushing in tight. The jet-black hair, simple braid, creased features. A single tear inches down his cheek. "People start pollution," a voice importantly announces, the tear filling the screen. "People can stop it."

The Crying Indian became one of the era's stars: printed, broadcast, stamped everywhere: Posters, newspapers, gazing damply from classroom walls and billboards. (A quandary for advertisers: how to convey the passion of a single tear by radio? This challenge left one executive "deeply worried.") The producers were dabbling in a paintbox of national myth: the Indian was the land, from before we arrived and fouled things up. He didn't accuse—the land doesn't speak; we'd simply let him down.

The spot has been viewed, in various formats, something like fifteen billion times. One media scholar calls it "perhaps the most famous visibly shed" tear in history. President Nixon endorsed the ad, it collected awards, got ranked by *Advertising Age* one of history's top campaigns, by *Entertainment Weekly* in the all-time top fifty. The spot was so popular that by the mid-seventies local stations were sending requests for new copies: they'd rebroadcast so often, their tapes were stripped.

The ad—fresh thinking about the environment—was the product of an antilittering outfit called Keep America Beautiful. Asked about its success, their president nodded to the miracle of casting. The Indian himself, Iron Eyes Cody, a Cree-Cherokee. "It was more than advertising," President Roger Powers said. "What we found—it was a stroke of luck—was a man who lived it and believed in it."

Iron Eyes Cody was committed to heritage. He declined, for example, to perform a Chant to the Great Spirit in the campaign. "I chanted it once on the *Tonight Show*," Cody explained. But the audience had been disrespectful. And when the moment came, he refused to weep on camera. "Indians," he said, "don't cry." So the actual tear was glycerin—a product of the factories.

The Keep America Beautiful ad made Cody, one Native American publication wrote, "quite likely the best-known American Indian in the world." He sat with presidents, caused stirs at assemblies, met Pope John Paul II, marched in Rose Bowls. He received a star on the Hollywood Walk of Fame, where he is to be found next to Errol Flynn, recording artist Eartha Kitt, and the man who played the Skipper on *Gilligan's Island*. (History is shelving.) In 2007, Keep America Beautiful was still describing Iron Eyes as "the embodiment of our entire organization."

He was also a fake. He grew up in New Orleans, a child of Sicilian immigrants; his birth name was Espera de Corti. As a boy, his sister recalled, "he had his mind all the time on the movies." (This domestic situation receives a touchup in his memoir. "I fell asleep instantly," Iron Eyes writes, "and dreamed of my forefathers racing across the plains after buffalo.") He graduated from playing the Indian in the schoolyard, testing and perfecting, to acting the Indian in Hollywood, crossing soundstages and deserts. He appeared in titles like *The Paleface* and *Gun for a Coward*. He rode beside John Wayne, Roy Rogers, Ronald Reagan. Offscreen, he traveled the streets of Los Angeles wearing buckskin and feathers, hair tucked beneath a thick braided wig. In 1995, a New Orleans reporter confronted him with his birth records. "You can't prove it," Iron Eyes said. "All I know is that I'm just another Indian."

And his organization was a fake, too. Keep America Beautiful was founded in 1953, by representatives of the beverage and packaging industries. The aim was to battle litter. But the true aim, the bullseye inside the bullseye, was litter regulation.

Before the Second World War, beverages were sold in hard glass bottles—"returnables." During the war, beer shipped to GIs in a new package, cans. Soldiers came home, and the whole industry went nonreturnable.

"Convenience packaging"—use it once, chuck it away. (One immediate problem: people were chucking cans out of automobiles.) By the early seventies, the industry's seventy billion annual nonreturnable containers, stacked end-to-end, would have made a tower five million miles high; about ten round trips to the moon. So Keep America Beautiful's objective was to head off a new concept, recycling.

They would do this by blame-avoidance, tactical ducking. "This whole environmental movement began a few years ago," President Roger Powers told the *New Yorker* in 1974. Prior to heading Keep America Beautiful, Roger Powers had worked as a lobbyist. For the beverage and packaging industry. "But unfortunately emotionalism crept into it and we began to point fingers at one another. Emotionalism like that can't solve problems," Powers said. "Our whole program is [to] stimulate a sense of individual responsibility for environmental improvement."

You could hear this in the slogan, by what it doesn't say: "People start pollution. People can stop it." That is, not industry. It was the gentlest way of saying, "Handle this yourself, you're on your own."

It was also, as a movie swindler might have whistled in appreciation, a beautiful con. You hear echoes of the National Rifle Association's famous motto, "Guns don't kill people, people kill people." Which made sense—the NRA sat on the advisory board. The board of directors was a beverage and packaging industry reunion. Presidents and officers from Coca-Cola, Pepsi, the Brewers' Association, Reynolds Metals, the National Soft Drink Association. American Can had advanced the money for the Iron Eyes Cody spots.

The advisory board also seated a number of environmental groups—plus the Campfire Girls, Boy Scouts, and League of Women Voters, who were included in the spirit of bulletproofing.

In 1971 and 1972, the national perimeters, Oregon and Vermont, passed the first recycling bills. Then the squares between began to fill in. Keep America Beautiful contested a 1974 California recycling bill. Roger Powers testified at a public hearing—and suddenly the jig was up. The Sierra Club and the Wilderness Society pulled out. As *Advertising Age* reported, critics were now denouncing the campaign as "a powerful political decoy."

Keep America Beautiful held a full, calamitous board meeting at New York's Biltmore Hotel. How could an environmental organization take down recycling? The corporate heads had the sour stomachs and sense of injustice of men with an investment going rotten: they had, after all, put up good money. In an emotional speech, the president of American Can described advocates of recycling as "communists." The Audubon Society, the Nature Conservancy, and the National Wildlife Federation resigned, taking with them the League of Women Voters.

Keep America Beautiful held the line against recycling for two decades. In 1988, Roger Powers was still telling the *New York Times* that politics was no way to tackle pollution. Let industry be. "You're not going to solve it by pointing a finger and scapegoating. Set an example." In 1998, the outfit decided to give Iron Eyes Cody, now near death, one last shot—a bid to reactivate what executives called "the spirit of the tear." ("The tear," a marketer told the *Times*, is "our equity.") Commuters wait it out at a bus shelter: a representative for each of us. The bus arrives, loads, departs. And of course, we've left a mess. The camera moves to the poster inside the shelter—Iron Eyes Cody. The poster sheds one tear. Look closely, and you see the litter is Kleenex, paper cups, sports sections; everything but a bottle or can.

Greenwash. Keep America Beautiful had marketed one opinion—we should all clean up—while secretly selling the opposite: large companies should not be on the hook for that cleanup. It was just what Dr. Louis

McCabe had warned about back in 1949; the "co-operative programs" whose true aim was "delay and defeat."

And this was only recycling, bins and curbsides. Not power plants, airports, factories, cars. If the climate scientists had glanced a few miles down the road, at the tactics and armaments waiting for them along the shoulder, they would have swallowed and said, "Uh-oh."

THE *YAMAL* AND THE FENCE—
Come North with Me

THE HALT IN WARMING HADN'T SHAKEN THE CLIMATE researchers. It just kept them out of the newspapers. If this were a Bible movie, here would be the sequence where the heroes' faith gets tested.

Roger Revelle began his career as a public scientist—advisor to President Kennedy's Secretary of the Interior, then head of the Center for Population Studies at Harvard. (Where he'd practice some internal climate-shifting on the young Al Gore.) Trips and pronouncements. He deemed the national water supply "adequate." He inspected an Egyptian flood zone. He said the American woman—because of "modern, mechanized society"—had become the "loneliest, hardest-working, most isolated woman who has ever lived."

Keeling kept outdoors, testing air at Mauna Loa. He flew to Stockholm, measured some carbon dioxide, and learned the different ways Europeans and Americans see energy. His office was chilly; he asked for a space heater. Then he took off his coat, rolled his sleeves. A couple of Swedes stuck their heads in the door and laughed: "Look at that Yankee."

His graph kept climbing. It was a locator service: the new carbon dioxide was finding its way to the air. (These are the figures that tapped Al Gore on his shoulder, fixing him on a course to the Nobel. "The implications," Gore wrote later, "were startling.") Strong numbers in hand, Roger Revelle chaired a panel for President Lyndon Johnson's Science Advisory Committee. Global warming first felt the bright, crowded light of a presidential speech on February 8, 1965.

"This generation has altered the composition of the atmosphere on a global scale," President Johnson said, through "a steady increase in carbon dioxide from the burning of fossil fuels."

The committee published. Their recommendations have the clench and excitement of something controversial you might read on Vox—the big change your body tells you will never come.

Since the path wasn't followed, the ground is as fresh today as it was then. (Tax emissions. Develop fuel cells. It needed to be done "now.") With Keeling's graph in hand, the panel could make specific predictions. "By the year 2000, there will be about twenty-five percent more carbon dioxide in our atmosphere than at present. This will modify the heat balance of the atmosphere to such an extent that marked changes in climate, not controllable through local or even national events, could occur."

Revelle had an airline pilot's gift for suave, haunting understatement. "Possibilities of bringing about countervailing changes," he added, "may then be very important."

Same message, different loudspeaker. The warning was coming not just from government-payroll scientists. It was coming from government itself—the difference between hearing rumors about a quiz in the lobby and cracking open the test booklet at your desk.

The *Washington Post* didn't shelve the recommendations in the science section. It ran them on page one, with a follow-up editorial. Here was a positive development, carbon dioxide and climate at last receiving science's "unprecedented attention." But the newsmen knew politics, knew the routine, and were able to make a wry, fedora-wearing prediction of their own. "Mankind's ability to protect itself," the editorial ran, "is less likely to be limited by inadequate knowledge, than by its notorious inability to turn knowledge into effective public action."

WARNINGS BY GOVERNMENT and military personnel came in a clump at the end of the decade. There was Colonel Joseph Fletcher, an expert with the Rand Corporation. (The military's planning arm; its name is a squeeze-together of research and development.) In 1969, Fletcher became one of the first scientists to bring global warming before the United Nations. A solemn, creeping before-and-after: A 10 to 15 percent rise in carbon dioxide already, a larger greenhouse effect to follow. "The purposeful management of global climatic resources," Colonel Fletcher predicted, "will eventually become necessary." He offered the annual meeting of the American Geophysical Union a grim timetable: science had "only a few decades to solve the problem."

There was Daniel Patrick Moynihan, President Nixon's domestic counselor, his "favorite new staffer." That year, Moynihan flew to Belgium, addressed NATO's Committee on the Challenges of Modern Society. He brought the government figures—Keeling's graph, Revelle's schedule; 25 percent more carbon dioxide by 2000, ice caps, sea level. Moynihan wrote the White House, "This very clearly *is* a problem."

And the ice was shrinking. Colonel Bernt Balchen was an air force consultant and a polar explorer: in the classic sense of the term, an adventurer. Blond-haired and square-faced, with a cockpit tan: usually pictured in uniform, wings on his lapel, stories in his head. He'd run polar air raids during World War II, flown low and quiet into occupied Norway. He was good friends with the sea-adventurer Thor Heyerdahl, and gave Amelia Earhart flying lessons. His autobiography is called *Come North with Me*. Balchen became the first pilot to cross the South Pole by wing, to spy it from the stick of an aircraft. The North Pole is what's called pack ice. Floating ice—all freeze, no soil. Now, as Balchen studied, he saw heat and melt. The declining ice didn't belong to some vague, faraway future, a thought bubble with letters trailing into dot-dot-dot. It was happening now, he wrote in a 1969 report; the ice could disappear within decades. An Arctic Ocean of open water. To a man like Balchen, this wouldn't just be signal. It would be the loss of a place he knew—winds, smells, experiences.

Balchen's report got people talking in Washington, started arguments around faculty tables. The *Times* noted that a pack ice disappearance "would probably have cataclysmic effects."

Here's why. World climate—the currents and heat streams—was then understood as a kind of endless bank shot: the Gulf Stream curling from the tropics to the North Pole, where it rebounded off a frozen wall into Europe, was one thing that kept England warm. It's the same latitude as British Columbia. It should be yarn sweaters, hike-minded dogs. No cold north, no warm Europe. (A later famous expression of this comes from

Wallace Broecker, who calls it "the Great Ocean Conveyor Belt." Picture the globe crisscrossed by a giant climate assembly line.) But the *Times* noted that even with some "evidence" of increased carbon dioxide, temperatures had cooled in recent years. Nothing to worry about.

HERE'S WHAT HAPPENED with Balchen's prediction. In 1971, a United Nations report, cosponsored by MIT, found the ice thinning. Just extremely slowly; the complete melt would take a century.

It's nice to imagine this as a movie montage: couple at the breakfast table—toast, bathrobes, hairstyles lowering and rising with the years—as the camera pans across their newspapers. 1971: "Study Says Man Alters Climate—U.N. Report Links Melting of Ice to His Activities." 1982: "Some Polar Ice Melting Linked to Global Heating." 2006: "Climate Change 'Irreversible' as Arctic Sea Ice Fails to Re-Form." 2017: "The Arctic as It Is Known Today Is Almost Certainly Gone."

By the nineties, a commando squad of climate change skeptics had been fielded by industry. These men were fire-jumpers. Within moments of a global warming report, they were ready to leap into the guest chair at any talk show, the columns of any newspaper. In 1994, a skeptic named Patrick Michaels introduced the poles as anti-evidence on ABC News *Nightline*. "The polar regions should be so warm now as a result of the enhancement of the greenhouse effect that we wouldn't even be talking here," he told host Ted Koppel. "It would be obvious. That's the region where the most warming is supposed to occur. There isn't any."

In 1995, the Arizona State geography professor Robert Balling reassured the boardrooms-and-golf-shoes readership of the *Wall Street Journal*: "The Arctic area, where most of the warming should be taking place, has not warmed over the past 16 years or over the past fifty years, as measured by standard weather stations."

As it turned out, both poles had been warming since Bernt Balchen's report. Signal came first from the animal world, from the nests and landscapes. Krill, the classic whale and seal entrée, were breeding in something like 20 percent of their previous abundance. During the eighties, Antarctic penguins began to expire in large numbers. (Because such episodes have joined the features of our era, there's now an official term: *die-offs*.) As the Alaskan permafrost thawed, once-flat streets drooped and sagged, became "roller coaster roads"; pine and spruce, listing at crazy, cock-eyed angles, are called "drunken trees." Some Alaskan homes sank up to their window frames. Navy subs had sampled polar ice thickness throughout the Cold War. Simple undersea self-preservation: if you need to surface, you want a map to the likeliest spots. Between the fifties and nineties, pack ice had thinned 40 percent.

In 1997, a polar research veteran read a new report—*Newsweek* printed his reaction this way: "an astonished 'Holy s___!'" Air temperatures on the Antarctic Peninsula had warmed five degrees in fifty years; ten times the global average. It was the camera starting to run at fast motion.

TWO TRIPS, in 2000. If you've noticed something, odds are somebody else has, too. That May, a private ice expedition set out for the North Pole by air. Water, sky, puffball clouds giving way to thin hard slices. The pilot set down. His wheels crunched, then sank; passengers quickly exited the aircraft. The owner went back to pick up his plane; it had cracked through the North Pole's ice.

A Russian icebreaker called the *Yamal* set out for the Pole from Norway that August. The *Yamal* is nuclear-powered, a behemoth, a set of fanged teeth drawn into the prow. It's a tour boat now—the Cold War over, the Soviet fleet for fun—a hired icebreaker for a certain type of experience-collecting tourist. Cruisers who want to attend lectures, fit cold binoculars against their cheeks, feel the thrill of standing at the top of the world, head home with a vacation story to take on all comers. Passengers were lofted aboard the *Yamal* by helicopter: a kind of deluxe, dramatic gangplank. Waiting aboard (an ice mirror of Jean-Baptiste Fourier's voyage to Egypt) was a deluxe academic crew. Professors from Cambridge, MIT, the American Museum of Natural History.

Dr. James McCarthy—balding, beard the thin color of winter reeds— was among the *Yamal* lecturers. He's a Harvard professor of ocean biology; he was also leading a United Nations panel on climate change. This lent the trip some geopolitical dash.

Four days of lectures, as the *Yamal* headed as far north as you can go. (Is there a picturesque word for traveling by nuclear power? The *Yamal* reacted north.) The lecturers probably started with polar etymology: *Arctic* comes from the Greek for north. So what *Antarctic* means is the opposite—anti-north, south. They talked about warming (in 1997 Dr. McCarthy had told the *Washington Post*, "There is no debate among any statured scientists of what is happening. The only debate is the rate at which it's happening"), about Arctic evolution and history, the difference between multiyear ice and seasonal ice.

It was like a long overture—then the star won't come out of the dressing room. As the *Yamal* pushed north, cruisers got to snap pictures of polar bears, sea birds, sunsets; everything but thick ice. Instead, thin ice and open water.

"None of us," Dr. McCarthy explained later on TV, "had ever seen anything like what we observed in this particular journey."

McCarthy and the academics had logged about two dozen polar visits between them. It's a five-hundred-mile route. The ice pilot is supposed to be an icebreaker's most edgy, non-relaxed man. Adjusting course, eyes peeled. As the *Yamal* reached ninety degrees north, he didn't even have to be on deck: the ice pilot was in his office, at his desk, playing computer solitaire.

No ice. Instead, a mile-wide lake, gulls keening. (Bonus for ornithologists: this was the first report of ivory gulls so far north.) Cruisers snapped deckside pictures. Tourists are tourists. The *Yamal* detoured six miles, found freeze thick enough for passenger weight. They debarked, everyone collected their boots-on-the-ice experience.

When they returned, the cruise offered a final extra: they got to be passengers on the inside of an international news story. Headlines, anchormen (ABC News: "Santa's Workshop Underwater"), columns, accusations, cartoons. McCarthy found the cartoons "perhaps the most interesting of all the responses. . . . The editorial cartoon is a metaphor for a deeper feeling." The North Pole had melted: What did it mean?

WHAT'S INTERESTING from here is the story of Dr. Mark Serreze. As a scientist, Serreze is a pole distant from James McCarthy—a kind of AntMc-Carthy. McCarthy was Harvard, rarely pictured out of a tie, with eyes that seemed to register that you've skipped class. Serreze gets photographed in open-necked shirts with a hangover gaze, heavyset and deliberate—the bulky style of the undercover cop. An upstairs, internal affairs guy, and a street guy.

Serreze was a senior research scientist at the government's National Snow and Ice Data Center. (He's got the director's chair now.) Compiling statistics on the cryosphere—poles and ice, the world's frozen parts—is his cold job. Here is the story of one scientist's evolution on global warming.

When the *Yamal* returned, with everything it implied about carbon dioxide, Dr. Serreze became the journalists' correction. The credentialed person who would explain that here was no big deal.

"There's nothing to be necessarily alarmed about," Serreze told the *New York Times*. "We have no clear evidence at this point that this is related to global climate change." Serreze reassured CNN viewers, "We've seen features like this before. We'll certainly see them again."

Serreze's own story, as narrated by the scientist himself in a series of broadcasts and interviews, is the story of a fence. In 1990, then again in 1993, Arctic ice posted record lows: developing thinner and slower, disappearing sooner. In 1995, the *Denver Post* framed the questions: Were temperatures rising, the packs melting? The Serreze answer held a scientist's caution. "I'm not sure." Was there a carbon dioxide effect? "It's probably too early to make any rash decisions."

On the other hand, Serreze argued—and what you hear is probably the sound of a man climbing onto a fence—"If you wait until it happens, it will probably be too late."

Serreze did his first public fence-mention in August 2000, the month of the *Yamal*'s return. "I've been a fence-sitter for a long time," he explained, meaning global warming. "I'm definitely leaning," the scientist added, but "I haven't toppled yet." Ice hit another record low in 2002 and Serreze told NPR, "Something weird is going on in the Arctic." In 2004, the Arctic Climate Impact Assessment study—a research supergroup, the combined international results of three hundred scientists—found the far north warming at "almost twice the rate as that of the rest of the world." (This same year an Arctic marine expert was contacted by the Bush White House to join the Arctic Research Commission. Her interview's first question landed far from qualifications or theory. "Do you support the president?" The expert said she did not. Interview over, no job.) In 2005, *Newsweek* was predicting open Arctic water within eighty-five years, the first time "since prehistory." Keeling's graph had climbed all the way to 380 ppm. If you started with 270 ppm in the 1850s, it made for a two hundred times speedier rise than any other period. Half a million years had passed since the air contained more carbon dioxide—there were now trees and robins in the Arctic.

"At this point," Serreze said, "I'm definitely on the other side of that fence."

In *The Sun Also Rises*, Ernest Hemingway offers a terrific description of how a person goes bankrupt: "Two ways. Gradually, then suddenly." First you're slow paying off one bill, then another. You take out loans to pay down previous loans. There's a pause. Envelopes stamped "Dated Material," hard looks from the landlady, voicemail you don't return. Then every bill comes due, and you're swamped.

Gradually then suddenly also describes feedback in the Arctic. Gradually: some ice melts. More exposed ocean, which absorbs more heat. (Water has one of the lowest albedos.) Less sunlight gets reflected back, now you've got warmth from upstairs and next door—warming goes fast. In March 2006, winter ice hit an all-time low. North Pole ice seemed to be melting even quicker, scientists were talking about tipping points, the Arctic had reached the suddenly phase.

"People have tried to think of ways we could get back to where we were," one Serreze colleague told a British journalist. "We keep going further and further into the hole, and it's getting harder and harder to get out of it." Mark Serreze said, "Even five years ago, I was kind of a fence-sitter on this whole greenhouse warming thing. We saw no reason to be raising these huge warning flags. And that's changed. Now we've got the flags at full staff."

Six months later, Serreze told reporters, "I for one, having studied this for twenty years, have never seen anything like this. . . . We're seeing an overall pattern of global warming." One last fence-mention: "If you asked me five years ago," he said. "I was a fence-sitter." The scientist continued, "What we see in the Arctic is part of a much larger picture. We hate to say, 'We told you so.' But we told you so."

The 2007 sea ice had contracted to half its 1957 area. You could find crows and salmon in the Arctic, beech trees and grass growing on the Antarctic peninsula. The melt "will be within our lives, not our grandchildren's," another Serreze colleague told reporters. "Things are on more of a hair trigger than we thought."

It was as if the plot had been branched—the same story being treated in two different genres. Down in Washington, it was senator against senator, idealist versus lobbyist, a still-friends beer after the vote: it was *The West Wing*. At the poles, it was a disaster movie. "A decade ago," the *Washington Post* reported, "melting at the poles was predicted to play out over a hundred years. Instead, it is happening on a scale scientists describe as overnight." Serreze told reporters, simply, "The Arctic is screaming."

In 2001, the United Nations panel had predicted sea ice would remain until 2100. In 2007, Serreze estimated it could be gone by 2030. "Climate change is upon us, and it has arrived well ahead of schedule," journalist Michael Pollan wrote in the *Times*. "The warming and the melting is occurring much faster than the models predicted. . . . Have you looked into the eyes of a climate scientist recently? They look really scared."

Serreze talked about emblems. He was discussing just what the post-*Yamal* James McCarthy had been struck by, in the editorial cartoons—"the visual image and symbolism" of a no-ice North Pole. The upstairs guy and the evidence guy had become the same person.

"The North Pole is just another point on the globe," Dr. Serreze said. "But symbolically it is hugely important. There is supposed to be ice at the North Pole, not open water." Eight years earlier, reflecting on his experience at the hot center of a news story, James McCarthy had mildly observed that an ice-free North Pole provided "an unusual opportunity to see the future." Forty years earlier, the *Washington Post* editorial board had guessed that when it came to carbon dioxide and climate, the real problem was not likely to be inadequate knowledge.

THE GLOBAL COMPUTER MODEL

SCIENTISTS WERE HAMMERING DOWN THE REMAINING questions. Looking over their shoulders and out toward the horizon. Past and future—how much carbon dioxide before, how much heat to expect?

At the 107th annual meeting of the National Academy of Sciences, the chairman of Johns Hopkins' Earth and Planetary Science Department announced that human activity would be shaping the atmosphere by the year 2000. "We are entering an era," he said, "when man's effect on his climate will become dominant." It was spring 1970: the scientist was flashing the hall a ticking clock. Research now, understand now. "It behooves us," he said, "while we still have time."

There was already one big change, a move away from Arrhenius' pencil. Along with interest rates and NFL point spreads, climate change is forecast by computer. The programs are called GCMs—the full name, general circulation model, is so blandly memory-resistant it's fruitless to try. (You'll correct the "C" to stand for "climate" or "computer," the "G" for "global.") The first GCM was programmed by Dr. Syukuro Manabe, at Princeton's Geophysical Fluid Dynamics Laboratory. Another government program, funds circulating in a stream through the Weather Bureau.

Manabe had grown up on the Japanese island of Shikoku, a dot so small American bombers ignored it. (They wanted splashier targets.) The postwar Japanese economy, to a young scientist's eye, was all pits and ruin. Manabe arrived in America in 1958, small and trim, with the scientist's knack for making every official photo look stiff and out-of-date the instant it's printed, an experiment that's already old hat.

Manabe took up the next question. He was jogging on the heels of Keeling and Revelle. Carbon dioxide: not staying in the water; okay, collecting in the air. So now: How fast would the Earth warm?

Arrhenius and Callendar had worked in pencil and paper. (Research has a great phrase for the moment you toss house keys into the bowl and the numbers hit you: they're called back-of-the-envelope calculations.) The idea was to create a program that contained, somehow, every earthly variable. Manabe wrote equations for continents, deserts, ice, ocean, rain, carbon dioxide. Sims World. He was in and out of the Princeton library, boning up on soil and water.

The first big GCM run took place in 1965, on a UNIVAC 1108. This is the kind of proto-computer you see in the old spy movies: an entire white room devoted to cabinets and tape, a technician slapping the console, "That's what this baby right here is going to tell us." (The machine had half a megabyte of storage. It could be beaten up by your cellphone.) Crunching twenty-four hours' worth of atmosphere took twenty minutes; some runs lasted fifty days. There's a photograph of Manabe, slumped beside his monitor, looking like a member of Sinatra's crew, Sunday morning at The Mirage—two-tone polo shirt, dark slacks, black shoes, white socks. His shoes barely graze the floor. He looks exhausted: he's created a planet.

Manabe's GCM results showed that carbon dioxide would indeed warm the atmosphere. Doubling the parts per million would raise temperature three to four degrees. His 1967 paper has become one of the most-quoted in climate literature. "I just happened," he modestly says, "to hit the jackpot the first time." (His streak continues. In late 2021, Manabe received the Nobel Prize, "for reliably predicting global warming.") It was Manabe's numbers, the geophysicist Wallace Broecker said, that first raised his eyebrows. The work "convinced me that this was a thing to worry about."

Wallace Broecker looked, in the kindest way possible, like a highly evolved frog. Flat cheeks, wide mouth, bulge forehead. Which makes a type of sense: his specialty was water.

He was a professor at Columbia, where he would eventually demonstrate how the seas cycle warmth, sweeping tropical air from the equator to the Arctic, heat leaking down, rolling with the deep waters south. He'd call this "the great ocean conveyor belt." Broecker had been following results from the first Greenland ice core studies. Manabe was using his program to peer ahead, training his computers on the future; ice is how climate scientists went about frisking the past.

Let's say you want a picture of the atmosphere from a thousand years ago, or a hundred thousand. Where can you go? Head for a polar ice cap, Greenland or Antarctica; position your drill, chew down. Extract a long rod of ice. Study the trapped air bubbles, and you get captured weather, skies and days, millennia of climate. (You can read the atmosphere from the decades of George Custer, or the Caesars; a famous polar researcher calls ice cores "the two-mile time machine.") This was ingenious, *CSI* stuff, investigators lifting a climate fingerprint from an unexpected surface. When a scientist explains there's more carbon dioxide in Earth's atmosphere than there's been for eight hundred thousand years, they're quoting ice cores.

In 1975, Broecker plugged Manabe's GCM numbers into what he'd learned from the Greenland ice cores: an unsettling mix. "By the first decade of the next century," Broecker wrote, "we may experience global temperatures warmer than any in the last 1,000 years." (Three decades later, in his book *Fixing Climate*, Broecker would note that, right on schedule, eleven of the previous twelve years ranked among the twelve hottest since record-keeping began.)

Broecker published in the journal *Science*—"Climate Change: Are We on the Brink of a Pronounced Global Warming?"—and pointed out in the *Wall Street Journal* that human-made carbon dioxide had multiplied twelve times over the century, from 21 billion tons in 1900 to 242 billion. (In the process, Broecker helped popularize the term "global warming.") An ominous line from his paper has become famous: "We may be in for a climatic surprise."

With the theory buttoned up and presentable, predictions got scarier. If you were reading the news, it's the kind of thing that would make you lower the paper, as if you'd just heard a gunshot by the mailbox. In 1977, a program director with the Energy Department did a survey of climate researchers. "The consensus," he told the *Christian Science Monitor*, who ran the story on page one, was "that carbon dioxide has potentially one of the largest impacts on the environment man could make short of an all-out nuclear war."

It wasn't just the West. In 1976, a group of US scientists flew to the Soviet Union. Good will: equations, exchange, vodka. Dr. Edward Epstein, a weather expert with the National Oceanic and Atmospheric Adminis- tration, was shocked to discover the Russians had adopted a 1974 climate change study put out by the CIA. The analysts had predicted massive disruption: food shortages, refugees chasing fair weather, borders going to holes.

Epstein wrote up a report for the Commerce Department, since good will is also spying. The CIA paper had not attracted a lot of domestic attention—it was the band that must tour overseas to gain a following. The Soviet government had gotten ahold of it, then issued a catch-up order to their own experts: research or else. Those scientists, Epstein wrote, were now "firmly convinced that a rapid climatic warming is occurring."

THE WOOD CHIPS
AND THE MALAISE

THE SEVENTIES WERE THE ENERGY CRISIS YEARS—THE first big fuel sock to the gut, which the country remembers the way a driver might recall dozing behind the wheel, then bouncing off a guardrail: the first alarming time you touched a limit. For six 1973 months, OPEC twisted shut the oil spigot. Price jumps, gas lines, rationing. (A traveler in a *New Yorker* short story, just home from London: "Things sounded so much worse here—people shooting people in gas lines and everybody freezing.") Two years later, the situation had grown so tight the secretary of state made a military threat.

One day Henry Kissinger is being interviewed by *Business Week*; the next, his words are being typeset under headlines. "I am not saying that there's no circumstances under which we would not use force," the secretary said. "But it is one thing to use it in the case of a dispute over price, it's another where there is some actual strangulation of the industrialized world."

He had powerful backing. The secretary of defense called the idea "feasible"; the press secretary said Kissinger spoke for the president. Plans were drawn up: Kuwait was to have been the target—400 miles of coastline, no tree cover, easy pickings for a combined air and sea force. The president of Egypt said that in any event the Arab world, faced with invasion, would blow the oil fields.

Peaceful solutions to the energy crisis were funny, extreme. The difference between an immediate problem (any fix) and a long-term one (let's make sure we've got all the angles on this deal). The crisis pinched so hard people looked for energy under every rock. Articles about creating fuel from spinach, from sugar cane, cattails ("It's not a panacea," warned a botanist), from cow dung. In Pennsylvania, investors with the Wood Power Energy Corporation unveiled a pickup truck powered entirely by wood chips. "We think the equivalent of five two-by-fours should take us 100 miles," they explained. "That's only theoretical.... Our group wants to make wood the primary source of energy by the turn of the century."

A more standard, lunch-bucket solution was coal. During the seventies, it became a booster's response to the Middle East; the sort of boast that bandages a wound. So what, Saudi Arabia had oil? America had resources too—it was "the Saudi Arabia of Coal." An industry slogan tried to recast the situation as groovy: "America has more coal than the Middle East has oil. Let's dig it." More than a quarter of the world's coal reserves were buried under America's hills and plains—enough kick to power the country for 250 years. (And it's funny to note this is just the situation—shortages, the military, lustful glances—that Svante Arrhenius had warned about in his glum final book.)

In February 1977, two weeks into his presidency, Jimmy Carter appeared on TV wearing a sweater. That cardigan was on purpose—a symbol of all the living rooms where thermostats would have to be turned down. "We must face the fact," the president said, "that the energy shortage is permanent." (He sounded like the decent and ashamed father in British melodrama: "We can find ways to adjust." A Wall Street moneyman sniffed to *Time*, "I don't like a president in a sweater.")

Two months later, Carter offered his energy plan. And there it was, right at the center, the black lump. "We can't continue to use oil and gas for seventy-five percent of our consumption.... We need to shift to plentiful coal."

He wanted to see coal in power plants and furnaces, pureed into something that could fill the gas tanks. Wish list numbers for Congress: increase production 66 percent, shoot for one billion annual tons. The Swiss army knife approach—use for everything. The "strategy will be," Carter said, "conversion from scarce fuels to coal wherever possible."

So when the first giant reports came in from the scientists, they were running smack against Carter's energy policy. It was two trains headed toward each other on the same track. One was really big and carried the president; the other was skinny and filled with formulas and scientists.

There's a scale in fossil fuels, a kind of least-wanted list. Natural gas burns the cleanest. Oil comes second. In sooty last place there's coal. One coal ton means three tons of carbon dioxide. Switching to coal was politically foolproof—an energy source with your home address. From a carbon dioxide standpoint, you couldn't do much worse.

And when the data started to arrive, it offloaded even to the investor press, where *Business Week* had some of the scariest quotes. This comes from August 1977: "For years, government and industry had hoped that the problem of carbon dioxide pollution would somehow disappear. This week they faced up to the prospect that CO_2—emitted by every factory and utility that burns fossil fuel—may be the world's biggest environmental problem." The head of the Energy Department's fuels division told the magazine, "The consequences may be horrendous. . . . If CO_2 proves to be the problem people think it is, we'll have to restructure our entire fossil fuels program." He added, "We'll make changes—and fast."

These reports are as much a symbol as the president's sweater—an emblem of global warming's bad dramatic instincts, its poor political timing. The findings arrived during an energy crisis.

ALL THE SCIENCE came together at the end of the seventies. Four government reports in two years.

It had been two decades since Roger Revelle published, announcing the big geophysical experiment was underway. There were few punches pulled. The scientists—having theorized, discovered, studied, predicted—now expected the next step, action. The first report came out of the National Academy of Sciences, in July 1977. It was the product of thirty months' research, and ran 281 pages. You could settle down to it like a long-reading, climate-change novel.

It was also a get-together: Keeling's numbers, Manabe's computer, and, chairing the panel, Roger Revelle.

The report made headlines around the country, front pages and editorials in the *New York Times*, the *Washington Post*, the *Los Angeles Times*. (The nation's three holy cities: money, government, fun.) It's interesting to imagine what lawmakers felt, opening their papers. Did they think, "Somebody just dumped a big load of rocks in my briefcase?" Did they say, like the sheriff in a monster movie, "Somebody better warn the town!" Did they decide, "Science—no constituency?" Two years later, one of the scientists would brief lawmakers in person. Picture a Washington office. Ties, good furniture, the solemn Capitol faith, in power advancing surely across difficulties. A lawmaker asked, "So when will these effects happen?" The scientist said, Forty years. The lawmaker replied, "Well, get back to me in thirty-nine."

The National Academy report was dire. It foresaw "highly adverse consequences"—especially from coal. It recommended a "lively sense of urgency." It said the country needed to act now, since changing energy sources—which the panel expected would become necessary—takes a generation. If action got postponed, "then for all practical purposes the die will already have been cast."

In February 1979, at the World Climate Conference in Geneva, American and European scientists warned that government policy would come to "a crossroads" within a decade. Roger Revelle told reporters the world faced "a Faustian bargain. . . . Whatever you do is bad."

The next report arrived out of the Pentagon—from the JASON Committee, a secret organization of Nobel winners and hopefuls. In the 1940s, physicists formed up to battle the Nazis: this gave us radar, the Manhattan Project, and the atomic bomb. In the late 1950s, Manhattan veterans formed up again as JASON. The organization is so secure that for years no one could agree on what the acronym stood for. (It's actually named after the Greek hero, who traveled under arms through strange lands; part military, part explorer.) Security clearances, members tapping new members, who are known to the defense community as Jasons. They became the lab voice whispering in the military's ear—what journalists call "the brain trust for the nation's defense community." A secret history of the last sixty years, the masked number in the equation. Jasons advised against nuclear weapons in Vietnam. They pioneered the electronic battlefield and stared down the Russians with fancy nuclear delivery systems. Since 2001, the Jasons have worked in counterterrorism.

In the spring of 1979, the Jasons weren't pulling any punches either; they called the potential shifts "ominous"—since any climate change would "produce stress and possibly disaster in some parts of the world." (The Department of Energy's own study, released the same month, declared carbon dioxide "potentially the most important environmental issue facing mankind.") The chairman of the report was Dr. Gordon MacDonald, a Jason—and the scientist who, on a Santa Barbara beach, had once kicked oil on President Nixon's slacks, to demonstrate the effects of a fuel spill. Now he was muddying another president's trousers. MacDonald predicted some greenhouse effects would be perceived by 1990. "The government," Dr. MacDonald said, "must start dealing with this problem now."

A few months later, the Council on Environmental Quality—a White House advisory group—released its own report. Man, the report stated, had set in motion a chain of events that "seem certain to cause a significant warming of the world climate over the next decades unless mitigating steps are taken immediately."

The council chairman told the *Times*, "The report is an extremely important, perhaps historic statement." He expected it to become "very influential in government decision making." (He was in a sense correct. After assuming office, President Ronald Reagan would quickly try to shut down the Council on Environmental Quality.)

The White House, presented with these reports—the Jasons' especially—followed the course charted by the Los Angeles oil companies: it asked for a second opinion.

So in the summer of 1979, the National Academy of Sciences formed an additional, follow-up panel. It was headed by Dr. Jule Charney—a professor at MIT, and a climate change skeptic. "I don't think we can predict climate now and I wouldn't trust anyone who said he could," Dr. Charney told a *Times* reporter in 1974. "Anyone who says he can tell you more than a few days ahead of time what the weather is going to be is practicing necromancy."

So Charney assembled a team of fresh eyes. No Keeling, Broecker, Roger Revelle. The panel retired to the Academy's summer research facilities on Cape Cod. (Which is probably a very nice place to work.) It took five days. They were the jury that returns a quick verdict.

The Charney panel had taken care to find "unbiased viewpoints on this important and much-studied issue," they reported. "The conclusions of this brief but intense investigation may be comforting to scientists but disturbing to policy makers. If carbon dioxide continues to increase, the study group finds no reason to doubt that climate changes will result and no reason to believe that these changes will be negligible."

They understood—it's the undertone to their conclusion—just what sorts of results the White House might have preferred. "We have tried," the panel reported, "but have been unable to find any overlooked or underestimated physical effects that could reduce the currently estimated global warmings."

The president didn't wait. On July 15, 1979, he gave a live address from the Oval Office. This broadcast has since become known as the Malaise Speech. (Carter himself never used that hazy word.) It was meant to be an energy talk. Instead, he spoke for thirty-two minutes about America being a place of the ballot and not the bullet. He wore a striped tie and stared directly into the camera. The most unsettling leadership mix—he seemed both angry and exhausted. He talked about a crisis in the American spirit. He talked about facing the truth. He sat at a shiny desk, flags on either side, curtains drawn. He said it was time to stop crying and start sweating. He massaged one hand with the other, sawed both hands through the air, he pointed, pounded on the desktop. It is among the strangest appearances ever by a US president. Packaged as a one-man show—as performance art—it could tour the festivals for years.

He called it "the battlefield of energy." He said it was time to end our "intolerable dependence on foreign oil." He glared at the camera. "Beginning this moment," the president said, "this nation will never burn more foreign oil than it burned in 1977—*never*." He said, "I am asking Congress to mandate, to require as a matter of law, that our nation's utility companies cut their massive use of oil by fifty percent.... and switch to other fuels, especially coal." He asked people to park their cars, seek public transportation, which he called an act of patriotism. He signed off with a humble, sad, neighborly request: "Whenever you have a chance, say something *good* about our country."

It's difficult to know what viewers heard. (In his 1981 Pulitzer Prize-winner *Rabbit Is Rich*, John Updike lets some characters reflect on the broadcast. "I thought it was pathetic. The man was right. I'm suffering from a crisis of confidence. In him.") What the climate scientists heard was a message aimed squarely at them. A policy in one word. For better or worse, the nation was rumbling toward coal. Carter had faced just the choice Nixon once laid out—smoke or jobs. Two days later, an administration official made the thinking plain. "The President apparently felt he had to fish or cut bait between energy and environment," he told the *New York Times*. "He chose to cut environment." JASON report chairman Gordon MacDonald later said Carter had been "a disaster."

THE PRESIDENT'S PLAN, of course, could never have worked. To stand against oil imports in 1979 was like declaring yourself anti-wind in the middle of a hurricane. In 1971, just before signing on with the Middle East, we spent $3.7 billion importing oil. In 1979, after six years as a Middle East customer, we spent $50 billion. In 2013, $427 billion. President Carter's plan failed.

It worked with coal. (In the late sixties, France had faced a fuel situation that was even bleaker: a lack of domestic *anything*. France bit the bullet and went nuclear. About 80 percent of their electricity comes from fission—from steam and fuel rods. When an American reporter asked how they'd managed it, a director with Électricité de France explained, "The nuclear issue in France was never a political one: the left side and the right side were both, for different reasons, convinced about the necessity of nuclear energy." The journalist asked whether reactors were inherently safe, and the director made the sort of beret-wearing, French observation—rich and weary with common sense—that used to be called *continental*: "We are not in paradise, we are here on Earth. And all technology may have some problems.") In 1977, when Carter took office, the nation produced

685 million tons of coal. By 1990, we'd crossed the president's milestone figure: a billion tons—two trillion pounds—for the first time. As of the next
decade, more than half our electricity was being generated by coal. Every
American consumed about twenty pounds per day: for the flatscreen, the
wifi, the fridge. Coal moved the subways and lit the cities at night. The
president's plan succeeded.

And, in a way, it also succeeded with science. More Research: more
panels, more dollars to print more unnerving charts and studies. In the
quarter century after the Charney panel, the National Academy of Sciences
would just by itself release about 200 reports on carbon dioxide. A second
opinion every forty-five days.

The end of the seventies brought global warming to the Capitol, where
it's loitered ever since. Hanging in the cloak room, on tiptoe in the back row
of every energy decision. Lab budgets bloomed. So in a way every interest
was satisfied. Decades from now, historians might say the president, and
presidents since, had arrived at an appeasement strategy. It wasn't a foreign
power getting appeased—some strongman who'd raised an army and filled
his head with colorful books. It was the cars and storefronts, the senators
and scientists, the burners and the drivers; the strategy at home.

THE FROG

THERE'S A FAMOUS ANALOGY FOR GRADUAL CHANGE. You've probably heard it; it's called the Boiling Frog. Vice President Al Gore splices it into his Academy Award-winning documentary *An Inconvenient Truth*, where it's presented as a cheerful animation: beaker, boiler, disturbingly relaxed frog.

"If a frog jumps into a pot of boiling water," the vice president explains, "it jumps right out again. Because it senses danger." He's strolling under the kind of starry blue light that suggests a museum's Hall of Minerals.

"But the very same frog, if it jumps into a pot of lukewarm water, that's slowly brought to a boil, will just sit there. It'll sit there, and it won't move. Until ... until ..." Boiled frog.

Climate change, in this analogy, is the worst kind of danger. One that dawns *too* gradually—until suddenly you're outside, and the sun's at noon, and you're cooked. In the movie, a vice presidential arm scoops out the frog pre-disaster. (No animated creatures were harmed in the production of this motion picture. "It's important," Gore deadpans, "to rescue the frog.")

"But the point is," the vice president explains, "our collective nervous system is like that frog's. It takes a sudden jolt sometimes before we become aware of a danger. If it seems gradual, we're capable of just sitting there."

In 1995, the magazine *Fast Company* decided to test out the analogy. They phoned the American Museum of Natural History. "Well that's, may I say, bullshit," said the Curator of Amphibians and Reptiles. "If a frog had a means of getting out, it would certainly get out."

They got on the blower to Harvard. "If you put a frog in boiling water, it won't jump out," explained a professor of biology. "It will die." He went on, "If you put it in cold water, it will jump out before it gets hot—they don't sit still for you."

The magazine engaged (heartless) experts, conducted time trials. Frog A was installed in cold water above a moderate heat. Frog B entered water that was lukewarm and heating. Frog A jumped away after 4.2 seconds. Frog B exited at 1.57.

Another way to test this would be by substituting news reports for the slow pot. Two concepts came out of the seventies scientific panels: "Urgency" and "Too Late." Urgency meant the problem could probably do with a little present-tense attention. Too Late meant that even with some warming uncertainties, action now represented bargains in the future. If the sofa and drapes start to smolder, better to hose down the living room than wait and see what develops.

Here's how Urgency made out, in five different decades, in the *New York Times*.

One day after the National Academy of Sciences' 1977 report, the *Times* editorial board explained that experts "counsel not panic but 'a lively sense of urgency'... the penalty for doing nothing until a climatic change is actually observed seems too high to risk."

In 1989, the paper reported that "world political leaders" were "rushing to broad accord on the urgent need to halt global warming." The following decade, in 1997, the paper asked, "Just how urgent is the problem of climate change?"

A decade later, and a 2008 editorial: "Another Failure on Climate Change." And it was as if a used car had got its mileage clock turned all the way back to zero, and it was still 1977. "We hope the next president will have the necessary conviction and stamina—and a real sense of urgency. Too much time has been wasted." New president, next decade, and a 2015 front page, "Citing Urgency, World Leaders Converge on France for Climate Talks."

Here's how Too Late did. In 1979, the chairman of the National Academy of Sciences' Climate Research Board was quoted in the *Times*: "A wait-and-see policy may mean waiting until it's too late." A year later, the editorial board expressed the idea in the paper's own voice: "By the time climatic changes can be unequivocally detected, it might be too late to reverse the carbon dioxide effect."

In 1987, a scientist from the National Center for Atmospheric Research passed along the notion to readers of *Time*. "By the time we know our theory is correct, it will be too late to stop the heating that has already occurred." (Another climate expert had this to tell the magazine about the ninety-year-old concept: "The greenhouse effect is the least controversial theory in atmospheric science.") A year later, in their 1988 cover "The Greenhouse Effect," *Newsweek* brought the phrase before its readership: "If we wait for proof beyond any doubt, it will be too late to take preventative steps."

In 1988, the *Times* warned, "Each decade of delay may commit the atmosphere to future warming." But it was 1989 that became, for the paper, the Year of Too Late. January editorial: "There's every reason to take action immediately, and not wait until that debate is concluded." February: "[Scientists] do not have precise answers about global warming, but if they wait for the answers and they are too long in coming, it will be too late to take effective action." March: "The EPA warns that action must be taken now, not after all the uncertainties are resolved." November: "Experts warn that if the world waits for incontrovertible evidence of major climate change, it will be far too late to do much about it." Same week: "Some experts, moreover, express concern that the industrialized countries, following a well-documented human tendency, will not take adaptive action until disaster is near, when it may be too late." Ten more days, this coda: "When proof of the warming finally arrives, the climate will already be locked into a significant temperature rise."

In 1995, *Time* did the broadsheet one better: Too Late had already come and gone. (Their piece is also a demonstration of the power of the question mark: "Heading for Apocalypse?") "A more realistic strategy, some scientists argue, is to spend what research money there is figuring out how best to deal with global warming when it comes. It's already too late, they say, to do much else." In 1997, the head of the National Oceanic and Atmospheric

Agency explained to the *Washington Post* just how firmly scientific opinion had solidified around CO_2: "There's a better scientific consensus on this issue than on any other I know." In 2001, *Science's* editor-in-chief wrote, "Consensus as strong as the one that has developed around this topic is rare in science."

As of 2006, the tone around Too Late had entirely shifted; shouldered aside by a new idea, Adaptation. England's chief science advisor, Sir David King, told a reporter from *Vanity Fair* that even if carbon dioxide emissions could be brought to an absolute halt, it would hardly matter. "Temperatures will keep rising and all the impacts will keep changing for about twenty-five years." The magazine continued, "The upshot is that it has become too late to prevent climate change; we can only adapt to it. This unhappy fact is not well understood by the general public; advocates downplay it, perhaps for fear of fostering a paralyzing despair." (A Princeton professor of geosciences and international affairs told the reporter, "For 25 years, people have been warning that we had a window of opportunity to take action, and if we waited until the effects were obvious it would be too late to avoid major consequences." He went on, "Had some individual countries, especially the United States, begun to act in the early to mid-1990s, we might have made it. But we didn't, and now the impacts are here.")

In 2007, *Newsweek* offered its own Adaptation story—the far side of the river, the shore you hope never to see. "Learning to Love Climate 'Adaptation.' It's too late to stop global warming. Now we have to figure out how to survive it." Some effects would "be so costly," the magazine reported, "that we'll look back on the era beginning in 1988, when credible warnings about climate change reached critical mass, and wonder how we were so stupid as to blow the chance to keep global warming to nothing more extreme than a few more mild days in March."

(All of which gave a *Time* cover from the period—"Earth Is at . . . the Tipping Point"—a touch of bleak comedy. "The climate is crashing, and global warming is to blame. Why the crisis hit so soon.")

In 2008, *the Times* was still playing the same music. It considered studies on limiting emissions—"The same studies also make clear that the costs of inaction will dwarf the costs of acting now." And addressing the 2015 question "How much trouble are we in?" the *Times* explained, "It is already too late to eliminate the risks entirely." Whatever new direction frog research takes, it may be only human beings who have the stamina and determination to stay in the same pot.

THE UNWARRANTED
AND ALARMIST REPORT

FROM HERE ON OUT IT'S POLITICS: STUDIES AND HEAR-
ing rooms. The turn of the seasons in the capital—faces on the news, headlines you forget, replaced by fresh headlines and other faces.

Global warming had become a Washington fact—that system where
all data is looked at the same way: an edge for your side, a boost to the
other guy. (Conservatives didn't see a new threat: they saw progressives trying to smuggle the old anticorporate agenda into the auditorium under a
fresh blanket.) If you tell me your politics, I can predict your thinking on
global warming. Which is a strange thing to say about science. Your attitude toward quarks, your policy on the electron, won't tell a pollster much
about which way you'll go in the voting booth.

The first parts of global warming are a detective novel. Politics changed
this to a story of non-detection—it became a movie where the hero cop
keeps getting yanked off the case.

Ronald Reagan came to Washington from Hollywood. He had a back-
lot past and studio mission, to plug up holes in the story. Keep the attractive
hero—America as charismatic and optimistic nation—ditch the subplots:
regulation, taxes, doubt. He cleaned house on the environment, because it
was complicated and where was it heading?

The chief of the climate research program got demoted. Then Dave
Keeling received word: For two decades, he'd been let alone, measuring
carbon dioxide to his heart's content. Now his government ride was at
an end. (Keeling's memoir gives a clipped explanation: "Ronald Reagan

had become President of the United States.") Climate scientists were instructed to button their traps. "The new administration laid plans to cut funding for carbon dioxide studies in particular," writes historian Spencer Weart. "Everything connected with the subject became politically sensitive."

The aim was "a total gutting." The White House science advisor had identified a potent new source of the national ills. Not the Soviet Union, the oil cartels, or smoke. Science journalists. "We are trying to build up America and the press is trying to tear down America," he said. The science advisor announced that, staffing up research panels, the qualification would be "competent scientists who understand what President Reagan perceives as the role of government."

The Department of Energy was a hangdog subplot. It wore the cast's long face—fixated on shortages, the home of warming research. (Where there'd been so much bureaucratic activity a staffer griped, "If anything has been meetinged to death, it's carbon dioxide. . . . If conferences could solve problems, it should be solved by now." A month before the new president's inaugural, the department released another study: global warming's effects would be "beyond human experience.") It was like the Hollywood red pen, hovering over a script element. President Reagan decided to scratch out the entire department.

The first energy secretary had a deep, tanks-and-espionage résumé: tours as secretary of defense and head of the CIA. Reagan's pick was a South Carolina governor whose longest professional experience was as a dentist. The new secretary cheerfully told Congress he aimed to shutter the agency. He was the new boss who flies in under corporate order to negotiate pensions and sell off the photocopying equipment. "Yes sir," announced the secretary, "my job is to abolish the department. I hope to be back in South Carolina hunting and fishing in April."

A FEW DAYS BEFORE Halloween 1983, President Reagan got his Carter moment. Every chief executive since has had one—when the different strands of the story knit together, and global warming rises to the top of the pile. By a freak of scheduling weather, two massive reports blew across

Washington the same week. One came out of the EPA. The other was from the National Academy of Sciences.

The EPA report landed first, ran 200 pages, and was, according to the *Times*, a "bombshell." As the paper explained on page one, a landmark, "the first warning by the Federal Government that the 'greenhouse effect' is not a theoretical problem but a threat whose first effects will be felt within a few years."

If you'd been following the story, this is the report you could have written yourself. Warming was imminent, inevitable, and would set in by the 1990s. By the late twenty-first century, changes could be "catastrophic." (Tom Brokaw, the NBC News desk: "The EPA said today that our earthly paradise is heating up at an alarming rate, and there's very little we can do to stop it. . . . It could have a big effect on life on this planet.") The EPA recommended immediate action. "We are trying to get people to realize," explained their director of strategic studies, "that changes are coming sooner than they expected."

The report had been test-read by one hundred scientists. Their most common complaint: "too conservative."

The second report tore through three days later—it was 496 pages (a double-decker) and welcomed by the White House, which triggered what *Time* called "a media brouhaha." (Ringing phones, more TV and headlines.) It was as if the scientific community had shown the administration two suits; the White House decided the National Academy study was the one it looked good in.

In many ways it was lots scarier. The carbon dioxide problem, the Academy found, had "no solution." (A later explanation was even more demoralizing. It made warming sound unrepentant and recidivist. "Viewed in terms of energy, global pollution, and worldwide environmental damage, the 'CO_2 problem' appears intractable.") The issue would involve "virtually every branch of science" and impinge on "virtually every area of human activity." It was caused by cars, power plants, airplanes—every piece on the modern-society game board. Changes would "carry our planet into largely unknown territory." This was because it promised "to impose a warming of unusual magnitude on a global climate that is already unusually warm."

Warmth would crowd the world stage with bad news—where it could prove "a potent source of international conflict." All in all, the Academy found itself struck by "a profound uneasiness about inducing environmental changes of this magnitude. . . . We may get into trouble in ways that we have barely imagined."

But this was the study the White House applauded. And by Friday, it was the EPA's work the science advisor was condemning as "unwarranted and unnecessarily alarmist."

Here's why. To the extent it could influence things, the White House had installed its own man on the panel. "I started this National Academy study with Bill Nierenberg chairing it," the Science Advisor recalled decades later. "We kept track of it." Dr. William Nierenberg had served on the Reagan transition team. Individual scientists wrote individual sections: Nierenberg handled recommendations, executive summary, press release—which put him in the catbird's seat, in charge of interpretation. Dr. Nierenberg used the potent word "uncertainty" sixty-five times in seventy-eight pages. He wrote of the potential "benefits," of keeping "an open mind."

And he had good news. Warming might be inevitable but it would not be unmanageable. About the same as renting a U-Haul, chucking the Dakotas for the Sun Belt.

"For the individual, in contrast to the environment, the idea of climate change in a generation or two is far from novel," Dr. Nierenberg explained. "A change in the climates where people live may not be altogether different from moving to another climate." The scientist added, "Not only have people moved, but they have taken with them their horses, dogs, children, technologies, crops, livestock, and hobbies. It is extraordinary how adaptable people can be."

The White House liked his overall counsel even better. Stick to the books. "The best single investment strategy for coping with the CO_2 issue," Dr. Nierenberg said, "is more research." (It makes you think of Louis McCabe and Los Angeles smog: "Research properly used is a valuable tool. It is not a panacea, and it is a dangerous sedative.") The science advisor spoke with reporters. "At this time," he emphasized, "there are no actions recommended other than continued research."

That's the message that went out to readers. *Time* repeated that warnings were "unduly alarmist." (And allowed Dr. Nierenberg to lodge appearance number sixty-six: "The issue," the chairman told the magazine, "is full of uncertainties.") The *Times* called the findings "far less urgent. . . . Academy scientists advised a wait-and-see attitude." Over at the *Wall Street Journal*, readers received the all-clear: "A panel of top scientists has some advice for people worried about the much-publicized warming of the Earth's climate: You can cope." Urgency drained out of the newspaper reports. Then the stories drained out of the papers. The issue went away for years.

The eighties would sign the register as the warmest decade in recorded temperature history. 1980, 1981, 1983, and 1987 checked in as the hottest years—then 1988 re-broke the record. That spring, England's lead climate researcher spoke with the *New York Times*. If the next decade was as warm or warmer, it "would be very hard to deny the greenhouse effect." The scientist added, "It is very hard to deny now."

Years later, an ice core researcher named Richard Alley would chair a much darker National Academy panel, which would reverse a lot of the 1983 recommendations. In a book, Dr. Alley offers a kind of implicit answer to Nierenberg's good words re extraordinary human adaptability.

He's writing about Greenland: what happened to people and livestock during a period of rapid climate change—a sudden cooling. "The settlers brought farm animals into their houses during the cold winters," Dr. Alley explains. "Eventually, the settlers ate their farm animals, then their dogs, then disappeared themselves."

THE UNDOING OF
THOMAS MIDGLEY

JUNE 23, 1988, IS AN ANNIVERSARY, A DATE GLOBAL WARM-
ing people remember, the way you would a graduation, a wedding, the birth
of a child. Dr. James Hansen, the head of a NASA study group, became
the first scientist to announce that everything had started, the theory had
become a fact. He told Congress the warming had begun.

It was the scientists' after-hours talk pressing into the nine-to-five—
what they said privately near the sign-in tables at conferences, in the parking lots headed for their cars. "Dr. Hansen," the *Times* wrote, "sounded the
alarm with such authority and force that the issue of an overheating world
has suddenly moved to the forefront of public concern." (Their headline
was "His Bold Statement Transforms the Debate.") Two decades later, the
Times could still track the landscape's wobble and shift: "Twenty years ago
Monday, James E. Hansen, a climate scientist at NASA, shook Washington
and the world."

DURING THE EIGHTIES, warming had been nudged aside as the environmental headliner. Its place was taken by a full-on, wait-and-see experiment; you got to watch what happened when Urgency and Too Late
slipped right over the falls.

In the late 1920s, Thomas Midgley Jr., a scientist in the labs at General Motors, invented a whole new class of chemicals. Midgley had hard,
kind eyes, the successful man's slightly combative smile. General Motors
needed a coolant for its Frigidaire line of refrigerators. Something safe

and nontoxic—it'd be working around food all day—to keep the milk and eggs fresh.

Midgley came from a family of inventors. His grandfather helped create the band saw, giving a universe of shop classes and Home Depots the smell of warm sawdust. His father helped perfect the automobile tire. Midgley was a big, trusted name in the industry. He'd already earned General Motors vast sums of money by inventing leaded gasoline. This took the knock out of car engines, and in the process spread lead everywhere a car could go. Salesmen called the no-sputter gasoline "a gift of God." (By the seventies, most Americans—and most Russians, Japanese, Europeans, Canadians—had elevated lead trundling around in their bloodstreams. The fuel was banned.)

Leaded gasoline took Midgley six years, a house-to-house search through the periodic table. His refrigerator solution took three days. He mixed chlorine, fluorine, and carbon—CFCs, chlorofluorocarbons. Was it safe? Midgley had a touch of the showman. At a meeting of the American Chemical Society, he stood behind a table. Then, like a dashing movie scientist, he swallowed a deep breath of his own compound: nontoxic. He exhaled over a candle: not flammable. Safe.

Marketed as Freon, the gas opened the gates to a climate-controlled world. To summer AC and condos in the desert—and malls, supermarkets, and offices with the bite of autumn. (In 1933, the *Washington Post* said air-conditioning had finally "made office work conditions ideal." The assistant with a July cardigan around her shoulders is a monument to Thomas Midgley.) Freon was adaptable. It shipped for the tropics as the GI's bug bomb, wiping out whole colonies of mosquitoes. In the fifties, it pushed into the bathrooms—where it made an excellent propellant for aerosol cans, delivering underarm deodorant and hair spray, joining the morning's noises. Alarm clock, coffee machine, shower, pfft! Production doubled, then redoubled.

Thomas Midgley never got to see any of this. In 1940, age fifty-one, he was crippled by polio. He was an inventor. He built a rope and pulley system to hoist himself out of bed. In 1944, depressed and alone, Midgley hanged himself with the cords. Which may have been for the best. Midgley was a generous, thoughtful man who aimed to improve things. A historian later wrote that he turned out to have "more impact on the atmosphere than any single organism in earth history."

F. SHERWOOD ROWLAND grew up as a kind of reverse Midgley. He was born the year before chlorofluorocarbons and raised in an academic household. In such homes, science becomes the family farm, a chore list you inherit. "I do not honestly remember when the decision that I would go to graduate school was made," Rowland wrote, in his Nobel Prize autobiography. "My father had studied for his Ph.D., and all of us took it for granted that I would, too."

He studied chemistry at the University of Chicago. An exciting, sobering place to be young: Giants at the blackboard, examples of science stepping off the green to do big work. During the wartime scramble for the bomb, under the football bleachers, Enrico Fermi had built the world's first atomic reactor. He used a squash court—which Russian journalists mistranslated, leaving their nation the odd image of Fermi splitting the atom in a pumpkin patch. "I see you made all A's in undergraduate school," Rowland's faculty advisor told him. This advisor, too, would eventually win the Nobel Prize. "We're here to find out if you are any damn good!"

Graduated, Rowland followed the postwar path. Some government research—Brookhaven National Labs. Some federal contracts—the Atomic Energy Commission.

By the early seventies, he'd become chairman of chemistry at the University of California, Irvine. A skinny six-five, with muttonchops and the grave patience tall people show a world that's on a much smaller scale. Rowland had the look of a local TV weatherman—owl glasses, weedy brows, friendly and concerned; a man who would use complicated phrases, then take the trouble to explain.

Rowland was naturally curious. To a scientist, any day's newspaper is also a menu of potential experiments. Reading about the scary mercury rise in commercial tuna, Rowland hunted down museums that preserved antique specimens of fish. He twisted open, tested. The level of a century ago matched the current one. Nothing to get worked up about, there'd always been mercury in tuna.

He published in 1972. So that's how Rowland entered the environmental record: as a pollution apologist, the ocean dumper's friend.

The episode neatly illustrates a divide: environmentalist and scientist. Environmentalists follow a cause, scientists are loyal to facts. Scientists often mistrust environmentalists—who are, in many cases, anti-science. The next year, his attention snagged on the spray cans.

About eight hundred thousand tons of Freon were exiting the nozzles and air conditioners every year. Where did that gas go? As a scientist, Rowland knew the strange truth: that no trip ends, Arrivals and Departures are really the same gate. (The writer Vladimir Nabokov has a haunting, funny gloss on this. He's talking about a splinter removed from his eye in 1911: "I wonder where that speck is now? The dull, mad fact is that it *does* exist somewhere.") Rowland understood the spray cans were a way station, an overnight hotel: Midgley's Freon would continue its journey, rise to the atmosphere. He took another grant from the Atomic Energy Commission. In September 1973, he began research with another man, a laser chemist named Mario Molina.

Things happened very fast. First two months, boring. Third month, terrifying. It turned out the CFCs would collect in the stratosphere. They would break down—and their freed chlorine would begin munching through ozone. The ozone layer, twenty miles up, acts as the planet's sunblock; it umbrellas the world from about 98 percent of the sun's ultraviolet radiation, which would otherwise blur you with cataracts, bump you over with skin cancer. CFCs turned out to be super-soldiers: a single chlorine atom could take out one hundred thousand molecules of ozone. And every minute, more CFCs were spraying, rising, drifting, arriving.

Rowland and Molina stared at each other. Then, very slowly, over three days, they rechecked every calculation. (This is, eerily, the exact number of days it took Thomas Midgley to create CFCs.) Rowland drove home. Palm trees, onramp, driveway, lawn. His wife asked if anything at the lab was bothering him. "The work is going very well," Rowland said. "But it may mean the end of the world."

They published in June 1974—and in September addressed the same

American Chemical Society where Tom Midgley had once taken his non-poisonous gulp of Freon. (Midgley likened the sensation to a cocktail—except that "these fumes do not arouse a desire to sing or to recite poetry.") Rowland and Molina called for an immediate ban. They anticipated an ozone loss between 7 and 13 percent—which meant an eighty thousand rise in annual skin cancers, a number that would itself be revised upward, to "several hundred thousand."

Immediate response. In antiperspirant terms, on drugstore shelves, this began the golden age of the roll-on. Consumers sent more letters to Congress than over any issue since the Vietnam War. The *New York Times* described the dark, giddy impact of Dr. Rowland's discovery: the news that by hitching a towel around your waist, pressing a nozzle, you were mucking with the future. "It was like finding out," the paper wrote, "that eating candy causes earthquakes."

CHEMICAL GIANT DUPONT (the full name is like a sweeping courtier bow: E. I. Du Pont de Nemours) had become the world's largest CFC manufacturer. Rowland's paper presented them with two options. One, they could have quietly lowered the curtain on a billion-dollar industry. They chose to fight instead, in the process perfecting nearly every strategy industry and lawmakers would use against climate change. A brilliant pollster named Frank Luntz later certified every one. Peeped at over the hedge of ozone, the campaign turns to rerun, a case of déjà argued.

The industry floated advocacy groups: The Aerosol Education Bureau, the Council on Atmospheric Sciences. Nonpartisan and trustworthy sounding, fact-finders summoned reluctantly to the barracks from the fields.

The Aerosol Education Bureau pointed out that the chlorine could have innocent origins—it might be entering the atmosphere from ocean spray, or volcanoes. In the next decade, the main fuel and automobile trade group opposed to climate science would call itself the Global Climate Coalition—which certainly sounds as if they're working the environmental side of the street.

Second, the industry funded its own scientists and called for More

Research. (In a political campaign, these would be the Opposition experts: steely-eyed and long-memoried, crawling over the other guy's record.) Decades later, in his 2002 strategy memo, Frank Luntz would promote this strategy especially, as a strong, late-quarter move against global warming.

"The scientific debate is closing (against us) but not yet closed," he'd write. "There is still a window of opportunity to challenge the science." Luntz's advice: become "more active in recruiting experts who are sympathetic to your view, and much more active in making them part of your message. People are willing to trust scientists. . . . For example, you should argue that America should invest more in research and development."

Which meant more days on the docket, more months on supermarket shelves. As a science journalist wrote in *Discover* magazine, "No thoughtful action could be taken until research was completed, which always seemed to be three to ten years away."

Next, bring the fight to your critics: question character, approach, motive. DuPont painted Sherwood Rowland as a goony-bird, head-in-the-clouds scientist—long on theory, short on proof. His work was all just simulations on a computer anyway: no more reliable, really, than thinking you could train for Wimbledon by playing Nintendo. (This would become a big argument against warming.) The words industry spokesmen used came from the defuser's section of the thesaurus: "Hypothetical," "assumption," "uncertain," "unwarranted," "evidence," "hard facts," "rational science."

This was another sharp instinct Frank Luntz would later codify, for use against global warming. "The most important principle in any discussion is your commitment to sound science."

Luntz is the man who suggested replacing "global warming" with "climate change," because it sounded less frightening. Luntz is a focus-group artist—a man who's put in the time under fluorescent lights. Coffee cups and opinions, trotting different words past voters, seeing what makes them jump. Luntz's 2007 how-to is called *Words That Work*. He is an ideal talk show guest: a man excited by his own specialty. Luntz explained to the New Yorker, "Perception is reality. In fact, perception is more real than reality."

Here's what focus groups would tell Frank Luntz about global warming: uncertainty belonged in the playbook, too. "Should the public come to believe that scientific issues are settled, their views about global warming will change accordingly," he'd write. "Therefore, you need to continue to make the lack of scientific certainty an issue in the debate." In 1976, the CFC industry's science advisor told the *Times* he would never presume to claim there was no ozone connection—only, that the effect was "uncertain." DuPont's R&D chief lobbied the National Academy of Sciences. Any legislation based upon "unproven theory," he maintained, was "the road to the rule of witchcraft."

Last, put your hands on those working words. Move down the political cafeteria line, loading your plate with values and symbolism.

Luntz again, turning notions and success into formula: "We have the moral and rhetorical high ground when we talk about values, like freedom, responsibility, and accountability." Here's the Luntz explanation, as summarized in the *New Yorker*. A lot of it has to do with sound, mental triggers. We turn out to be susceptible to a certain band of moral noise. "In general, words starting with an 'R' or ending with an 'ity' are good," the magazine reported. So " 'reform' and 'accountability' work, and 'responsibility' really works." ("Listening" and "children," on the other hand, offer a lock on female voters.) "Americans want a free and open discussion," Luntz would write. "The words on these pages are tested—they work!"

Luntz's effort has since become notorious: an example of what's lost when science and politics get gulped by marketing. It's thought to have been especially helpful to Exxon and the second Bush White House. In 2006, cornered by the PBS show *Frontline*, Luntz replied, "Look, you want me to say it? It was a great memo. It was great language. I busted my ass for that memo."

In 1975, a CFC spokesman told reporters that, what with all the "shortcomings and uncertainties," action on the chemicals would be "utterly against American tradition." At a press conference, he said, "We would like a fair trial, not a lynching party."

DuPont's chairman leased a full page in the *Times*. One of the most high-priced and high-profile executive lounges: a place where a man can loosen his collar, mix a drink, deliver an argument as calm and undisputed as in a dream. Rowland's work was downgraded to a "controversy." Which meant nothing worse than opposing positions. And "nothing," the chairman wrote, "is settled." He spoke of CFCs as "these useful, inert gases"—as though they were an unfairly marginalized group—he talked about fairness, responsibility.

AND NONE OF IT WORKED. Maybe DuPont expected Rowland to fold, back down, sue for peace. If so, they'd pegged Rowland as the wrong kind of pasty, wobbly-kneed scientist. Rowland had played baseball in the navy; he hitchhiked home two thousand miles after his bit. He earned extra money in graduate school as player-manager with a minor-league ball club, the Oshawa Merchants, and led his team to a pennant. These were the middle innings, when you stick.

"We know that a ban is inevitable," Rowland told reporters. "The question is merely when." In 1975, Oregon passed the nation's first aerosol ban. (There's something about Oregon. In 1974, the state passed one of the first recycling bills, helping send Keep America Beautiful around the bend. In 1993, Portland would become the first American city to regulate its own carbon dioxide.) A state senator called Freon "simply an unconscionable method for handling underarm spray." A year later, the FDA, EPA, and the Consumer Product Safety Commission teamed up to support a nationwide ban. Exemptions would be granted for asthma inhalers, contraceptive foam, insect sprays: for personal health and public good.

Otherwise, no more CFC spray cans as of 1979. The seventies featured iron-on decals, regrettable hairstyles, a version of "Sgt. Pepper's Lonely Hearts Club Band" as performed by the Bee Gees, but the political system worked. (Imagine a state of affairs where the seventies looks like a more reasonable era than our own.) The entire process, from discovery to preventive measures, had taken just four years.

THE UNDOING II—RED DAYS

EXCEPT. THE CFCS STILL PERCOLATED IN THE AIR CONDI-
tioners and refrigerators. They still squared the Styrofoam for Big Macs and
supermarket eggs. They were in the car's AC—also the seat cushions and
the dash. This represented half the market.

Manufacturers did not even consider the seventies a setback. One public relations chief later described them as "a smashing success." Advocacy
groups, paid editorials, hired scientists—all of it managed to "buy time" for
the industry, still taking in a billion dollars per year.

In the eighties, the industry took off on a new tack; a heading that
would become very popular in our own era. Why act without the rest of the
world? You'd just be handing the market to foreign suppliers, a case of help
there, harm here. After all, the sky recognized no borders. A smoky India, a
Freon China, ate just the same ozone as a polluting Ohio.

This strategy would also receive the Frank Luntz blessing. "The 'international fairness' issue is the emotional home run," Luntz explained. "Given
the chance, Americans will demand that all nations be part of any international global warming treaty. Nations such as China, India, and Mexico
would have to sign."

Dry pens in India and China would provide the Senate with an official
excuse for turning its back on the first international climate change treaty,
the 1997 Kyoto Accords. China was so important—full of power plants and
aspiring motorists—that two months before negotiation, Exxon Chairman
Lee Raymond would travel to Beijing and make the case personally.

"I hope that the governments of this region will work with us," the CEO said, "to resist policies that would strangle economic growth." Raymond gave the developing world his assurance: their "most pressing environmental problems" would be solved by "increasing, not curtailing, the use of fossil fuels." (As of our era, the leading cause of environmental death in China is increased use of fossil fuels.)

Science and politics were speed bumps. In 1986, EPA released a draft report. Ozone loss could mean eight hundred thousand deaths and forty million extra skin cancers in the next century. (The sun would be biased about this. Affecting, the *Times* reported, "people with fair skins," blue and green eyes, Celtic heritage, or "poor tanning ability.") Ten days later, the industry was still touting the "enormous benefits"— still describing Thomas Midgley's invention as "among the most useful compounds ever devised."

And lots of markets had opened up, far from the bathmats. "The CFC industry managed to find other uses," Sherwood Rowland said. These mid-eighties markets, Rowland sighed, "now put the same amount of CFC gases into the atmosphere as before."

A *New Yorker* writer named Paul Brodeur visited the scientist. Rowland had stopped being a researcher who scans the headlines for lab ideas. He'd become the kind of man you see at the library in the afternoons, reading the newspaper with a rooting interest, a grudge. (It's unsettling to wrestle an industry. Your best punches go into a void; the return jab comes from a thousand arms.) Rowland handed over clippings, discussed reports. Back in 1974, a DuPont spokesman had sat down before a congressional subcommittee, leaned into the microphone bouquet, and made a promise: If "credible scientific data" showed the gases were a threat, DuPont would halt production.

"The CFC industry," Rowland said, "appears to have decided that it does not intend to consider any evidence credible." He called this approach "blatantly cynical." He said, "From what I've seen over the past ten years, nothing will be done about this problem until a significant loss of ozone has occurred." What Rowland meant was, the only convincing evidence would be "a disaster."

THE ANSWER CAME FROM the quiet part of the class. From Halley Bay, on the Antarctic coast. (Winter temperature minus 20 degrees Celsius; summer minus 10, which means zipping up fewer parkas.) In 1977, a team with the British Antarctic Survey began to record a tremendous plunge in ozone. The figures were troubling, puzzling, and embarrassing. What researchers thought it meant was that their equipment was busted. They had new instruments flown in, same result.

In 1984, they confirmed the data with a second British team stationed in Argentina. The sun's UV radiation—the cancer-causing band—had jumped by a factor of ten; the ozone layer had declined nearly 40 percent. Overhead, a NASA satellite called Nimbus 7 had been logging the same figures. But the satellite was programmed the same way as the scientists: for years, the machine had tossed out ozone numbers it considered impossibly low. A hole twice the size of the continental United States had opened in the ozone above Antarctica. (In satellite images, the hole resembles a giant eyeball; in others, the hero of the video game Pac-Man, mouth parted, headed straight for you.) The hole was larger than Antarctica itself. It could have served as a road sign for astronauts: it was big enough to be seen from Mars.

The hole sparked what one researcher calls a "global scientific frenzy." Within two years, world ministers had met, hashed out their differences, signed an international freeze. President Reagan's secretary of the interior argued you could do the job just as well with a public service campaign. Under his plan—called "Personal Protection"—industry would continue to manufacture CFCs; everybody else would just work around them. The public would be encouraged to clap on sunglasses, lotion, celebrity-size sun hats. "People who don't stand out in the sun," the secretary explained, "it doesn't affect them."

DuPont rented another full-page in the New York Times, welcoming the ban. Because when you switch sides, you switch sides all the way. "Progress been made," their ad announced. Also, "Here's what you can do now" and "The sooner the better." Sherwood Rowland received the Nobel Prize for Chemistry. DuPont revealed it had only ever earned about 2 percent of its income from the chemicals anyway. Pocket change. (Accountants did a tour of the company books: "To cease manufacture of CFCs would have no meaningful impact on our financial results.") Which raised a question: How long and nasty could a fight get, how many teeth and claws would an industry show, if the other 98 percent were at stake? Sherwood Rowland, with a rueful head shake, posed one of the great questions of the age. "What's the use of having developed a science well enough to make predictions," he asked, "if in the end all we're willing to do is stand around and wait for them to come true?"

THE OZONE DROP has caused between one and two million excess skin cancers worldwide since 1975—in the winter of 1995, eighty thousand in Western Europe alone. The ozone hole has also given us a preview of what Adaptation might look like. Punta Arenas, at the tip of Chile, is the world's southernmost city; 120,000 people live under the ozone hole. Walkaround mascots—men in penguin suits, in the tradition of the theme park—appear at school assemblies, to demonstrate the proper application of sunblock. Alerts are published in the papers. They tell what the *agujero*, the hole, is going to be doing on any given day. A stoplight system. Green means go about your business normally. Yellow, sunglasses and a hat. Orange, wear sunscreen. Red, keep to the shade. On red days, the sky goes white, the ocean turns to glare, sun bounces off fenders and windows like a searchlight beamed into a soup can. In one seven-year period, skin cancer jumped 66 percent.

The director of the city's Ozone Education Program encourages people to imagine themselves in a personal relation with the *agujero*, to welcome the hole as a friend. "There is nothing else they can do," the director says.

In the United States, the ozone hole touched different people different ways.

DuPont became a model corporate citizen, the reformed parishioner who ushers at church. In 1991, the company saw "sufficient evidence" of global warming, its industry affairs manager said, "to warrant what in our judgment was prudent action." (Thinking about DuPont after CFCs, I remember what Martin Amis wrote about KGB agents under Stalin. "When it was their turn to be purged, former interrogators...immediately called with a flourish for the pen and the dotted line.") The company has since cut its greenhouse emissions 72 percent, a better record than almost any government.

For citizens, ozone laid conceptual track, pioneering a once-incredible thought. Gases, even safe and invisible ones, might one day circle back around to harm us. It changed the feel of things. In some moods, you looked at the sky, and instead of clouds and blue you saw menace and overuse.

For lawmakers, the hole became a model of government action. A GPS line showing the passage between there and here. "It was the first time," an assistant secretary of state told the *National Journal*. "There's no question that set the form institutionally."

And for hired scientists, ozone set the ground conditions for the next decade. Dr. S. Fred Singer, an atmospheric physicist, wrote a long piece in 1987 defending CFCs. He used carbon dioxide as his club. The thinking is tricky to follow: one step, second step, crazy swerve. CFCs were not only a threat to ozone, they could also "lead to climate warming. . . . However, their effects are small compared with those of carbon dioxide from the burning of fossil fuels," he explained. "And no one is suggesting that we stop energy production and freeze in the dark." Two decades later, this same Dr. Singer was telling ABC News that warming science was "all bunk." Ozone taught such men flexibility. Dr. Singer would become one of the world's most prominent climate deniers.

And for government opponents, ozone provided a chance, from their spot on the line, to alter the way climate treaties reached the books in the first place. In 1990, President Bush's Chief of Staff John Sununu opposed an international plan—$100 million to finally bring China and India into the ozone family. The sums debated were paltry: the US share was $25 million over three years, the national equivalent of buying an exercise bike in installments. But China and India were also a climate-change sticking point. As the *Washington Post* reported, John Sununu was "adamantly opposed to the plan out of concern that it might become a precedent" for treaties about global warming.

THE HOME OF DONNA REED

THE EIGHTIES WORKED THE NERVES OF THE CLIMATE scientists. President Reagan never spent more than fifty million dollars a year on warming research—roughly the budget of the first *Ghostbusters*, a Hollywood comedy about scientists produced under Reagan. (The figure is 2.4 annual billions today.) A British study ranked history's six warmest recorded years in descending order: 1988, 1987, 1983, 1981, 1980, 1986. Christmas Eve in the capital, 1982. Outdoor cafés, shirtless joggers, 66 degrees. "Man, there's only one explanation," a White House employee told the *Post*. "Washington is one messed-up kind of town."

In 1987, warming made the cover of *Time*. An emotional moment for a national issue, resembling the announcement when a Hollywood journeyman like George Clooney lands his first starring role. "We play Russian roulette with climate, hoping that the future will hold no unpleasant surprises," Wallace Broecker wrote that year in *Nature*. "I am less optimistic about its contents than many."

In fact, not very optimistic at all. The scientist testified before the Senate Subcommittee on Environmental Protection. "I come here as sort of the prophet," Dr. Broecker said. "There are going to be harsh changes."

As the Pulitzer Prize-winning science journalist Jonathan Weiner writes it, climate researchers have the sound of intelligence analysts in the summer of 2001: drumming on doors, pushing files in your hand, running from office to office with their hair on fire. "By the second half of the nineteen-eighties, many experts were frantic to persuade the world of what they thought was about to happen," he writes. "Yet they could not afford to sound frantic, or they would lose credibility. They were placed in a curious position as human beings."

In 1987, Jim Hansen brought charts, figures, and predictions before a Senate committee. It was November, football weather; no one noticed. "We had a very, very distinguished panel," a senator recalled. "And who was in the cavernous hearing room? Six or seven people, and two or three of them were lost tourists."

Dr. Hansen asked for another shot. The senator—Tim Wirth of Colorado—had learned the great lesson from President Reagan's White House. Everything plays better with a special effect or two. Wirth phoned the National Weather Bureau: What, traditionally, was the city's warmest date? June twenty-third. "So we scheduled the hearing that day," Wirth said. "And bingo: It was the hottest day on record in Washington, or close to it."

And the national weather map, obligingly, became a giant supporting player. 1988 saw the worst drought in half a century. Federal emergencies in thirty states. On the grain fields, they were calling it The Big Dry. On the Mississippi, barges beached and wallowed in hot mud. The day of Hansen's testimony, eight cities broke heat records.

In Washington, Senator Wirth's staff didn't leave matters to chance. A fraternity stunt, like a sequence left over from *Animal House*: Aides snuck into the hearing room, Dirksen 366, the evening before, and pushed open all the windows. This let the heat gust in, setting the Senate ventilation a nearly impossible task. By the time Hansen sat down to speak, comfort had evaporated from the room. It was nearly a hundred degrees, a bank of TV cameras (Senator Wirth had tipped reporters) making it even hotter. Coats came off, sweat glossed foreheads.

I wonder what the moment felt like to Hansen. He was in Washington—for a scientist, a strange, hostile land, filled with loud authorities and bewildering customs. Roger Revelle once explained the difference between lawmaker and researcher. "The politician," Revelle said, is "gregarious, garrulous, and has a strong gambling instinct." The scientist, on the other hand, "is publicly modest, introverted, relatively inarticulate, and seeks certainty rather than risk."

Hansen had to squint down the part of his brain that naturally said, "Well, maybe not." He was overcoming a career's worth of training—he was forty-seven years old, he'd never been anything but a scientist. He told an interviewer he'd mulled his testimony for weeks: "I weighed the costs of being wrong versus the costs of not talking." Hansen told another journalist, "I think you just have to do what is right—that's what I learned in 1988." I've gone back and rewatched the moment many times. Hansen tucks in his chin; he reads from notes, the words scholarly and official. Maybe this helped: It was the figures talking, not Hansen. He says, "The greenhouse effect has been detected, and it is changing our climate now."

It had taken almost a century, from Svante Arrhenius in 1895 to 1988 and the Dirksen hearing room, to get it said. Another Hansen remark from that day has also become famous. It shows what Revelle meant about the scientist's love of certainty; how the skills that outfit you for life in the lab—rinsing an experiment with doubt over and over in your mind—are the ones that misoutfit you for public life. "It is time to stop waffling so much," Hansen said, "and say that the evidence is pretty strong that the greenhouse effect is here." It's like the old identity-politics satire: "Shy Rights—Why Not Pretty Soon?"

By August, Vice President George Bush was raising the issue at campaign stops. "Those who think we are powerless to do anything about the greenhouse effect," he said, "are forgetting about the White House effect. As president I intend to do something about it." Hansen's testimony made him, as *Newsweek* later wrote, "weatherman to the world."

A JOURNALIST ONCE DESCRIBED Hansen as looking—in his office clothes; khakis and sweaters—like a PTA dad. But that's slightly wrong. He has the catcher's mitt face of a farmer, someone you might see giving reluctant directions to a sportscar at the side of a long, flat highway, or perched in the metal saddle of a tractor, shaking his head over a bum crop. Hansen grew up in Denison, Iowa: birthplace of the movie star who played Mrs. George Bailey in *It's a Wonderful Life*. It's still the biggest thing to happen in town. Signs announce "HOME OF DONNA REED," and the town slogan is "It's a Wonderful Life"; there's also a Donna Reed Museum, festival, foundation, and scholarship. (One hotel maintains the Donna Reed Suite; which—I'm quoting from its website—is "decorated with photographs of Ms. Reed in various screenplays.")

His childhood was a mix of Capra and Steinbeck, a child you don't much get to see anymore. Hansen was born in 1941, the fifth of seven children. His father was a tenant farmer. Then he became a bartender. So Hansen got to be poor in the fields, then poor in town. Colleagues and reporters made a lot of Hansen's Iowa style, his "shambling, gee-whiz, farmboy" manner. But Hansen is also a product of the taprooms and arguments, the knuckled parts of America. He played sandlot baseball, earned college money with a newspaper route, walked for miles along rail lines with a dog named Skeeter. He enrolled at the state campus, University of Iowa, and graduated summa cum laude.

December of his senior year he had a sort of origin moment: the event that makes your skin prickle, fills your head with plans and ideas. Hansen and some other astronomy students gathered on a hilltop for the lunar eclipse: winter coats, plains wind, necks straining up. Hansen expected to watch the moon glow. Instead, as the eclipse began, the moon simply vanished. An Indonesian volcano had gone up that March, the usual spray and show. Nine months later, the residue was still blocking light in Iowa City. Hansen ran science's two-step, its if-then. If smoke could do that to the sun's brightness, what might it do with heat?

Hansen took a master's, then a PhD; Iowa, then Iowa. He had a mentor, Dr. James Van Allen. (Famous as the discover of our planetary mantle of radiation, the Van Allen Belt. "I was so shy and unconfident," Hansen later recalled. "I didn't want him to realize how ignorant I was.") Allen modeled how a scientist might act and live: a "guy who walks slowly and is very thoughtful and is very nice. Not aggressive at all as good scientists often are, and have to be." Van Allen arranged a job for Hansen with NASA's Goddard Institute of Space Studies. For a PhD, NASA is the head office, the call from the big leagues. Hansen made the entire trip in one touch of the gas pedal. "I drove all the way to New York," he said, "without stopping to sleep."

Goddard sits a few blocks from Columbia University; in a gray building at the corner of West 112th Street and Broadway, six stories above Tom's Diner. This is a big X on tourist maps: the *Seinfeld* coffee shop. One way to measure Hansen's life is that distance: born in Capra, relocated to *Seinfeld*. The series debuted a year after Hansen's testimony—so throughout the nineties, two of the largest products of American culture were situated in the same building. The charm of being alive in the modern world on the ground floor, some complications accumulating six stories up.

In 1976, he was principal investigator on NASA's Pioneer Venus Orbiter project. Hansen read the numbers: runaway carbon dioxide equals uncontrollable heat. He compared our planetary neighbors. Venus and Mars "provided the best proof at the time of the reality of the greenhouse effect," Hansen later said. "Mars, which has only a thin atmosphere of greenhouse gases, is only a few degrees warmer than it would be if it had no atmosphere. Venus, on the other extreme, has a thick atmosphere of carbon dioxide and such a strong greenhouse effect that you could bake pizza on its surface." In fact, that pizza would crumble to pizza ash, "because the temperature is about 800 degrees Fahrenheit. The Earth has an intermediate amount of these gases, which keeps the surface a comfortable sixty degrees warmer than it would be without." It didn't take long, Hansen said, "until I was captivated."

His interest looped home—the pragmatic Iowa streak. Earth made an even better subject. More exciting "to study a planetary atmosphere that was changing—scientifically exciting," he said. "And also of practical importance."

Hansen and his team put together a GCM, a global computer model. They tinkered, experimented, perfected. His model became the world's most reliable, issuing accurate prediction after accurate prediction. Historian Spencer Weart calls his work "absolutely groundbreaking." (As Weart told the *New Yorker*, "It does help to be right.") In 1981 Hansen became head of Goddard. That August, he and his team published. The planet had been warming for a century. This was the exact type of report the Reagan White House did not want to see. New warming would arrive within the decade, with more in the nineties, and heating of "almost unprecedented magnitude" in the century to come.

It would have been bad enough had Hansen confined this to the academy—launched his report to float on the backwaters of the scientific press. He leaked his findings to Walter Sullivan, a science reporter at the *New York Times*. Who ran them on page one.

Headlines and editorials. Hansen calls this his "original sin." (Hansen had met Sullivan at a local Chinese restaurant, Moon Palace; the touch-and-go of a life. No restaurant, no Sullivan, maybe no Hansen.) The Department of Energy pulled his funding. Five people had to be laid off from Goddard.

I think this is the moment that made him. When people around you get fired, how you respond tells who you are. Army people call this a meet-yourself experience. (When people around you get fired, it's like the card FedEx leaves on your door. *We're Sorry We Missed You.*) Six months later, Hansen traveled to Washington to address the annual meeting of the American Association for the Advancement of Science. Maybe it's difficult to intimidate someone who grew up on a tenant farm, in a bar. They've seen how hard things can get, they know it's really not that bad. He was standing a few blocks from the White House. He repeated the same warnings as before.

He testified for committees Al Gore started up. He testified in November 1987, when skies are cool and warming seems not just manageable but preferable. No response. He testified in June. Giant response. As the papers said, he'd had "a stunning impact on political thinking."

Climate change went national, international. By the end of the summer, Princeton atmospheric scientist Michael Oppenheimer was marveling, "I've never seen an environmental issue mature so quickly, shifting from science to the policy realm almost overnight. . . . He felt comfortable saying clearly and loudly what others were saying privately." Hansen had piled family on top of job: Years later, and a thousand miles from Iowa, it was still Donna Reed. "I don't want to be in a position," Hansen told a TV interviewer, "where a few decades from now, my grandchildren say 'Opa understood what was going to happen, but he never tried to make it clear.'"

A life has repeating patterns, the background in a cartoon. Tree, boulder, Bedrock. An Indonesian volcano had drawn Hansen into atmospheric science. In 1991, global warming resumed its usual rotten timing. A Philippine volcano went up, dust cooled the world, and for years the line between media and global warming went dead. And—more bad luck—a smaller volcano was kicking up dust and obstructions in the West Wing.

THE PIRATE

IN CLIMATE TERMS, THE FIRST BUSH ADMINISTRATION
got off to a flying start. George Bush on the stump was not shy about making his feelings known. "My first year in office, I will convene a global conference on the environment at the White House." He wanted to roll up his
sleeves, get right down to it. "All nations will be welcome—and indeed, all
will be needed. The agenda will be clear. We will talk about global warming . . . and we will act."

This turned out to be one of the last times George Bush said the words
"global" and "warming" in the same sentence. As if saying them turned the
problem real; avoidance kept solutions in the hazy attic.

And it must have been a kick in the pants for Ronald Reagan. One
day everything is agreement. The next, he's leaving office and his former
vice president—the tall, peppy fellow down the hall—is trashing his policy
on the natural world. "In the past few years we have not done enough to
protect the environment," George Bush told reporters. "I am an environmentalist. Always have been . . . and I always will be." Reagan returned to
California under a shower of retrospective honesty. *New York Times*: "The
angry anti-environmentalist polices of his predecessor, Ronald Reagan."
National Journal, re warming: "The Reagan approach—that is, stalling in
the guise of doing more research." *Washington Post*: "The Reagan administration, which ignored the global warming problem." The Bush family was
big on nicknames. The first President Bush even nicknamed himself: "the
Environmentalist President."

Every gesture had the feel of a promise. After his victory, Bush became the first president to tap an actual environmentalist to head the EPA: William Reilly, who had steered two organizations with Siberian tiger posters in the waiting room—the World Wildlife Fund and the Conservation Foundation. EPA was on board for doing something about climate change. The Departments of Energy and Treasury were on board. Most importantly, James Baker was on board too.

James Baker was the first President Bush's David Axelrod, his Karl Rove—the man who steered him to office, what David Mamet calls "the guy behind the guy behind the guy." He was also George Bush's best friend. (And acted additionally as family retainer, its break-glass-in-case-of-emergency. Jim Baker is the man who would neatly pocket, on behalf of the second George Bush, the 2000 Florida recount. A quiet man in a flawless suit, with a set of keys to unlock every door.)

Baker had served in Reagan's White House; he'd read the science, understood global warming. If you had Jim Baker, you had the new president. Bush made his friend secretary of state. Baker made global warming the text of his first speech. (Aides put the word around: this was on purpose.) Baker welcomed members of a new United Nations task force, the Intergovernmental Panel on Climate Change. The clearest gesture—as the *Times* ran it, a "signal that the Bush administration was placing the problem of global warming and climate change high on its agenda."

Baker took in the climate delegates, spoke with his soft Texas accent, its mix of saddle leather and the club room. "The truth is, as I don't need to tell those of you who are here, we face some very difficult problems. It is also true, though, that we now recognize them to be problems. And in my experience in government that is at least half of the battle," Baker said. "We are all in the same boat." No senior member of any administration had ever spoken this way before. Among the scientists and advocates, there must have been damp eyes, husky throats. "So, if I may borrow a phrase from the environmentalists, the political ecology is now ripe for action. We know that we need to act, and we also know that we need to act together." The secretary closed with a benediction: "Godspeed."

Excitement and relief. William Reilly, the new EPA head, said the speech was "terrific, succinct and clear, and established priorities that I think people heard." All those years—and sensible people had at last arrived in the White House.

You can feel the word coming. There's always an *except* with global warming, the string of cans knotted to its tail. It's like dieting, or giving up cigarettes, or joining a gym. *Except* in this case was 5 foot 9, had tight rhino eyes and a Jimmy Olsen haircut and was named John Sununu.

He was Bush's chief of staff. And would become his administration's dominant figure. *Time* ran him on its cover in May 1990, under the tagline, "Big Bad John Sununu." They quoted an admirer's description of Sununu—who weighed 190 pounds—as "this fat little pirate." Sununu had 180 points of IQ; he responded to queries he didn't like with "Small minds ask small questions." He made the blunders—underlistening, overreach—committed mostly by people who know themselves to be very smart. Long-time Republican strategist Ed Rollins called Sununu "living proof that you shouldn't give children their IQ test results."

He arrived in DC with an interesting twist on the whole better-to-be-loved-or-feared question of governance. "I don't care if people hate me," Sununu said. "As long as they hate me for the right reasons." *Time* called him "Bush's bad cop" on the environment. People he frightened called him "King John."

And Sununu did not believe in global warming. Decades of research had orbited one word: Urgency. Sununu, on the other hand, told an interviewer, "There's no urgency." This was Bush's crux moment. His campaign had promised to act; his environmental chief, his best friend, both said "act." His chief of staff said "don't." It's Sununu who convinced President Bush the threat was overblown, a case of huff without puff. Each time the pressure built, Sununu would manage to get in front of the issue, like the man at the fair who blocks cannonballs with his gut. And so, from that flying start, it was back to the ground-bumping Reagan approach on climate.

BUT BUSH'S CHIEF OF STAFF is also the story of how America ended up in the first global warming treaty. We drew there inadvertently, a series of decisions and Sununus.

This started in the spring of 1989. But it really started in fall of 1988, with the formation of the Intergovernmental Panel on Climate Change. (At the United Nations, and sounding very like the panel Sweden's Hans Ahlmann once proposed, four decades earlier. "I hope an international agency can be formed to study conditions on a global basis," he'd said. "That is most urgent.") You know the panel from news stories, where it's referred to by monogram: the IPCC. It's the world's ranking authority on climate science. In 2007, Al Gore and the panel would split the Nobel Prize.

European governments never faced America's trouble establishing a beachhead on global warming. The anthropologist Willet Kempton has studied response to climate politics for nearly thirty years. (It's that kind of topic; it leaves a long, broad wake. You can follow it for miles.) In 1991, Kempton wanted to know about the differences—what "motivates European policymakers to press for strong action now."

One diplomat made the torch-passing case; centuries of torches. "Next year, it will be a thousand years old, this country. So, there is certainly the idea that it should go on.... So we have to take care of the threats to the environment." Longer history meant wider experience of every type of politics. "To put it crudely," another official said, Europe was passing air quality regulations "while the States were fighting the American Civil War."

And the fact that Europe had never been a diverse place offered an especially shabby way to make the case: immigration. "Why is global warming dangerous?" one diplomat asked Kempton. Sea-level rise. "If this ever happens, there will be huge fluxes of people from those low-lying areas—millions and millions and millions." So if you couldn't persuade a voter with altruism or science, you could always make them squirm with the sidewalks of the future. "You simply appeal to the self-interest," the diplomat explained. "And tell him, 'Well, dear friend, you are against anything being done about global warming. OK, but then your children and grandchildren will have to live with millions and millions of people moving up from Asia, moving up from Egypt, and so on.'" An awfulness contest. To win over the wrong kind of voter, Europeans had learned to fight global warming with racism.

And then there was England. Prime Minister Margaret Thatcher had been President Reagan's most reliable ally—the tonier female version. In a police drama, they would have been mismatched partners, arriving through misunderstanding and mutual rescue at silent devotion.

Global warming would've split the team. It took a day. Thatcher invited cabinet officials, industrialists, and researchers to the ministerial residence on April 26, 1989. They were briefed on the current material by British experts. That was it. Examine data, reach a conclusion. Thatcher stepped outside, and Britain called for "rapid action" on warming. Before April, England had been known as the Dirty Man of Europe—the coughing island, with dusty clothes and flies in its hair. In the fifteen years after 1990, Britain's economy grew by 40 percent; its carbon emissions decreased 14 percent.

The IPCC was meeting that May in Geneva. First day, Margaret Thatcher called for actual progress: the group wanted to begin moving toward an international treaty. Jim Baker was in favor; the EPA was in favor. (Not just that. They wanted to drive the pace car—they were, as the *Times* said, "urging that the United States take the lead on a convention to meet the threat of global warming.") John Sununu stood against it.

Here's where Jim Hansen comes back in. That same week, Dr. Hansen was called by the Senate to regather his results and retestify. If you're a government employee—Hansen's ID had the round blue NASA logo that in the trade is called The Meatball—you submit all testimony in advance. Hansen had cleared testimony under President Reagan. When he suggested bigger investment in climate science, the idea came back deleted; thanks, don't tell us how to spend our money.

The Bush administration had made an advance. Hansen intended to say his NASA research showed more warming coming sooner. (Which is in fact what happened.) Hansen checked his testimony. The White House revision didn't just have stuff taken out; it had new stuff added in. He was now expected to say his own work shouldn't be counted on: it was "estimates," not "reliable predictions." Hansen had written that new greenhouse gases were primarily of human origin; the edit stated that the cause of global warming was a mystery—and climate change "remains scientifically unknown."

"They put words in my mouth," Hansen said. "They even put it in the first person. . . . I can understand changing policy, but not science." What this meant was that if you listened to Hansen in the Senate, or read about the scientist in the news, you'd hear results opposite to the ones the government had funded him to obtain.

Hansen went to the committee chairman, Al Gore—who instantly gave the hearing a new subject: the White House was distorting science! It became, the papers said, a circus, a furor. "They are scared of the truth," Gore said. (Gore includes this in his movie, which is like the profile photo you post on Twitter; it's a moment he must have especially liked himself in.) Hansen was on the newscasts, with NBC now calling him, "the government's best-known scientist researching the issue." The nation's TV sets became a courtroom: there was Gore, sharp-profiled, a crusading prosecutor. "If they force you to change a scientific conclusion, it's a form of science fraud—by *them*."

This was Monday. Tuesday, the British were agitating at the United Nations for a greenhouse treaty. Hansen stayed in the news for days—a week-long White House embarrassment. (Jim Hansen has the charm of directness. In the press scrum, a reporter asked, "Doctor, are you worried about any retaliation?" Hansen flexed his eyebrows. "Well, I am *now*.") Toward week's end, Sununu summoned the EPA head to his office. It was nighttime; the request probably cost him something. The next words must have been especially hard for a pirate to say. Sununu asked, "What do we do?" The EPA head suggested going along with the climate negotiations. So Sununu sent a telegram—this was the last twilight decade before email—to the delegates in Geneva. Make the strongest efforts to get a treaty process started: a workshop on setting the terms would be held in the United States.

The workshop convened a year later. Sununu kept busy. The President was supposed to address the IPCC in February. Jim Baker, the EPA, the Department of Energy all approved a speech—a White House message rolls through many levels of quality control. President Bush was to outline specific steps to stabilize carbon dioxide.

When Sununu read this, he got mad and tore it up. He rewrote the president's speech, to emphasize doubts about climate science and recommend More Research. The story broke. The director of one environmental group griped to the *Times*, "The American people did not elect John Sununu." But Sununu couldn't help himself—he'd become, another official said, "obsessed with the issue."

The treaty workshop convened on April 18, 1990, Sununu crouching behind every weed. Hiring a pricey conventions expert, picking speakers, editing press releases, demanding updates by the hour. The name of the conference has the Sununu ring—and here is a good test to see whether a news event is designed to mislead you. If you'd spent the last six months out of the country, or up on a mountain taking a news fast, then read the headline, would you have any idea what was going on? The event was called "The White House Conference on Science and Economics Research Related to Global Change." A near-perfect conductor: All euphemism, zero information. President Bush took care, during the two days, to give the phrases "global warming" and "greenhouse effect" a wide berth. (Reporters jumped all over this. "How can you call a White House conference on global warming—and then not even use the words?") Talking points were distributed to delegates. Here's what to say, more importantly what not to. They read like a battle plan, an outline to be followed by anyone combating a decision on any issue. The Sierra Club got hold of the points, leaked them to the media. Here was the White House policy on global change.

> *It is not beneficial to discuss whether there is or is not warming, or how much or how little warming. In the eyes of the public, we will lose this debate. A better approach is to raise the many uncertainties that need to be better understood on this issue.*
>
> *Don't use specific numbers—degrees, dollars, rates, etc. [Emphasize] this conference is accelerating the international discussion and understanding of these issues.*

*Don't get into an advocacy position of the merits of vari-
ous policy proposals.*

*Don't let reporters position this conference as an attempt
to delay serious decisions on this issue.*

The *Times* gave a sour nod to the president's campaign promise: "Some
White House effect."

Within a month, scientists from the IPCC had recommended an
immediate 60 percent cut in emissions. Margaret Thatcher joined the
chorus of basically every European leader. "The problems do not lie in the
future," the prime minister said. "They are here and now." On warming,
America had become isolated, cut off, a policy ocean away from its allies:
the solitary territory where it would reside for most of three decades.

JOHN SUNUNU LASTED another year and a half. His career illustrates a lesson: In the schoolyard, bullying works. Noisier, more aggressive, and better-connected usually means getting your way. Once people
put on suits and exit the classroom, adulthood seems to offer a kind of
natural bully insurance—there's a limit to how much pushing around
grown-ups will tolerate. By 1991, the *Times* could run this piece: "The
answer: John Sununu. The question: Why are administration officials
like Secretary of State James Baker, William Reilly, head of the E.P.A.,
afraid to speak their minds on the greenhouse effect." The paper continued," Baker, like most pros, backed away from the greenhouse cause when
he saw Mr. Sununu's ferocity."

Sununu fired a Jim Baker deputy who tried to move ahead with climate
proposals. He yelled at senators, glared at staff. (Sununu was a marathon
glarer. If he didn't like what you'd told the president, he could glare for
an hour.) He ticked off Democratic and Republican senators. How they
got mad tells a lot about the differences between the parties. When he was
dismissive of thirty-year veteran Robert Byrd, the Democrat gave a dignified speech: "Your conduct is arrogant. It is rude. It is intolerable." Sununu
called Trent Lott—later the Republican majority leader—"insignificant."
Lott's response: "He just stuck the wrong pig."

At the end, the president was seen as punch-drunk, a figure woozy with infighting, bulldozed by his chief of staff. In summer 1991, *Newsweek* was calling Sununu "the tyrant of the West Wing." The joke image from the Capitol: the White House run by "John Sununu and a thousand interns." When a reporter called an administration official, nosing for Sununu dirt, the lawmaker turned the tables. "Please tell me you have more. We have to do him in."

What finally got him were the perks. He was a serial privilege abuser; a free-rider. In 1991, it came out that during the previous two years Sununu had hitched seventy-seven Air Force rides—one for a family ski trip in Colorado, another for a visit to the dentist. You know your career is approaching the terminal when they get the joke down to two words. Here, it was Air Sununu. The president ordered Sununu to keep out of the airways. A few weeks later, the chief of staff commandeered a White House limo for the four-hour shoot to New York, where he spent $5,000 at an auction of rare stamps. (He bought a Ben Franklin and a series of three Zeppelins. Postage Sununu.) Sununu was forced to resign. There's another way to look at it: John Sununu, who'd done so much to hold the line for the coal and oil industries, was brought down by overuse of fossil fuel. Carbon did in John Sununu.

George Bush, unplugged, de-Sununued, ended up at the 1992 Earth Summit in Rio di Janeiro. This was the next stage of the treaty process the government had locked itself into, after adjusting Jim Hansen's testimony. The Earth Summit was the largest gathering of heads of state ever: 117 leaders and thirty-five thousand visitors descended on Brazil. (The Palestine Liberation Organization and the self-esteem dance troupe Up With People both sent representatives to Rio.) George Bush was able to make himself utter the science's gentler name. "We come to Rio with an action plan on climate change," the president announced. "And I'm happy to report that I've just signed that Framework Convention on Climate Change."

This is the framework that, in 1997, would become the Kyoto Accords—and, under President Barack Obama, the 2015 Paris Agreement. So it was George Bush and John Sununu who brought us there.

It's like one of those weekend-gone-wrong comedies. The gestures and

excuses you make to cover up mistake number one grow so big they eventually set you on a whole new course. As the *Times* wrote about Jim Hansen,
the treaty was "part of his legacy." It's the way things happen: the administration ended up in the last place it ever wanted, a global climate negotiation, because in 1989 it decided to sand a corner and amend the research of
one NASA scientist. They might have arrived there by a hundred different
routes; in our world, that's how they made the trip.

In the fall of 1992, George H. W. Bush found himself in another unforeseeable place. The president who said he was and always would be an environmentalist ran against Bill Clinton and Al Gore on the environment.
He heckled Senator Gore as "Ozone Man." He warned that victory for the
ticket meant "we'll be up to our neck in owls and out of work for every
American." George Bush had ended up a far distance from where he began.
Maybe the job does funny things to presidents. News events give you a jos-
tle, staff reposition you, and you look out a different window on a whole
new view; maybe that's part of what he meant by the White House effect.

THE PILOT LIGHTS AND
SOMEBODY'S WORLD

ON JUNE 6, 1991, THEN FOR ABOUT WEEK, THEN ESPE-
cially from the 12th to the 15th (an eruption is like a person who hasn't
spoken for a long time starting an argument; there's stuttering, throat-
clearing), the Philippine volcano Mount Pinatubo exploded. It sent an ash
cloud twenty-eight miles into the air. From far away, the cloud resembled
a frozen nuclear detonation; spikes of smoke, immense power. At ground
level, it backed the landscape like a puffy mountain range. In fleeing cars,
windshield wipers cleared tiny arcs from the spatter: mud, ash, pieces of
earth fell in a kind of clumpstorm. General William Studer decided to evac-
uate Clark Air Force base when one of his subordinates told him, "General,
you better fill your pockets with jam, because we're about to be toast."

A quarter of a million Filipinos abandoned their homes. They returned
to a hushed gray landscape—smashed roofs, the morning-after of a barking
dog. Volcanic mud had flowed through some towns, rising ten feet, then
hardened. It made returning residents feel like giants—at shoe level with
the rooftops, eye-to-eye with town statuary. Volcanoes are stunning things;
people revert to deep training. When one of last geologists to leave Clark
opened the door to one of the last transports, the driver was crossing himself. The geologist found another ride.

Pinatubo also injected about twenty million tons of sulfur dioxide into
the air. After three weeks, the cloud made a ring that circled the globe: a
Pinatubo belt. By early winter, the dust had gone everywhere. So for a good
part of America, 1992 was another Year Without a Summer. (In Massa-

chusetts, this meant salad bar shortages—a Year Without Tomatoes.) The winter of 1993 was powerful. All-time rust-belt lows: twenty-two below in Pittsburgh; Indianapolis (the town Edison knew as Railroad City) hit minus twenty-seven. Chicago closed its schools, simply due to cold, for the first time in history. In New Jersey, stranded motorists were rescued by snowmobile.

Issue amnesia crept in. "Whatever happened to global warming?" *Time* asked, in early 1994. Then decided, "It might be more to the point to start worrying about the next Ice Age instead. After all, human-induced warming is still largely theoretical."

Science journalist Jonathan Weiner wrote in the *Times* that friends had begun avoiding the topic with him, out of simple commiseration, workplace tact—they assumed he'd gotten the story wrong. Skeptics took to the columns and airwaves in a festival mood: "The greenhouse effect isn't doing a very good job of what it's supposed to do best."

Jim Hansen—this is why scientists are happy in the world—saw the explosion as an unanticipated plum. A good way to check the equipment. The Pinatubo sulfur would reflect about 2 percent of the sun's radiation back into space. Global warming was heating the Earth by around two extra watts per square meter. Which doesn't sound like a lot. You always wonder if scientists doubt—and there's a lovely moment about this, Jim Hansen sitting on a beach near Manhattan: "It was getting dark. There was a strong sea breeze coming in. It seemed very powerful, nature. And it seemed incredible that the small force due to this very tiny change in atmospheric composition, now about two watts per square meter, meant anything to those crashing waves."

One standard way of describing the two watts is to say that warming is like adding a pilot light to every square on the grid. The Earth is about 500 trillion square meters. So, a half-quadrillion steady blue flames. (Testifying before Congress in 2006, the president of the National Academy of Sciences broke down the figure: "About one hundred times larger than all of the energy usage by humans worldwide on the entire planet, from all sources—fossil fuels, nuclear, wind, you name it," explained Dr. Ralph

Cicerone. "It's a big number.") The computer said Pinatubo would block out about four watts. So: net global cooling.

Hansen might be shy in a living room, but he's not a shy scientist. Six months after Pinatubo, he and his team went public with their figure. Temperatures would drop one degree Fahrenheit; then warming would pick up right where it left off. "We should see it," Hansen said, "or there's something wrong with the models."

As *Newsweek* wrote, Hansen had just "put his reputation on the line," by making "a testable prediction." The skeptic Pat Michaels went after the idea on CNN. "When the Earth does not show one or two degrees of cooling—which would be a massive amount—it's going to be said, 'Oh, look, in spite of the volcano the globe is still warm,'" Michaels explained. "This thing is going up as a smokescreen."

By fall of 1992, the cooling trend was in. "It's not a home run yet," Hansen told climate writer Bill McKibben. "The temperature still has to start coming back the way we've said it will." 1990 had set the record as the warmest year in measurement history. In 1992 and 1993, the planet cooled one degree; 1994 logged in as history's fourth-warmest year, and 1995 re-broke the record. But by then it was too late to take back the effect. As the *Washington Post* explained, "Mount Pinatubo not only temporarily cooled down the climate, it blotted the issue off the national media map."

AT THE SAME TIME, a set of refinements to the greenhouse effect was being gouged out of the world's top and bottom. Vostok Station used to be the Soviet drilling project in Antarctica; it's now a jointly run Franco-Russian-American shop. Antarctica, in pictures, is like a photo-shopped memory. An ideal January post-blizzard morning. Blanketed landscape, chilly sunlight, everywhere powder. Descriptions of the continent tend to end in the syllable *est*. Antarctica is the highest, windiest, driest, frozenest; also the loneliest. (The enchantingly named Edwardian explorer Apsley Cherry-Garrard called Antarctic work "the cleanest and most isolated way of having a bad time which has been devised." His *est* words are *bravest, farthest, hardest*.) Total summer population climbs to about 4,400—then sinks to 1,000 diehards during the rough Antarctic winter. It's like an ant farm in sole tenancy of a shopping mall.

The nearest landmark to Vostok is what's actually known as the southern Pole of Inaccessibility—the point farthest from the Antarctic coasts, and so one of the most difficult spots to reach on the globe. Things hang around like stray frozen thoughts. A bust of Lenin erected by the Soviets is still here, when Lenin busts have been recalled virtually everyplace else in the world. A team of British-Canadian adventurers used the bust as an orienting point, to reach the pole by foot in January 2007; their feat drew international attention. The whole continent is a dare.

At a tiny site like Vostok the look is of an underdog construction crew riding out an impossible winter. Vostok is where the planet's lowest-ever ground temperature was recorded: minus 128 degrees Fahrenheit, on July 21, 1983: Hemingway's birthday. (Gradually or suddenly, he probably would've liked that.) Another name for Vostok is the Pole of Cold.

So it was the coldest spot on Earth that proved out global warming. Ice coring started at Vostok in the late seventies. By 1987, the team—twenty-five researchers in high season—had drilled more than a mile down, for an atmospheric record stretching back 160,000 years. Jonathan Weiner calls this work "probably the single most important piece of greenhouse evidence beside Keeling's curve."

No one had a precise idea of what carbon dioxide was up to in the pre-industrial, pre-human era. Vostok gave us those numbers. What Arrhenius arrived at by pencil and paper: It turned out carbon dioxide matched temperature fluctuation in every era. Cold temperature, low concentrations. High concentrations, heat. The pioneer Swiss climatologist Hans Oeschger always accepted greenhouse theory; when he examined a Vostok chart, saw the effect played out across hundreds of centuries, he shivered.

By 1998, drills had chewed down two miles and 420,000 years. Keeling's ppm, the parts-per-million number, is 421 now; the old high-water mark, 330,000 years ago, was 300. (For a sense of what parts per million can do: ozone, for all its sun-blocking properties, exists in the atmosphere at only 15 ppm.) As Elizabeth Kolbert writes, in her calm and frightening *Field Notes from a Catastrophe*, "the last time carbon dioxide levels were comparable to today's was three and a half million years ago." To find an era when they were higher, you'd probably have to hike back fifty million years. At that time, Kolbert points out, "crocodiles roamed Colorado, and sea levels were nearly 300 feet higher than they are today." By 2008, the Antarctic record reached back 800,000 years and researchers had found another CO_2 low, 170 ppm. As Jonathan Weiner writes, seeing those results "would have meant a great deal to Arrhenius."

EVIDENCE OF A DIFFERENT, unsettling effect was being extracted from Greenland. The geologist Richard Alley writes about the great visual beauty of this work: Sunlight passing through years and years of crunched snow. "Water," Dr. Alley explains in *The Two-Mile Time Machine*, whether "liquid or ice, absorbs red light" so that "only blue reaches your eyes." At the Greenland pits, that's the color you see. "The snow is blue, something like the blue seen by deep-sea divers," he writes. "An indescribable, almost achingly beautiful blue."

Compared with Vostok, Greenland is Caribbean. When Elizabeth Kolbert deplaned there (dressed in thick boots, a snow-fighter parka), the ice core field director met her wearing an unzipped coat, tennis shoes, dainty icicles strung from his beard.

There was another startling thing about the Vostok record. Graph temperatures in the United States for a given year, and you'll get a slowly rising and descending hill—cold to warm back to cold. The kind of slope you can take on a weekend afternoon, wearing Rockports. The Vostok record isn't like that; it's alpine. Sudden ascents, deep plummets, like a mountain range drawn by a child.

But it's the Greenland cores that did away with the idea of gradualness. Kolbert writes that the work led to a "wholesale rethinking of how the climate operates." The Greenland record doesn't go as far back as Vostok; it taps out at 110,000 years. But the picture is more crisp, a section of landscape and history shot through a zoom lens.

Behind Adaptation is one idea: that climate change will be steady, predictable. A guest you can prepare for, turning down the guest room sheets, stocking the fridge. The word Greenland researchers have settled on is *abrupt.* (Other favorites are *drastic* and *nonlinear.*) The last ice age didn't end with slow warming: Glaciers trickling, grass seeding a valley. It seems to have happened over three years.

In the Greenland record, there are eras where temperature changes fifteen degrees in a half century, periods when heat leaps eighteen degrees in a decade. Here's Richard Alley:

> *For most of the last 100,000 years, a crazily jumping climate has been the rule, not the exception. Slow cooling has been followed by abrupt cooling, centuries of cold, and then abrupt warming, with the abrupt warmings generally about 1,500 years apart. . . . At the abrupt jumps, the climate often flickered between warm and cold for a few years at a time before settling down. One can almost imagine a three-year-old who has just discovered a light switch, flicking it back and forth, losing interest for a while, and then returning to play with it again.*

Predictable climate, researchers stress, is at root level the requirement for everything. History—everything—began when climate became something you could plan for. Elizabeth Kolbert sat down with the Greenland field director, the man with the beard icicles. (Kolbert is funny about her own motivation. Which she describes as "an interest—partly lurid, but also partly pragmatic—in apocalypse.") "Now you're able to put human evolution into a climatic framework," the field director told her. "You can ask, Why did human beings not make civilization fifty thousand years ago? You know that they had just as big brains as we have today."

The answer was temperature. "Ten thousand years of very stable climate," he said. "If you look at it, it's amazing. Civilizations in Persia, in China, and in India start at the same time, maybe six thousand years ago. They all developed writing and they all developed religion and they all built cities, all at the same time, because the climate was stable. I think that if the climate would have been stable fifty thousand years ago it would have started then. But they had no chance."

When the Greenland data went public in 1993, the *Times* called it remarkable. Five years later, the data becoming clearer, the paper explained, "The implications for Federal and international climate policy are enormous." ("There has been a widespread assumption that if humans are changing the earth's climate, the effects will be felt gradually and smoothly, making it easier to adapt to the change. But a growing accumulation of geological evidence...") One paleoclimatologist described the present situation—not knowing where or when a line might be crossed—as "walking the plank blindfolded." Another paleoclimatologist pictured the toll—the surprise pummelings—early cultures must have taken. "Climate changes that we thought should take thousands of years occur within a generation or two," he said. "It's certainly something that would have rocked somebody's world."

It's believed that changes in climate finished off Greenland's Vikings, the Bible's Assyrians, the Anasazi in Arizona, Utah, Colorado. Too cold, nations never form. Too warm, water evaporates, survivors drift, cities turn to relics. You're left with wind against walls, a mystery.

The 1983 National Academy of Sciences report had a soft title: *Changing Climate*. Something you might see in an ad supplement for a golden-years community. Leave Buffalo, resettle in Lauderdale. The 2002 report, chaired by Dr. Richard Alley, was called *Abrupt Climate Change: Inevitable Surprises*. A blank traffic sign before a patch of bad road. Another way of looking at abrupt climate change is as something like the ozone hole. Effects link up—and suddenly, the lurch forward, something nonlinear, an inevitable surprise.

And this was at the same time that warming reached maturity as a political issue: a question of voter trends, pin maps, roll calls. In 2007, the former Republican chief of staff for the House Science Committee got interviewed by *Newsweek*. His name was David Goldston, this was for another cover story. "In the House," Goldston told *Newsweek*'s Sharon Begley, "the leadership generally viewed it as impermissible to go along with anything that would even imply that climate change was genuine." He explained, "There was a belief on the part of many members that the science was fraudulent, even a Democratic fantasy. A lot of the information they got was from conservative think tanks and industry.... there was a constant flow of information, largely misinformation."

The same year, William K. Stevens, the science reporter who'd broken stories on Greenland, Antarctica, and Pinatubo in the *New York Times*, came out of retirement to write a kind of farewell essay for the paper. He was, in the manner of David Goldston, a veteran of the information fights. But his nicks and memories came from the journalism side.

> *In the decade when I was the lead reporter on climate change for this newspaper, nearly every blizzard or cold wave that hit the Northeast would bring the same conversation at work. Somebody in the newsroom would eye me and say something like, "So much for global warming."... Such an exchange might still happen, but now it seems quaint.*

Stevens reeled off the data—the scientists, the decades of research, national and international. Then, with a nod to Roger Revelle's famous 1957 sentence, he signed off.

> *It has been pointed out many times, including by me, that we are engaged in a titanic global experiment. The further it proceeds, the clearer the picture should become. At age 71, I'm unlikely to be around when it resolves to everyone's satisfaction—or dissatisfaction. Many of you may be, and a lot of your descendants undoubtedly will be. Good luck to you and to them.*

MARK MILLS

MARK MILLS IS ONE OF MY FAVORITE PEOPLE IN THIS story—a man, and a name, I often think of. (He's also a person it's difficult to find information about: his organization existed just long enough to get its figures read into the record.) Mark Mills is an illustration of something I've always wondered. Where do the numbers come from?

During the nineties, and maybe also like you, I didn't know very much about climate change. The big news events are an airport you're rushing through, where CNN plays above the gate: you overhear what you can. The issue went away, then it came back. Here's what I thought: In the late eighties, scientists decided warming was coming. Then they changed their minds and decided it wasn't really—good news. Then, for whatever reason, they'd decided it was on again, but were maybe not totally sure. Then Al Gore won the Nobel Prize.

In 1991, about two dozen coal-burning electric utilities—plus some big coal producers, and the National Coal Association—devised what they called a "creative strategy." Their goal, as expressed in a planning document, was to "reposition global warming as theory (not fact)."

Everything about the campaign was artful: a matter of market segments and custom bullseyes. The group engaged a public relations firm to create an acronym that fit the letters ICE—the opposite of "heat." The firm was judicious. It recommended Information Council for the Environment—which says an interesting thing. According to experts, people trust blandness; we prefer a name that's nonstick. One name they vetoed was Informed

Citizens for the Environment; this had an "activist" sound. You saw beefy concerned fathers, mothers with glasses, raising their hands in a meeting at the school gym. Informed, Citizens, Environment—together, it was everything you didn't want.

The campaign wouldn't be focused on the Informed, people boning up with lots of news. Like someone sizing up particular guests at a party, ICE understood where its best chances lay. The campaign would go after "younger, lower-income women" and "older, less-educated males." It picked cities and broadcast markets with equal care. Electoral districts with coal-burning power plants, representatives who served on the House Energy Committee. ICE ads would say things like: "If the Earth is getting warmer, why is Minneapolis getting colder?"—then show a man and his snow shovel.

In fact, as one journalist pointed out, Minneapolis had warmed something like two degrees over the previous century. Another ad read, "Some say the earth is warming. Some also said the earth was flat." (The president of one utility publicly broke ranks: the issue, he told reporters, was too potentially severe "to be dealt with in a slick ad campaign.") The head of the National Coal Association sent out memos, urging coal producers to kick in: "Public opinion polls reveal that sixty percent of the American people already believe global warming is a serious environmental problem. Our industry cannot sit on the sidelines in this debate." The ICE president cast the threat in existential terms; all or nothing. At issue was "the future of fossil fuels."

The group formed a Science Advisory Panel. It went out, located three scientists who had their doubts about global warming. (Two of whom—because there wasn't a longer list—promptly asked to have their names deleted from future elements of the campaign. They didn't want be the only ones out there on stage.) They were Robert Balling, skeptic Patrick Michaels, and Sherwood Idso. In December of that year, expenses paid by a trade group, the three testified on behalf of coal at a California hearing. They were joined by a fourth scientist—and this is one of the first times I saw him in the paper—named Mark Mills.

It ended up an absorbing, rewarding, long-term gig. High profile, travel. A few years later, a journalist in *Harper's* magazine revealed that, all told, Dr. Robert Balling had received about $200,000 from coal and oil interests. Dr. Balling disputed this figure in the *Arizona Republic*. "Actually," he said, "I've received more like $700,000 over the past five years."

Skeptic Patrick Michaels has been on the road for a quarter century. Giving interviews, publishing op-eds, sitting in at CNN, MSNBC, NPR, Fox News. One of Dr. Michaels' best credentials was that he was the official climatologist of the State of Virginia—until it was revealed that, in 2006, he'd received $150,000 from some coal-burning Colorado interests, at which point he was directed by the governor to cease identifying himself as official climatologist of the State of Virginia.

Dr. Michaels has toured as long and hard (small rooms, big venues) as Bob Dylan. In 1999, he spoke at a brochures-and-folding-tables event in Brewer, Maine. Addressing a group of older, less-educated males, Dr. Michaels didn't base his appeal on science. He used social discomfort, people his audience might be expected to dislike. Climate advocates, Dr. Michaels explained, were "some kid in a Phish t-shirt and an earring." Environmentalists were really a kind of labcoat welfare bureau, "just looking for an excuse to take the money you earned and give it to somebody who didn't." Dr. Michaels has headed overseas, turned up in a notorious British anti-warming documentary, on newscasts in Australia.

Sherwood Idso is a scientist the *Times* described as "an adjunct professor of botany and geography at Arizona State." (Dr. Idso is now president of the Center for the Study of Carbon Dioxide and Global Change. It's an organization founded and run by his two sons, Craig and Keith.) In the early nineties, Dr. Idso was hired by the $400 million coal outfit Western Fuels. Western was one of the ICE campaign's backers; its chief executive officer, Fred Palmer, gave PBS his motives for joining the fight. (People are shockingly open with PBS, the way you might spill secrets to the brainy, not especially popular kid in class: Who are they going to blab to?) Global warming was "a game-ending kind of issue for the American coal-fired electric industry." For Palmer, the actual argument ascended to a higher debating circle. Not economic or scientific. Theological. "Fossil fuels," Palmer wrote, "are a gift of the Creator." He gave an account of this dispensation to an international conference of coal manufacturers. "It is easy to conclude that, under a preordained plan, coal and oil lay in wait for exploitation by humans," Palmer explained. "To permit our creation of an environment on Earth conducive to the spectacular success of our species."

So in the early nineties, Western Fuels greenlit Dr. Idso. Gave him $250,000 to produce a documentary telling the story their way. The movie, *The Greening of Planet Earth*, became a Bush White House favorite—in heavy rotation on the chief of staff's VCR. It offered a kind of gardening and hiking reassurance message. Global warming was going to be—in terms of food, farming, anything to do with vegetation—pretty awesome.

The movie is a simple alternation: growing plants, talking heads. The narrator, a pleasant-looking woman, appears not just dry but perspiration-impaired, dressed for a perpetual autumn: chinos, upturned collar, sweater around shoulders. "As more and more scientists are confirming, our world is deficient in carbon dioxide," she explains. "And a doubling of atmospheric CO_2 is very beneficial." Crop-crowded fields, deserts yielding to lawns. There was Dr. Idso, with the crack-voiced, slightly pop-eyed delivery of a person who can't quite believe they are on television: "A doubling of CO_2," he promises, "will produce a tremendous greening of planet Earth." (In 2001, Duke University scientists ran the experiment on an actual forest. The big winner turned out to be poison ivy, spreading 77 percent faster.) Dr. Idso gave an interview to the power industry organ *Electric Utility Week*. "I believe that the worst thing we can possibly do to the planet," the scientist said, "is to try to limit emissions of CO_2."

In the manner of a successful Hollywood product, the movie spawned a sequel—*The Greening of Planet Earth Continues* (which, like most Part Twos, tells the exact same story), and an internet fan club, the Greening Earth Society. Among the society's science advisors: Robert Balling, skeptic Pat Michaels, Mark Mills. In 1998, the industry shipped a *Greening* copy to every member of Congress. This was the era of deep canopy chic: of Ben & Jerry's Rainforest Crunch ice cream in the freezers, Amazon-themed restaurants like Rainforest Café at the malls. The video arrived in a sleeve, picturing a monkey in lush, CO_2-invigorated foliage.

It's hard to get across—without dipping into a tone of comedy or exhausted snide—just how intimate, how homey, the skeptical operation was. During hearings, Senator Al Gore pointed out the long, plush international ride given to *Greening* by fossil-fuel concerns and OPEC. He asked Dr. Idso about the movie's provenance.

> SENATOR GORE: It was made by a company which you established on the side. Is that correct?
> SHERWOOD IDSO: It was—it was helped to be made by a company which I established on the side, but which I haven't been associated with for about a year now.
> SENATOR GORE: I see. Who's the head of that company, for the record?
> SHERWOOD IDSO: My wife.

Ross Gelbspan was a reporter and editor with the *Boston Globe*; in 1995 he wrote an exposé of the skeptics that became a lifelong project. Tracking the funds, the interconnections—making the types of charts you see on bachelor walls, arrows between the photos. (A career like Gelbspan's, which is a public service, is also a casualty of climate change. Sea level rose and swallowed him.) Gelbspan described the skeptic enterprise, for the PBS news program *Now*, as a "massive campaign of deception and disinformation." In 2006, he told *Vanity Fair*, "The goal of the disinformation campaign wasn't to win the debate. The goal was simply to keep the debate going."

You could see this in the polls: in 1991, *Newsweek* reported the number of Americans who worried about warming "a great deal" at 35 percent; half a decade later, with the science firming, that number had softened, to 22 percent. "When the public hears the media report that some scientists believe warming is real but others don't," Gelbspan said, "its reaction is 'Come back and tell us when you're really sure.'"

Spencer Weart is not the type of historian given to large pronouncements. He's a trained physicist; he brings lab caution to every paragraph.

Weart sighed to an interviewer about the results of the skeptic project. "It sowed confusion and doubt into the public," Dr. Weart said in 2008, "that is now irremediable."

IN THE EARLY NINETIES, if you were following the science, you could feel the swell—the uptick that means a consensus is forming, a new floor everybody can stand on. The 1990 draft report by the IPCC's scientists had called temperature rise a "virtual certainty." As the *New York Times* reported in 1991, "Truth is, it's hard to find a climatologist these days who doesn't believe in global warming."

Which is part of what made the number so surprising. In 1991, Gallup conducted a survey: it revealed that most American climate scientists *didn't* believe in global warming.

The number was so strong—its gravity exerted so much tug—it bent coverage. Fourteen years later, the former *Newsweek* environmental writer Gregg Easterbrook announced he was "switching sides"; flipping from "skeptic to convert." For nearly a decade and a half, the journalist had aired doubts in places like the *New Yorker*, the *New Republic*, *Washington Post*, Slate—and a fat book that Mobil liked so much it placed supportive ads ("The sky is *not* falling") in the newspapers.

Easterbrook columned about his switch in the *Times*; the Gallup number had helped send him down the doubting path. "A 1992 survey of the American Geophysical Union and the American Meteorological Society," he wrote, "found that only 17 percent of members believed there was sufficient grounds to declare an artificial greenhouse effect in progress."

It was a devastating number. First, Gallup: so, authoritative. Second, it was experts: saying the exact opposite of what you'd expect. This gave the number journalistic charisma—a plot twist, a narrative, two positions approaching each other on the field of battle, statistics rattling. The number came at a delicate social moment, as warming was working its way through the party. Now, handshakes were made in an atmosphere of confusion.

The number got released in February 1992—the debut product of a nonprofit, nonpartisan outfit called the Center for Science, Technology and Media. By spring and summer, it had made its way to hearings in the House and Senate. The first time the number hit cable news was that June. "The jury's still out," a conservative commentator told CNN. "You can look it up, with the Gallup Organization." By the fall, it had joined the world of electoral politics, the November campaign. The deputy secretary of energy used the number to sock Al Gore. Warming, the secretary explained, was "a problem that the experts are not sure exists."

For print journalists, it was a restless number. It never removed its coat; it kept its bags packed from the very beginning. It went on a visiting tour of conservative organizations: in spring 1992, the Cato Institute and the *Washington Times*. (It also stopped over with the big-time, plutonium-and-anxiety press: the *Bulletin of the Atomic Scientists*.) In September, the *National Review*. From there, it found its way to the bowtie-celebrating opinion-writer George Will, who put it in his column.

Then it went everywhere. It logged air miles with Rush Limbaugh ("only 17 percent are devotees of this dubious theory"). Poll numbers have a mayfly lifespan; this number proved a special case. By the end of 1994, it had touched down so often, in so many spots, it had become a fixture in the landscape. The Environmental Defense Fund shook its head over longevity and impact: "This poll has been quoted so many times it has become gospel for the proponents of the environmental backlash."

IT WAS ALSO INCORRECT. When the number was just seven months old, right at the start of its travels, Gallup did something rare. It tried to take back the number. (A product with their name on it was ravaging the countryside.) Gallup made a public statement. In the original 1991 survey, "67 percent of those scientists" investigating the problem said that warming had in fact begun. As Gallup's correction stated, "Most scientists involved in research in this area do believe human-induced global warming is occurring now."

But the number kept traveling. To papers in the Northwest, Midwest,

Southwest, Northeast, DC, *USA Today*, the *Christian Science Monitor*. It left our shores, made landfall in Canada, the UK, New Zealand.

In 1997, its sixth year, the number was repackaged by a conservative media group. (They papered over the date and labeled it "a recent Gallup poll.") Though now officially inaccurate—traveling under a junk passport—the number took off on another tour and re-went everywhere. Among others, to the *Providence Journal-Bulletin*, *Chicago Tribune*, *New York Times*, *New Orleans Times-Picayune*, *Arizona Republic*, *Wall Street Journal*, *Bismarck Tribune*, *Kansas City Star*, *Detroit News*, *Denver Post*, *Florida Times-Union*. Experts disputing global warming coast to coast, from the *Orange County Register* to the *Boston Globe*.

The ideal combination: reliable pollster, august institutions. Here's how the *Wall Street Journal* editorial board reran the number in December 1997. (Still calling it "Hot News.") "Let's bear in mind that according to a recent Gallup poll, only 17% of the members of the American Meteorological Society and the American Geophysical Union—people who surely know something about the matter . . ."

So Gallup issued *another* correction. The number was now six years out of date. Even back in 1991, 66 percent of climate scientists had said exactly the opposite. Gallup understood the power of the number. Climate-change opponents had "used the study to support their position," the organization explained. "These writers have taken survey results out of context that appear to show scientists do not believe global warming is occurring."

But it was a hardy number. It had been making its own way for years; the corrections didn't do much to slow it down.

It headed back to Washington and politics—five separate global warming hearings in two years, 1998 and 1999. Senator and future Attorney General John Ashcroft spent some time with the number. ("The truth is, there is great uncertainty among scientists.") The chairman of the Republican National Committee picked it up twice, for swipes at President Clinton and Vice President Gore. ("It is not certain that global warming is taking place.") Approaching the turn of the century, the number was still chugging along, arriving in newspapers, books, websites, the American Petroleum Institute.

———

AND THE ORGANIZATIONS SURVEYED?

,   When the number ran a second time in the *Providence Journal-Bulletin*, a scientist wrote in, perplexed. He was a fellow with the American Geophysical Union; he couldn't remember any such question being asked of the membership. And the number struck him funny. The union, after all, awards an annual medal named for Roger Revelle.

Both groups eventually released policy statements, voted by their councils, expressing science's belief in human-induced global warming. The geophysicists went first, in 1999, followed by the weather guys. The Geophysical Union later revised their statement, with an eye toward making it even stronger.

> *The Earth's climate is now clearly out of balance and is warming. . . . As of 2006, eleven of the previous twelve years were warmer than any others since 1850 [as of 2020, per NASA, that number was 19 out of 20]. . . . The cause of disruptive climate change, unlike ozone depletion, is tied to energy use and runs through modern society. . . . Members of the American Geophysical Union, as part of the scientific community, collectively have special responsibilities: to pursue research needed to understand it; to educate the public on the causes, risks, and hazards; and to communicate clearly and objectively with those who can implement policies.*

After the turn of the century, their information manager took up the educating-the-public role personally. He wrote to newspapers still retailing the number—in Canada, Pennsylvania, Oregon, Palm Springs. Sometimes he sounded prickly. Sometimes he sounded like a linesman calling a ball foul. Sometimes he sounded weary. He wrote, "This survey is frequently misrepresented in op-ed articles, usually by climate change skeptics." He wrote, "The survey (not a 'Gallup Poll') was not conducted recently but in 1991." He pointed out, "Even in 1991, 66 percent of the scientists surveyed said yes. In the ensuing 15 years, the scientific evidence . . ." He said, "The

actual position of the American Geophysical Union, an international society of 45,000 scientist members, is that 'Human activities are increasingly
altering the Earth's climate . . .' "

But a statistic is like a fish or a breath; easy to release, tough to recap-
ture. It had reached the largest-circulation English-language daily in Bangladesh by 2004. In 2006, the number was still in migration, popping up
on both coasts: the *New York Times*, the *San Francisco Chronicle*. In 2007,
Vanity Fair was again debunking it. In 2007 and 2008, graying now, plump
and winded, the number was still barnstorming through the smaller markets, dropping its bags in humbler accommodations: Pennsylvania's *Sunday News*, Kentucky's *Louisville Courier*, the *Augusta Chronicle*. (You could
also find it in *U.S. News & World Report*.) In March 2007, you could fire
up the remote control and hear it on Fox News. "Let me give you a couple
of other facts here," explained Sean Hannity. "You know, when you look
at the most recent data of scientists, the overwhelming majority of those
that are surveyed, over and over again, they say the same thing. That they
dispute—only 17 percent of members of the Meteorological Society and
American Geophysical Society think warming of the Twentieth Century
had anything to do with greenhouse gases." The most recent data. Not a
bad performance—in public opinion terms, not a bad haul—for a number
that just one month earlier had turned fifteen years old.

SO THAT'S THE STORY. Of a number that got disinherited, twice,
by its polling agency. That was cut off, twice, by the scientists surveyed.
An orphaned number that faced hardship, overcame obstacles, and, as the
philanthropist Jeffrey Lebowski says, "went out and achieved anyway."
Where did it come from?

In 1997, a journalist wrote Gallup. What he sent was, in effect, a word
balloon containing a question mark.

An executive wrote back. Gallup's answer was the sponsor: the Center for Science, Technology and Media. "Apparently, the survey sponsor
recalculated some responses and arrived at an erroneous conclusion." The
Gallup man noted wryly, "As far as I can determine, the survey sponsor is
no longer an active nonprofit organization."

Mark Mills is one of my favorite people in this story. For months, I talked about him with everybody I knew. For one thing, it's a great name to say, like a character in a boy's adventure novel. Not the hero—the side-kick who stands lookout at the cave entrance, the best friend who holds the ladder, whose backpack contains the map, an apple, and the flashlight. For me, Mark Mills is an emblem for how permeable the culture is. Every fence is also a door, every wall is a gate, information can enter anywhere. Mark Mills is the scientist who commissioned the Gallup study.

He earned his bachelor's, physics, from the Queens University in Canada. Began graduate work at Rutgers, felt the wind and restlessness that sometimes come for the older student, left without a degree. By the late seventies he'd arrived in Washington. He pitched his desk at the delta where science and business flow into politics: where an organization has aims it needs a good scientific backing to support.

He worked for the Atomic Industrial Forum—nuclear power—in communications. A job where you peel your ears, listening for the chance to nudge the office topic into the national conversation. Reading a headline about the energy crisis, Mark Mills would write the newspaper to say how many fuel barrels the atomic industry could spare us. "The development of nuclear power," he'd explain, "is certain to lead to a lessening dependence on oil."

It's also a spot where you audition for the larger Washington cast—people whose job it is to be useful to other people. He became head of the DC office of Scientists and Engineers for Secure Energy. (Another nuclear power outfit.) He became a staff consultant to President Ronald Reagan's Science Office.

Then Mark Mills lit out on his own. His company wouldn't be providing science, exactly. It would be opinions around science. He called his new firm Science Concepts, Inc. (A name the *Wall Street Journal* considered "rather unfortunate." I think of it in the manner of Velveeta, whose label says you aren't buying cheese, but a processed cheese product.) As climate change became a recurring character in the news, Mark Mills incorporated it. He pointed out that coal, by releasing carbon dioxide, was enhancing the greenhouse effect; even innocent-looking wood enhanced it, "because deforestation removes trees that absorb CO_2."

After 1988—the summer of drought and Jim Hansen—Science Concepts pushed deeper into the greenhouse. Mark Mills pointed out how, across fifteen years, nuclear power had spared the world 3.4 billion tons of carbon dioxide. He attended a global warming conference: Jim Hansen spoke, Sherwood Rowland collected an award, Mark Mills gave a talk on the "greenhouse implications of various nuclear energy strategies." He suggested the simple, local gestures any individual could make: fax a letter instead of calling in FedEx, you just shaved two pounds of carbon dioxide. In 1991, he generated good buzz by inventing a new climate idea, the ecowatt; tips for reducing carbon dioxide—as he wrote, "one of the chief 'culprits' implicated in global warming."

These Science Concepts date from Mark Mills' nuclear era; the nineties were the start of his coal age. In December 1991, Mark Mills traveled west, took his seat at the witness table for a California regulatory hearing. (His appearance was funded by Western Fuels. Western's CEO told reporters, "The importance of the California proceeding cannot be underestimated." Western promised to "meet the professional environmentalists head-on in California and elsewhere.") Mark Mills joined the three scientists from coal's ICE campaign. Together, they testified that global warming wasn't going to happen; that if it *did* happen it would be good news for plants; that if it was happening already, it was the answer to seasonal prayers—cool summer days, warm winter nights. Sweater weather.

Mark Mills would spend the decade writing reports for the coal industry, with titles like *Coal: Cornerstone of America's Competitive Advantage in World Markets*, and *The Internet Begins With Coal*. (This one would later receive a signal boost from George W. Bush.) And three months after California, he released the Gallup number.

In 1991, Mark Mills added a phone line, printed up letterhead, started the Center for Science, Technology and Media. Now he was in the non-profit business. He put together funding. (Some came from the Scaife Foundations, which made *nonpartisan* a bit of a tug on the eyebrows. Richard Mellon Scaife is the Pittsburgh billionaire who threw millions into a long pursuit of Bill Clinton, with Monica Lewinsky et al. waiting at the finish. The Scaife approach was gruff and direct, fists and elbows. A Columbia Journalism School professor once asked about his politics. Scaife replied, "You fucking communist cunt, get out of here." The philanthropist then added unflattering remarks about the professor's mother and teeth. "That's the funny thing about tone," Al Franken later observed. "It's so subjective. I usually find it's enough to call someone 'a fucking communist cunt.'")

Andrew Revkin handled the environmental beat for the *New York Times*. Years later, he'd explain on TV how climate change got all mangled up. "There was a concerted campaign by lobbyists and communicators for industry," Revkin said. "And scientists who had partnerships or relationships with either libertarian think tanks or with industry directly, to cast doubt—basically, to focus everyone on the uncertainties."

Mark Mills introduced the number in a briefing at the National Press Club. He took reporters' questions, relaxed and friendly. You might say this press conference had a theme, which was: global warming is uncertain. Mark Mills deployed the potent word ten times. List his statements in order and they become a kind of opinion poem, a ballad to ambiguity.

> Now, from a scientist's perspective, what this speaks to
> > Is uncertainty.
> What this says, in scientific terms, is they clearly have
> > A high degree of uncertainty.
> In short, we found a great degree
> > Of uncertainty.
>
> What that says is that there is
> > Uncertainty, not certainty.
> Illustrating the extent of the
> > Uncertainty.
> That says, again, uncertainty,
> > It doesn't say certainty.
>
> We're finding a high degree of uncertainty.
> > This says uncertainty.

But again, all I can say is that it's
   An intriguing, uncertain result.
And again:
   It's uncertainty

Another uncertain thing was the distinction between Mark Mills and the Center for Science, Technology and Media. A center is reliable, quotable. It makes you picture an actual, physical spot: interns, planning sessions, a library with one of those ladders on wheels, a soda machine. As it turned out, the center was located in the exact same office as Science Concepts, Inc. Same street address, same suite door. On a page detailing the center's objectives, the phone number provided was also the number for a new Mark Mills venture: Mills, McCarthy & Associates. This hit a more sober and professional note than Science Concepts, Inc., which was now quietly pastured. (The McCarthy seems to have been Mark Mills' wife, Donnamarie, the company vice president.) The three organizations shared a fax line—so if you faxed the center, you'd also be faxing Mills, McCarthy and Science Concepts, Inc., at the same time.

When Mills, McCarthy got its own office, the Center for Science, Technology and Media packed and unpacked too. Same address, suite number, fax line. Mark Mills started another business, a newsletter called "Breakthrough Technologies"; its fax number was also the center's. They must have burned through the ink cartridges.

In the late nineties, Mills, McCarthy relocated again—to Mark Mills' residence, on a trees-and-carports street in suburban Maryland. The distinction between the center and Mark Mills had by now become largely theoretical: if you tried the center's old fax number, the crunch and screech would sound someplace inside Mark Mills' home.

At the press conference, Mark Mills had shared big plans for his center: lots of studies, lots more Gallup. "We're going to ask this kind of question," he said, "on every Gallup poll we do of experts in other fields."

As it happened, the center released only two more studies. (For whatever reason, Gallup did not collaborate with the center again.) The first was about cancer coverage. The media spent too much time on chemicals and pollutants, not enough on stealth threats like diet, or sunlight. The second was about the environment. Only 19 percent of waste experts believed recycling did any practical good. (Somewhere, Iron Eyes Cody and Keep America Beautiful dabbed their eyes and smiled.) And that was the end of the center's productive life. Its last publication was 1994, around the time the number had become "gospel."

When I think about Mark Mills' career, I see Adaptation—an intelligent man learning to read and follow the trends, making the small adjustments a person must. It's the reverse of Dr. Jim Hansen, who's known the good and bad of saying the same thing in every room. That both men succeeded shows you advance by picking a single strategy; the middle course—sometimes flexible, sometimes adamant—won't take you as far.

In the nineties, Mark Mills contributed pieces to a skeptic publication funded by the coal industry. In 1998's "Want to Improve Your Nation's Health? Burn Coal," he wrote, "Environmentalists see increasing fossil fuel use to power electric power plants as a health risk."

Actually, it was the nuclear-age Mark Mills who'd seen that. Among the health risks he once noted for coal were: "routine radiation emissions typically higher than from nuclear plants"; "lung hazards from microscopic particulate emissions"; "dozens of potential carcinogens and toxins"; "Donora P.A. and London"; and, of course, the greenhouse effect. With a Science Concept, you don't so much have persisting facts, as a change of clients.

That's what I know about Mark Mills. His wife contributed a charming, openhearted essay to an anthology of parenting stories: she quotes the Bible and Joni Mitchell, and must be an interesting person. The couple donates to a local Shakespeare troupe; they've hosted a benefit for veterans at their home. And whatever else Mark Mills was doing for the past quarter century—eating dinner, playing catch, staring at constellations—his number was out there, in the columns and airwaves, changing minds.

That's how a number can enter the national discourse. Very little border security. Sometimes, when I read news stories and come across figures nobody can account for—health care, immigration, education, warming—I think, "There's a Mark Mills number." And after releasing it, Mark Mills didn't speak about the number again: not when Gallup and the scientists refuted it, not when it circulated across newspapers and TV, a Science Concept of his that traveled the world. He doesn't seem to list on corporate bios having once directed his own nonprofit. He later became chairman of a green company that produces environmentally friendly batteries. That's where the trends carried him; and that's where the numbers come from.

Part III

DENIERS

History is full of stories that aren't actually true.
—LORD CHRISTOPHER MONCKTON, 2001

OLD JUDGE, OR TOBACCO
KILLED A CAT

AND THEN YOU HAVE CHRISTMAS 1953. AND FOR THE CIG-arette industry, a year whose best quality was that it was narrowing to an end. Cigarette chiefs called a panic meeting. Because they were CEOs—in extremis, luxury—this was at the Plaza Hotel.

That November, *Time* had run a story about Evarts Graham, a doctor who'd used cigarettes to grow tumors on mice. In the *Journal of the American Medical Association*, physician Alton Ochsner had been the prosecutor who points an unwavering finger. *Time*'s title was legal, funereal, just about worst case: "Beyond Any Doubt." The only title less appealing would have been "Cancer By the Carton"—which had run on the *Reader's Digest* story that kicked off the year. *Reader's Digest* was then the most-read magazine in America; they ended "Cancer by the Carton" with four ominous, monster-movie words: "Alert the smoking public."

Historian Allan Brandt gets Victorian about this. Which makes sense: in business, politics, and science, it's among the most influential sit-downs ever. The crossroads—where you sign the wrong deal, in crimson ink, and come back strange. Brandt calls that December Plaza day a "fate-ful meeting."

The firms summoned John Hill, head of the public relations firm Hill & Knowlton. Hill & Knowlton has since become the world's go-to suit-and-crisis negotiator: The booming persuasive megaphone carried inside the trim leather case. Hill & Knowlton has escorted the Church of Scientology across some dicey landscapes; when the nation of Kuwait decided to invite the American military over for what became the first Iraq war, billing went to Hill & Knowlton.

Craning in from our century, we can guess what John Hill advised.

He told the cigarette chiefs to reverse field. Science was a problem. The physicians Evarts Graham and Alton Ochsner were problems. An amateur might try to block research; an expert knew to order up lots of the stuff—much more research. (This was the beatitude that becomes gospel. "If we have a crusade," tobacco's head lobbyist said a decade later, "it is a crusade for research.")

The first-draft idea was to call their organization the Cigarette Information Committee. John Hill pushed to make room for the key concept.

"It is believed that the word 'Research' is needed," he explained. "To give weight and added credence."

It's like spotting your own front door in a vintage news photo. At that Plaza meeting, John Hill helped set a course for the next seventy years.

The "ploy was unrivaled in its genius," writes Siddhartha Mukherjee in his Pulitzer-winning *The Emperor of All Maladies*. "Rather than discourage further research into the link between tobacco and cancer, tobacco companies proposed letting scientists have more of it."

Uncertainty; the canniest way to start an argument. "If more research was needed," Mukherjee writes, "the issue was still mired in doubt."

And that course was locked in: an itinerary that included news desks, press conferences, and laboratory gladiator matches, scientist versus scientist. "This strategy," Harvard's Allan Brandt writes, "would ultimately become the cornerstone of a long range of efforts to distort the scientific process in the second half of the twentieth century."

CLIMATE SCIENCE IS A Hall of Ironies. Climate denial is the broken bottles and roasted tire smell of the parking lot. It's the scientists never invited inside.

There's a cigarette industry recruiting document from the early nineteen-nineties. This lists best-case qualities, in scientists being fitted for team colors.

"Ideal," runs the memo, "are people at or near retirement, with no dependence on grant-dispensing bureaucracies." That is, without inboxes or nosy colleagues; institutional bachelors.

Climate change is a vine twisting through the decades and disciplines. It pushes through strange ceiling panels, lifts floorboards, props itself in surprising corners. In the nineteen-nineties, climate denial got all tangled with the men and women suiting up on behalf of tobacco.

It began with that panic meeting in 1953, four years before Roger Revelle declared the start of the great Geophysical Experiment. It accelerated in 1992, four years after the Summer of Jim Hansen. Climate science had its heroes—Hansen, Roger Revelle, Svante Arrhenius. Denial also needed personalities to quote and celebrate. Cigarettes built the model of the celebrity denier, in the same years that Roger Revelle stepped forward as the first prominent American climate scientist. Before we meet its full climate form, here's the story of how denial became a profession.

TWO LENSES FOR LOOKING at the cigarette. One is as the climate change story in a crushproof pack. A modern convenience. Use becomes general. The scientists arrive, with their rainy-day predictions. Such moments, as the *Times* wrote about spray cans and ozone, are like "finding out that eating candy causes earthquakes."

Second way is as a romance. Flowered at the newsstands, withered in the courtrooms. Some wonderful memories: Moonlit smokes, foggy cafés, the elegance of the lighter. Then six decades of messy divorce. The Harvard professor of the history of science and medicine Allan Brandt calls the past hundred years "the Cigarette Century."

Cigarettes showed up in the shops after the Civil War. Early brand names have a tipped-hat sound: Welcome, Opera Puff, Town Talk, Old Judge. (The story in one sentence. Greetings, outings, attorneys.) For cigarettes to take off, three things were required. First, the rolling machine, in 1881. Human hands could stuff only four per minute, nicotine crêpes. But the gears kicked out twelve thousand an hour, tobacco fast food. Second, portable matches. With the ability to light up anywhere, the world became your smoking lounge. (Henry James, 1891: "My hand, in my pocket, was already on the little box of matches that I always carried for cigarettes."

Early matches were not dependable: they shot sparks, fouled parlors. They were called Lucifers. Frank matchbox advisory: "Persons whose lungs are delicate should by no means use the Lucifers.")

Last was an occasion. A monumental trade show, to gather customers and showcase the features: World War I. The big change at the front was the cigarette. People who'd never smoked, smoked; men who had preferred the splat of chew, the ruminative pleasures of the pipe, took to the pale convenience. "You ask me what we need to win this war," said General John Pershing, leader of the American Expeditionary Force. "I answer tobacco, as much as bullets." The army instituted a ration: Four cigarettes per day per doughboy. Volunteers distributed packs at the front; many came home smokers themselves. And veterans brought the habit back like a medal or a story. As Allan Brandt writes in *The Cigarette Century*, "Soldiers returned home committed to the cigarette."

Pre-war industrialists had disliked spotting them on the factory floor. Thomas Edison and Henry Ford both swore they'd hire no man who admitted the practice. Edison believed something in the rolling paper turned smokers goofy. Henry Ford produced a book, *The Case Against the Little White Slaver*; this included scare chapters like "Tobacco Killed a Cat."

After the war, opposition collapsed. It's like a bedtime story for manufacturers. Within a few decades, annual production jumped from 25 billion to 255 billion of what men in uniform had called "dreamsticks."

The celebrity sky crowded with endorsers. Humphrey Bogart smoked through pistol scenes and farewells, every kiss and gunshot footnoted by a cigarette between the lips. Frank Sinatra was grand marshal at the first National Tobacco Festival. Ellsworth Vines won Wimbledon, made *Time*'s cover, served so hard he could knock the racket out of your hand. And there he was, 1934's number one, a pitchman. "Championship tennis is one of the fastest of modern sports," Vines said. "You sometimes feel that you just can't take another step." The cigarette was then viewed as a kind of perfectly legal performance enhancer: puffs made you *healthier*. "Camels have a refreshing way of bringing my energy up to a higher level."

For men, enhancement. For women, reduction. Pioneer aviator Amelia Earhart and Lucky Strikes: "For a Slender Figure—Reach for a Lucky Instead of a Sweet."

This made the confection industry so mad it hired a former Chicago health commissioner. (And helped set a pattern: In a fight, an industry will reach for an expert.) The commissioner prepared a pamphlet: "The Importance of Candy as Food." The cigarette men—from our hilltop, a surprising vantage—fought back on grounds of health. "The authorities are overwhelming that too many fattening sweets are harmful." Old Gold cigarettes stepped in with an advertising truce: "Eat a chocolate. Light an Old Gold. And enjoy both! Two fine and healthful treats."

For everybody, physicians. The famous Camel campaign: white coat, white hair, pale cigarette. "He's one of the busiest men in town," ran the copy. "The doctor is a scientist, a diplomat, and a friendly sympathetic human being all in one." Their slogan: *More Doctors Smoke Camels than Any Other Cigarette.*

At mid-century, if you'd forgotten yours, you could count on the pack riding in the next pants pocket, there as naturally as house keys. Here's John Updike, a courtship and Harvard poem set in 1952: "There were cathedral fronts to know / And cigarettes to share—our breaths straight smoke." (Five decades later, the writer died of lung cancer.) The average person smoked four thousand cigarettes a year. As Siddhartha Mukherjee points out in *The Emperor of All Maladies*, this came to eleven butts a day, three times the old war ration, "nearly one for every waking hour."

The *New York Times* had at the beginning given cigarettes a skeptical once-over: "One thing is certain—namely, there is an ever increasing subjection to the influence of this narcotic, whose soothing powers are requisitioned to counteract the evil effects of the worry, overpressure and exhaustion which characterize the age in which we live."

This was the exhaustion and overpressure of the 1880s. By the 1950s, tobacco experts were prescribing the pack as a specific overpressure cure.

"It is a very good therapy for a great many nervous people." These are the words of a doctor of biology named Clarence Cook Little—the first celebrity denier.

In the sixties, a *New Yorker* writer went on a notebook safari through the big cigarette offices. One Philip Morris exec said a famous physician had just occupied the very chair where the journalist now sat. And had held forth on the perils of *not* smoking. "He said that if people were to stop smoking, there had better be something pretty powerful to take its place," the executive explained. "Or there would be more wife-beating and job dissatisfaction than people's natures could tolerate." Cigarettes had helped win a war, and were keeping the nation unbruised.

You think of the prescription pads of our own time, with their drowsy, end-of-the-alphabet names—Prozac, and Xanax, and Zoloft. There must be some anxiety, some chill to modern life, that makes us want to fiddle with the thermostat setting inside our heads.

STOCKINGS AND CHAIRS

PEOPLE WERE ALSO MAKING ADJUSTMENTS A LITTLE FARther down. The Tulane physician Alton Ochsner was born in North Dakota, trained in Saint Louis, eventually heading the American Cancer Society and the American College of Surgeons. In 1919, a patient with cancer of the lung was admitted to the Washington University training hospital where Ochsner was a student. "As usual," Ochsner said, "the patient died."

But his disease was so uncommon, it meant: field trip! A professor gathered up students. "He had the two senior classes come out to witness the autopsy," Ochsner said. "He said the case was so rare, he thought we'd never see another as long as we lived."

Two decades later, lung cancer was no longer the mountaintop you traveled to see. It was the scenic vista that follows you home. This was 1936. "Nine cases in six months," Ochsner said. "Of a condition that seventeen years previously I wasn't supposed to see for the rest of my life." To Ochsner, this meant an epidemic. Which also meant a cause.

The surgeon drew connections. In the wards—the faces on the pillows—what was taking place was a sad-eyed veteran's reunion. "All the cases were men. They all smoked heavily," he said. "And they all began smoking during the first World War."

Ochsner brought his suspicions to Dr. Evarts Graham. Graham was the Platonic ideal of the doctor in the Camel ad. White coat, white hair, broad reassuring body. He was Washington University's chief of surgery.

The scene plays like an episode from the old medical drama *House*. You see gurney, nurses, two physicians conferring in a corridor.

Graham, a heavy smoker, would have held a cigarette; you could smoke in hospitals then. He was perhaps the nation's foremost lung man. He laughed off Ochsner. Sure, lung cancer and cigarettes had climbed the charts together. So had a lot of things. "So," Graham said, "has the use of nylon stockings." His overall verdict: "How dumb and stupid."

The classic paranoid trope: invisible because they walk among us. This is Siddhartha Mukherjee—in his non-writing hours, an oncologist—from *The Emperor of All Maladies*. "When a risk factor for a disease becomes so highly prevalent in a population," he writes, "it paradoxically begins to disappear." (This is because "our intuitive acuity . . . performs best at the margins." When it's all signal, it's also all noise.) Mukherjee quotes an Oxford scientist. "Asking about a connection between tobacco and cancer was like asking about an association between sitting and cancer."

Graham began a study. In a mood of snide. Science is omnimotivational. Graham aimed to disprove. After a few months, he wrote Ochsner: "I may have to eat humble pie." He cut out smoking. When he published in 1950, Dr. Graham sounded like his friend. Cigarette use and lung cancer were "approximately parallel."

Scientists and statisticians share a motto: Correlation does not imply causation. This was the point of the chair and stockings jokes. Once the carbon dioxide rise and the warmth became inarguable, it is a slogan deniers rediscovered en masse.

Demonstrating cause put Graham in a bind. Subjects would need to smoke in a locked room for decades. They'd have to remain under observation from arrival to autopsy, on easy terms with personal extinction. "I will say to those who wish to volunteer for such an experiment, 'Please form a queue to the right,'" Graham joked. "No crowding, please."

So Graham invented a smoking robot: a nightmare nicotine-addicted mechanical lung. It puffed all day; for the good of its figure, doctors fed it Lucky Strikes. The machine distilled cigarette tar. Then Dr. Graham painted the tar on the backs of mice.

Mice pay for our sins. The animals grew tumors. Dr. Graham published in the journal *Cancer Research*.

BY THE EARLY SIXTIES, lung carcinoma had become the leading cause of cancer death among adult men. Forty-one thousand deaths per year—up from four thousand three decades prior. Senator Robert Kennedy gave the sad roundup in a speech: "Every year, cigarettes kill more Americans than were killed in World War I, the Korean War, and Vietnam combined." These were numbers Evarts Graham never got to read. In 1956, he came down with the sort of flu that sends you to bed, then tugs you out with just enough energy to cough in a waiting room. For physicians who'd trained under him, Graham was a legend. They argued over who should deliver the news. It was a former protégé who told Evarts Graham he had lung cancer.

The doctor was as incredulous as any patient. He asked, "Are you sure?" He made a brave little joke. "That cancer must have been awfully mad at me."

To guarantee a non-tactful second opinion, Graham scrubbed his name, brought the slides to another colleague, who pronounced his tumor inoperable. Graham wrote a fellow doctor. "I suppose you have heard," he said, "about the irony that fate has played on me." He wrote Alton Ochsner that the cancer "had sneaked up on me like a thief in the night." That's what the malady was: the housebreaker who sacks up everything, all potential futures included. When he died two weeks later, Graham left his body to Washington University—to be examined on the same tables where Alton Ochsner had got his first look at cancer of the lung.

GENETICS

CIGARETTE CHIEFS HELD THEIR PANICKED 1953 CHRIST-
mas meeting. With Hill & Knowlton's John Hill, they settled on a name
for their new organization.

There's a nice joke contained in the minutes of that session: everybody
agreed "with the urgency of getting such a program underway immediately."

Here is Urgency sailing by its actual dictionary definition. One week
later, a full-page ad ran in 448 American newspapers: the message reached
forty-three million people. "We are pledging aid and assistance to the
research effort in all phases," the firms promised. "This group will be known
as TOBACCO INDUSTRY RESEARCH COMMITTEE."

NOW THEY JUST NEEDED a human face—a particular someone to
stand beneath the opinions, speeches, and word balloons. Advertiser and
theologian Bruce Barton had understood this in the twenties: we need to
see faces. This was because "you and I are interested most of all in ourselves,"
is how Barton put it. "Next to that we are interested in other people."

The tobacco chiefs' ad included a bold casting promise. Leading "the
research activities of the Committee will be a scientist of unimpeach-
able integrity and national repute." Now they just had to go out and find
that paragon.

A Hill & Knowlton memo later acknowledged this to be a "real challenge." They were offering a Manhattan office and salary; the money would

not be entirely free of taint and odor. The man they discovered was named Clarence Cook Little. A fellow geneticist once described him to *Time* magazine as a "handsome numbskull."

Clarence Cook Little set the mold. The first scientist whose job was to remain publicly and inconsolably unconvinced of something. He would develop the first uncertainty catchphrase: "Not proven."

He had the good numbskull genetic luck: born handsome, stayed handsome. As a young man—tall and dark—he resembled Jake Gyllenhaal. Later on, his mustache went gray, he developed the folded-umbrella chin of the older man. This lent him an air of probity. He looked like the actor portraying the bank manager in a sitcom.

His background was also perfect. Bloodlines all the way back to Paul Revere. Three degrees from Harvard. (Little had to take his doctoral exam twice, was the sole blemish.) His father was America's first breeder of Scots Terriers—if you admire the breed, this may be the family's signal achievement. (The adult Clarence Cook Little would become a popular judge at dog shows.) He was the youngest president ever at the University of Maine. Then he lit out for larger greens and enrollments, and became the youngest at the University of Michigan. Then he directed the American Cancer Society. In 1954, he was supervising a research laboratory in Maine—on the Daniel Defoe-sounding Mount Desert Island—and that's where Hill & Knowlton found him, at the sunset age of sixty-six.

This is the overstory. But it's the understory that shaped and kneaded Clarence Little for the role of first celebrity doubter. As a student, he'd gone head over heels for genetics. When a Harvard family with pedigree dogs on the rug and Paul Revere on the mantel starts talking about "breeding," there's going to be trouble.

Little became one of America's premier eugenicists. Racism, in its science-language spelling. He befriended Madison Grant, author of the famous white call-to-arms book, *The Passing of the Great Race*.

Grant is the figure F. Scott Fitzgerald is making fun of, early in *The Great Gatsby*. "Well, these books are all scientific," Tom Buchanan explains. "It's up to us, who are the dominant race, to watch out or these other races will have control of things." It's important to recognize this was not just the spirit of the times; Fitzgerald was enrolled at Princeton when Clarence Little studied at Harvard. But Fitzgerald's book looks on Tom with pity: "Something was making him nibble at the edge of stale ideas."

Here is Madison Grant taking stale racial attendance: "The cross between a white man and an Indian is an Indian; the cross between a white man and a Negro is a Negro . . . and the cross between any of the three European races and a Jew is a Jew." The past is a strange, vivid, storm-tossed place. Leaders of very different stripes bowed to Grant's erudition. Former president Theodore Roosevelt praised him: These were the "facts our people most need to realize. . . . All Americans should be sincerely grateful to you." And Grant belongs to a very select club. Americans who've received fan mail from Adolf Hitler: "The book is my bible."

Clarence Little stood with Madison Grant in the halls of the American Museum of Natural History, when Grant called the Eugenics Committee of the United States to order. He was also an enthusiastic member of Grant's Galton Society (named for a race-proud cousin of Charles Darwin), which offered the clubhouse bonus of being Jew-free.

Little became president of the American Eugenics Society, the Neo-Malthusians, and the Race Betterment Congress. He featured in headlines—"SEES A SUPER RACE EVOLVED BY SCIENCE"—and delivered pro-sterilization speeches: "We favor legislation to restrict the reproduction of the misfit." He spoke out against opportunity. "America is based on the false premise that all persons are born free and equal. This is an absolute absurdity. . . . We must segregate men according to their standing." He helped found the American Birth Control League. (Shucking off influences and associations, it would eventually evolve into Planned Parenthood.) Contraception and eugenics, he explained, were inseparably linked. Accepting the league presidency, Dr. Little gave thanks, for popularizing race science, to "the gentlemen who rule Italy, Japan and Germany."

His administrative career was a succession of misfires. He got into hot water at University of Maine: lavish budgets. The University of Michigan, being larger, offered broader avenues for misfortune. Seventy years later, a U. Mich journalist would write, "It's difficult to see Little's career as any-

thing but a blemish on our collective past." He got in trouble for outlaw-
ing student cars. Also for a crackdown on what was called "petting"—an
antique sexual activity whose exact parameters are difficult to fix. ("At parties," Little explained, bringing it all home, "the car parked outside is often
used as a drinking and petting parlor.") The last straw was an extramarital
affair with a student. The first straw was speaking out on eugenics and birth
control. This especially horrified Michigan's governor, born ninth in a fer-
tile and cheerful family.

Little was invited to resign. He received the same invitation next decade
from the American Cancer Society. Where he had not been doing enough
to fight cancer. "Dr. Little," an industrialist said, "I would like to conclude
by saying this Society is too small to have both you and me in it. I intend to
stay." As Richard Kluger writes, in *his* Pulitzer Prize tobacco history *Ashes
to Ashes*, "Events a decade later would suggest that the scar never healed."

He was a thwarted public man. War against the gentlemen who rule
Italy, Japan, and Germany had somewhat dampened the market for eugenics. Bad luck is an opportunist. Little holed up at the research island in
Maine—which in 1947 burned to the ground. When the news broke, former Cancer Society colleagues were surprised to see Clarence Little seated
in the tobacco research chair. They totted it up to bitterness.

To a scientist friend with misgivings, Little explained his rationale. And this offers us, right from the start, some ideas about denial
motivation.

It wasn't for scientific challenge. Or the stickler's pleasure—keeping a
discipline skeptical and honest from the outside. Instead, salary, travel, and
a sort of padded recklessness that comes to reputations later in life. "I need
this job," Little explained, "because I would like to go to New York as often
as I can, and I need the money." And at sixty-six, the scientist's accomplishments were already safe in the rear-view. Little was not really a numbskull;
the calculation shows his intelligence, and the intelligence of the industry.
"Above all," Little told his friend, "I can do whatever I want to do, because
my sarcophagus is built."

"That was his attitude," his friend later explained. "And that is what got
him into deep trouble."

So at this early moment of denial, we get an early appraisal of the value of sponsored opinion. (George Orwell boiled it down to a sentence: "A bought mind is a spoiled mind.") "My attitude didn't change as far as him as a friend and human being was concerned," Little's colleague said. "But I didn't trust his judgment in scientific matters anymore."

A SCIENTIFIC GYMNASTIC FEAT

THE INDUSTRY DISTRIBUTED PAMPHLETS. REASSURING titles. Laboratory: "A Scientific Perspective on the Cigarette Controversy." Lamppost: "Go Ahead and Smoke," "Smoke Without Fear." Phrasing was bold: "If you have tried to give up smoking a dozen times and failed, quit trying. . . . Clarence C. Little [states], 'If smoke in the lungs were a sure-fire cause of cancer, we'd all have had it long ago.'"

Just last season, doctors had been scientist, diplomat, and sympathetic human being rolled into one. Overnight, they became "alarmists," "publicity seekers," the promoters of a "health scare." Members of the Research Committee—as deniers would learn to do with climate change—attributed the cancer rise to other factors: better record-keeping, or the larger issue of pollution. (This was soon after Donora.) Just as deniers would later ascribe warming to causes such as solar variability, or the influence of cosmic rays.

Under Little's direction, tobacco received the changed winds coverage that would later ruffle through climate. "Doctors Disagree"; "Is It Smog?"; "A Scare Fades with a Puff of Smoke."

THAT FIRST DAMP-ARMPIT YEAR became known as the "1954 emergency." Next year, Hill & Knowlton was sending around dry-browed memos. "The lynching party seems to have been called off, at least temporarily." Credit went, the firm wrote, to the doubting and finicking of the scientists around Clarence Little.

"Thanks to his renown," the *Wall Street Journal* later reported, "the Council was able to attract an illustrious scientific advisory board."

The *Journal* is not a notably soft-stomached paper; but the backward glance made it queasy. Clarence Little's new job marked the start of "the longest-running misinformation campaign in U.S. business history."

Clarence Little barnstormed. And these would become the perks of the denial life, as it matured across the century. The town car to the event. The sit-down with the editor eager to ensure balanced coverage. The gratifications of the interview—watching, beyond the horizon of your crossed ankle, as your opinions are carefully taken down.

Little was happy. He had "rediscovered the public limelight," Allan Brandt writes, "at this late point of his career."

On CBS TV, he even got to sneak in some contraband from the bad old eugenics days. "We're very interested in finding out what *kinds* of people are heavy smokers," Dr. Little told viewers. "Is it a different nervous type of person?"

The widest trail the scientist blazed—clearing brush, dotting tree trunks—was uncertainty. Dr. Little's famous quote boiled down to those two words *not proven*. (These are the climate words we still live by; it has never *quite* been proven.) Not wrong train or track—just no arrival, no car disgorging passengers. The necessary fuel was "more research." Little put the cancer rise down to genes (of course). To hormones, to diet. Even mood. "A person who is unhappy," Dr. Little explained, "is more likely to develop cancer. . . . These bodily functions are controlled by glands." So along with the medical stigma came a darker American stain: you'd thrown away years on a bad mood. Cancer was being discharged from life on the basis of attitude.

He even got to stick it to the old outfit. In 1957, the American Cancer Society published findings. (The organization's head of epidemiology had wanted the results out fast: "These lives," he said, "are on my conscience.") Cigarette links to lung cancer—also heart disease, cancer of the esophagus, larynx, bladder, tongue, the *lip*.

Clarence Little lobbed his objections from the field of sport. (He'd been a shot-putter at Harvard.) This multi-causality, Little said, would be "a scientific gymnastic feat."

His annual report for the Research Council was unperturbed. It did note a risk factor: men who quit smoking, "showed a tendency to gain weight."

Alton Ochsner—the doctor told in 1919 that he'd never again see lung cancer—delivered a great anti-uncertainty speech to the American Cancer Society. A Russian count has become suspicious of his young wife. So he announces plans for a journey. Then the count holes up next door and spies. Sure enough, a sleigh arrives in the moonlight. The count's wife welcomes a handsome cavalry officer. Through the upstairs window, the count watches their embrace. The room goes dark; a candle has been snuffed out. The count beats his brow. "Proof! If only I had proof!"

FACES WORK; smoking made its comeback. "You and I are again puffing away," financial journalist Sylvia Porter wrote in the *New York Post*. "And we will continue to smoke." (Porter died of emphysema.) All those patients, studies, hospitals, Russian couples, tumorous mice. Flattened by Clarence Little and nine scientists saying "doubt."

Peak tobacco arrived *after* what tobacco companies called the "scare." In a way, this is a long-odds achievement to be proud of. A demonstration of what expertise can do. With illness darkening the hour, bad press murking the handholds, cigarette use continued its climb; 369 billion cigarettes in 1954 became 488 billion in 1961.

Hill & Knowlton held a sort of internal celebration over that year's numbers. "The importance of the Scientific Advisory Board," one executive explained, "cannot be overemphasized." A memo to Dr. Little congratulated his board on carrying their "share of the public relations load."

When Clarence Little died, in 1971, he was still in the saddle. Seventeen years. His longest away-from-New England job.

MORAL QUESTIONS sometimes borrow shading from political ones. What did Clarence Cook Little know, and when did he know it? It's clear what the cigarette makers knew. Historians have tried to fix the exact hour at which hints and evidence buttoned on the formal wear of knowledge. "Senior scientists and executives," write these researchers, "knew about the potential cancer risk of smoking as early as the 1940s, and most accepted the fact that smoking caused cancer by the late 1950s."

Those were the people paying Clarence Little his doubter's salary. What did the scientist himself believe?

Testifying at a liability case in 1960, Dr. Little answered, "My mind is open." He added, "So is the mind of the whole Scientific Advisory Board."

In 1958, three English tobacco scientists visited here. A British expeditionary force: to share data and strategies, exchange commiserations. The British scientists met with Clarence Little, sat down with his whole Scientific Advisory Board.

Whose minds were, in fact, not open. "The individuals whom we met with believed that smoking causes lung cancer," the visitors reported. "If by 'causation' we mean any chain of events which leads finally to lung cancer and which involves smoking as an indispensable link."

Clarence Cook Little knew. And still he said "not proven." Later in his testimony, you can hear the dollar hopes, the bank plans: It's a voice happily counting years and paystubs on the pillow. "We have got hold of a big enough question," Dr. Little said, "to use every bit of energy for all the rest of our lives and far beyond it." He enjoyed New York, he could do with a little extra money. (Two conditions frequently encountered together.) *Proven* would have ended the job.

When I read about people who take unhappy positions, my mind always jumps to families. In 1971, the legal counsel at one top firm quit. He'd been striving on behalf of tobacco for a decade. "I had to come home every night," the attorney told a reporter, "and face my kids saying, 'Daddy, why do you work for a cigarette company?'" A 1980s executive pushed for chemical makers to stop widening the ozone hole. Same living-room motivation. He'd been "suffering intense pressure from his family, when a *New Yorker* article attacked him by name as a destroyer of the environment." New decade, different idiom: A tobacco scientist who'd wandered down the path Clarence Cook Little made was confronted by his teenaged daughter. "Dad," she told him, "you've got to be careful. These guys are pimping you."

With Clarence Little, it was the reverse. No family got in his way; they just came round after to tidy up the memory. Little died at eighty-three. His wife and children outlined his life for *Newsday* and the *Times*. The trail halts at 1954. No record of tobacco service. This is one of the first denial stories; it turned out the work could be denied, too.

WALL OF FLESH

YOU COULD ALSO PRACTICE DENIAL BY LAW. AS OF THE late eighties, there'd been three hundred trials—smokers puffing, sickening, lawyering—and the industry had never paid a nickel in damages. It used the number to stare down future lawsuits, as a football team might intimidate you with an undefeated record.

This strategy became known as the Wall of Flesh. You sued—in Minneapolis, Mississippi, Florida. And the circus came to town. Strong men, barkers, lion tamers, fortune tellers with a 100 percent accuracy rate: You are going to lose.

"A legal machine," is the summary of one reporter. "Hundreds of attorneys, paralegals, researchers, scientific advisers, and private investigators, not to mention public relations consultants, that form the defense when any U.S. tobacco company is challenged in the courts."

Philip Morris had supplanted Reynolds Tobacco and Camel: Marlboro became America's number-one cigarette; in 1989, America's leading packaged good; then the planet's most successful brand. Not Disney, the Coca-Cola polar bears, Apple's apple. The Marlboro Man. An ad executive explained the effect to the Smithsonian Institute. "Why it works," the advertiser said, "is it captures the spirit of an alienated country and recaptures a lost time." Office Camus, cowboy Proust. "The pressures of society have made us something we don't want to be. The cowboy is an antithesis."

Actually, Marlboro began life as a lady's cigarette. It had a gimmick: the red tip wouldn't show lipstick stains. Its slogan was "Mild as May."

Ads ran pretty much the antithesis of cowboy for three decades. "More and More Femininity." Plus a word that would have made ranch hands push their hat brims up and scowl: "*Soigne* to the very last puff."

For a time, it was marketed at mothers—and not ideal mothers. Mothers ashing and yelling and maybe taking it out on the kids. "Before you scold me, Mom," says the ad's worried baby, "maybe you'd better light up a Marlboro."

In 1954, the company decided to rehang Marlboro. It simply takes titanic confidence, resources, and ability to bring a thing like this off. As the son of Philip Morris' CEO put it, the idea was to transition Marlboro into "the cigarette with balls."

Billing went to Leo Burnett Advertising of Chicago. Burnett—who did not resemble the Marlboro Man—examined the corpus. Marlboro was "a sissy smoke," he concluded. "A tea room smoke." The "new Marlboro wasn't this kind of a cigarette at all."

The rollout was done via tattoos; inked hands, on interesting and brawny men. Pilot, hunter, craggy veteran. "Follow any man with a tattoo," one of the ad boys remembered having read in Jack London, "and you will find a romantic and adventurous past." Copy developed a deeper, rougher voice. Hemingway, in a cold kitchen, reheating yesterday's coffee and muttering. "The filter works good and draws easy. . . . The Flip-Top Box keeps every cigarette in good shape and you don't pay extra."

Marlboro just missed being about something other than cowboys. The Burnett team met in the conference room—an impromptu round of the old *Saturday Night Live* game show *¿Quien Es Mas Macho?* "What's the best masculine image in America today?" A bunch of people suggested "Cab driver." For a moment, national myth teetered in the balance. We might have ended up with the worldly Marlboro cabbie: the beaded chair, the request to say your destination a second time, the wild damp territory in the backseat of the taxi instead of the purified flat squirt of the plain. Somebody else said, "Cowboy."

After that it was all saddles. Sales climbed 5,000 percent the first year. By 1975, the newly testicularized Marlboro had become the all-time bestseller; our planetary number one.

A British filmmaker approached the company the following year. He proposed a documentary for the UK news show *This Week*. Philip Morris agreed; they craved a larger slice of the British market.

The movie secretly proposed a nasty experiment. Let's take Marlboro at their word: Let's find ropers and riders who really had spent their whole lives smoking. The show featured six cowboys; five had lung cancer. The sixth—emphysema—was filmed on horseback, oxygen tubes in his nose. "I thought to be a man you had to have a cigarette in your mouth," one cowboy says, gazing into the campfire. "All I got out of it was lung cancer. I'm gonna die a young man."

Philip Morris hit the ceiling, registering their displeasure in cowboy. "We were had," said the director of corporate relations. "Sand-bagged and double-crossed."

Even the title was scare-year bad, armpit bad, 1954 bad. *Death in the West: The Marlboro Story*. It's still considered the "most effective attack ever" on the cigarette industry. Dr. Clarence Cook Little's old outfit the American Cancer Society planned to enlist the film in education drives. Even worse, CBS had entered negotiations to air *Death in the West* on *Sixty Minutes*.

It was the worst thing to happen to cigarettes for a decade. So what was the solution? Wall of Flesh. Philip Morris dispatched riders from their Kansas City law firm of Shook, Hardy and Bacon. With a mission: Track down those cowboys. Prove they were never ranch hands at all.

Two cowboys eluded this posse forever, by promptly dying. But that's the great practice of denial: it's the thumb on the remote, changing the channel, sending the fight to take place on a different network, against a more favorable backdrop.

Philip Morris took a documentary about cancer, turned it into a debate over job title: whether six dying men qualified as cowboys. As tremendous as changing the gender of a cigarette.

The documentary actually contains the hard incidental poetry of the West—the sound Marlboro copywriters labored to find. This is the former rodeo rider Junior Farris onscreen, discovering he is terminally ill: "I got to spitting blood and stuff around, my legs just went down. . . . And I laid there a bit and crawled in the barn where it was cool and took an invoice of myself."

Two Shook, Hardy lawyers tracked Junior Farris to his place of business at the Oklahoma stockyards. Here's that collision the ad man spoke of. The alienated office, the cowboy antithesis—who we were, what we've become. "He stated that he was a family man and had several children," the lawyers reported back to Kansas City. "He mentioned that just a few moments before he came to talk with us he was offered a drink of whiskey.... He said he was part Choctaw and did not drink to excess." They added, "The pens and alleyways are covered with excrement from the livestock. The odor is quite intense.... At the conclusion of our interview we thanked him for his time and parted on cordial terms." This really is denial in a very pure form. Junior Farris, dying of lung cancer. Offering proof of stereotype to two cordial attorneys from the firm of Shook, Hardy.

The Wall of Flesh crowded the London courts. Thames TV, with its cramped urban budget. Philip Morris, with Grand Canyon resources. *Death in the West* aired only the one time, in the fall of 1976. Per the settlement, the production company was allowed to retain a single copy; remaining reels were shipped to Philip Morris, with a memo reading "To be destroyed." Three of the remaining cowboys also were gone within the decade. The film never aired on *Sixty Minutes*. Denial, in its legal form, is about keeping the cost of honesty prohibitive.

SIMPLE ANNIHILATION

START WITH A TITLE. *ON THE FRONTIER: MY LIFE IN* _____.
What's grand enough to complete it? "Combat"? "Platform Shoes"? "Low-Earth Orbit"? Stephen Hawking kept his autobiography's title simple: *My Brief History*. (The grim kick finds you after.) Chelsea Handler, host of a raunchy talk show, also went unassuming: *My Horizontal Life*. Physicist Fredrick Seitz arrived at the stage where a person compresses the big events into memoir, and decided his presence had cast a long shadow. *On the Frontier: My Life in Science.*

There's a photo of Dr. Seitz. Sixty-seven, at a faculty party, beside his wife—Elizabeth Seitz also looking long-boned and hard-eyed. They stare together from the edge of a group portrait, other faces at ease. This was 1978, some months before the physicist accepted a consultancy with Reynolds Tobacco. Decades later, *Business Week* would call Frederick Seitz the "granddaddy of global-warming skeptics." With S. Fred Singer, one of the two graybeard prophets who launched a movement.

For industry, the scientist became a Rambo to train, toughen, then release. Tobacco executives met that fall in New York. They were themselves hard thinkers: baked and veteranized by a quarter century in the fights. American Tobacco's general counsel reminded attendees that no written record of the proceedings should ever be distributed. Among other things, they were looking for a Clarence Cook Little replacement.

Little's former outfit had lost a step with the years. It had endured a makeover, around the time of the surgeon general's famous report. In 1962,

President Kennedy and his top medical man must've taken a look at the calendar. High time to provide an answer. A domestic version of climate's IPCC: A scientific panel to review all seven thousand previous cancer studies, return a final guilty or innocent.

The panel was scrupulously balanced. Shirts and skins; five nonsmokers, five coughers.

It didn't *stay* balanced. The panel's medical coordinator later called the data "hair-raising." The surgeon general himself was a pack-a-day man; he fell back on pipes. Another panelist simply quit. Dr. Louis Fieser—Organic Chem, Harvard—filled committee ashtrays with eighty butts a day. (Fieser must have loved smoke; he developed the weapon napalm for use in Vietnam.) When deliberations were over, the scientist lost a lung to cancer.

"You may recall that although fully convinced by the evidence," Fieser wrote his fellow panelists, "I continued heavy smoking throughout the deliberations of our committee." He was like one of those politicians who join the pride march only after the personal shove of learning a son or daughter is gay. "My case seems to me more convincing than any statistics."

Armed guards were posted outside meetings. Tobacco was among the nation's biggest industries. (For one thing, it was America's largest single advertiser.) A wily stockbroker might have snuck in, determined wind direction, made trades, wreaked havoc.

In January 1964, the surgeon general's committee returned its verdict. Considerately, they did so on a Saturday, when Wall Street sleeps in. Historians use language familiar to us from climate headlines.

The link was "unequivocal." Cigarettes caused cancer.

It's important to recognize: like Du Pont with ozone, like the coal people and Exxon, cigarette executives were not a special subgroup of immoral men and women. They were good people who'd landed fine and demanding jobs. But, in a pinch, they forgot what they had in common with faces on the street; responsibility and community retreated to the faces on the office hall. In a sense, they became too skillful at loyalty to each other.

In the wake of the report, a certain hardness, a bordertown frankness, entered the denial field. They knew they were villains now. The report freed them up, allowed them to become wily guerillas, outnumbered fighters way behind enemy lines. By telling them where they stood, the surgeon general invited them to become *less* responsible, and stonier-hearted.

You can chart this by memos. In 1969, an executive with Brown & Williamson set down four of the most famous words in denial.

His strategy document describes a pinned situation: "We," the exec writes, are a "harassed and restricted industry . . . in conflict with the awesome forces of the federal government." Tobacco, he explained, could thus operate "with the confidence of justifiable self-interest." That is, the report had liberated them to become unscrupulous.

And then the executive set down the four famous words. "Doubt is our product," he explained. "It is also the means of establishing a controversy."

Not competing facts. Just doubt—a fog that creeps in, never lifts; a damp that lingers and stinks. Next decade, a vice president at the Tobacco Institute—the industry's lobbying arm—codified things. The industry had found ways of "creating doubt about the health charge without actually denying it."

Somebody must have remembered Bruce Barton's old advice to General Electric, about dropping that unfriendly third word, *Company*. Clarence Little's Tobacco Industry Research Committee had got blanded: it was now called the Center for Tobacco Research. This wasn't getting the job done either. At that 1978 meeting, a Shook, Hardy lawyer gave the sad rundown.

It's amazing how direct—how honest and clear—these professionals could be. When doors closed and every elbow rested on the same table. The center, the attorney explained, had been "set up as an industry shield." It had "acted as a 'front,'" with some "public relations value." It was now spilling radiator fluid, muffler banging, as it crawled into long-term parking. For one thing, the scientists who'd come aboard with Clarence Little had aged beyond courtroom use. They couldn't be trusted to withstand "the rigors of cross-examination." This was a "major problem."

It wasn't all bummers. The Shook lawyer detailed a "major public relations plus." ("Major" was itself a major 1978 word.) A scientist named Gary Huber had accepted industry support at Harvard Medical School. Tobacco money, flowing in the Harvard fundstream! Here was a good safe path for the industry to walk down. "It is extremely important," the lawyer said, offering advice that would become gospel for Exxon on climate, "that the industry continue to spend their dollars on research to show that we don't agree that the case against smoking is closed."

All of which made Dr. Frederick Seitz a spectacular, necessary get. The second coming of Clarence Cook Little.

Seitz was the most-decorated scientist ever to cross over to the leaf and denial side. Recipient of the National Medal of Science. Former head of the American Physical Society. (This sounds gym-related but is really the clubhouse for physicists.) Member, then president of the National Academy of Sciences. Seitz, when tobacco found him, was sunsetting; about to step down as head of Rockefeller University. Seitz wrote a friend he anticipated a "very active retirement."

AS WITH CLARENCE LITTLE, there was an understory. An unhappy love affair. Seitz had romanced the country. By the seventies, the nation decided it was ready to see other people.

Frederick Seitz was a super-patriot. He was born on July 4, 1911—the sort of coincidence that takes up residence in the head. His parents ran a San Francisco bakery. After college, Seitz headed for Princeton University with William Shockley—a scientist who'd later develop the transistor, win the Nobel, and still later found Silicon Valley. "Seitz and I drove East in my 1929 Desoto roadster," Shockley explained. "And I recall arriving at Princeton one evening, the moon shining on the buildings and so on." That's the scientist: romantic and humdrum in the same sentence.

Seitz wrote a foundational textbook on solid-state physics. He brought his expertise to bear on the war against Germany. In science circles, there's a swanky way to say you crewed on the atomic bomb: Seitz worked in the Manhattan District at the University of Chicago.

It's the post-war that transformed him into the man who'd stump for tobacco, then attack on climate. The Cold War raised a strange crop of veterans. Men who never saw action but fought in sour stomachs and midnight coffee and bravely facing facts—then resumed civilian life without scars or ribbons.

Seitz sat in a coat and tie beside three other Manhattan District scientists, for an NBC roundtable broadcast to millions of homes. This was 1950: in the shadow of the Russian bomb, researchers were calmly discussing the end of life on Earth. Following an atomic exchange, thirty to sixty million Americans would need to be relocated. One physicist put in a deadpan, nuclear-age joke: "The ironical conclusion in this respect," he said, "is that it becomes easier to kill all people in the world than just a part." Another nodded: "That is definitely so."

Journalists called the NBC show "the chiller to top all chillers." The former head of the Atomic Energy Commission accused the men of fomenting a "cult of doom," then asked, "Why scare the daylights out of everyone?"

As with Clarence Cook Little, it wasn't just a question of the spirit of the time. "To this sort of talk," *Time* assured readers, "equally informed physicists react with astonishment or distaste."

Apocalypse was Frederick Seitz's eugenics. Every battle was the last battle. Contested at dawn, with the banners of the enemy choking the valley. Three months later, he was making a recruitment speech in the *Bulletin of the Atomic Scientists*. The joyful clench of spreading the worst possible news.

Scientists, Dr. Seitz said, needed to drop their "sense of sin" when it came to the nuclear spear; they must step up, protect their neighbors from communism. After all, Seitz asked, "Who among us will feel sinless if he has remained passively by while Western culture was being overwhelmed?" His paper set off, journalists wrote, "an extraordinary stir."

Response this time arrived from the Western street corners Frederick Seitz was defending. *Time* magazine letters page: "If physicist Seitz wants to prostitute his scientific talents inventing weapons of destruction, that is an affair of his own conscience."

A couple years later he was at it again. Once more, the *Bulletin of the Atomic Scientists*. Which must have made for squirmy Cold War reading. Seitz actually deployed the uncongenial word *Armageddon*. (Really just Greek for the biblical city of Megiddo. Apparently a rough place.) Seitz was living inside the twists and turns of the sort of novel air passengers buy to soak up excess attention mid-flight.

It's funny because decades later—in his memoir *On the Frontier*—this is just what Seitz would hold against the environmental scientists. Such researchers served as prophets, "but more of doom than of salvation," he wrote. "Their tactics encourage universal fear."

Here is Dr. Seitz in 1953. Detailing Soviet plans for the world: "overthrow." Their blueprint for America: "simple annihilation." Their vision for the post-war: "There is no reason to expect them to display mercy." The whole sad affair to be climaxed by a total Western junking—"obliteration of that line of ethical and cultural tradition which traces its origin to Greek and Judaic sources."

Altogether, things would not be looking too rosy. To top it all off: "Exile to Alaska without adequate preparation for millions of individuals."

At this dark pass, ethics did not seem the most essential tool for the go-bag. "The issue of morality," Dr. Seitz wrote, "perhaps lies in the philosopher's domain."

So it came as no surprise. When Seitz got elected head of the National Academy of Sciences, in 1962, he told reporters he "planned to prod even more scientists into working for national security."

Romance is timing. Frederick Seitz and the National Academy during the 1950s: love match. Frederick Seitz and the National Academy during Vietnam: cage match.

If we'd never fought that war, I don't believe Fred Seitz would ever have worked for tobacco, denied on climate. It was the start of a chain. As Academy president, Seitz found himself "highly unpopular," he wrote, "virtually overnight." Working with the military—his longtime collaborator—was suddenly "in bad odor." The Academy became a mess: scientists resigning "with a brilliant display of fireworks."

Even the holidays were conscripted. His last year, Seitz sent a Christmas card, illustrated with a dove. Yuletide brows lifted. "It made you think," a colleague told the *New York Times*. "Because Frederick Seitz has never been regarded as a dove. You'd have to say he was rather hawkish." Fifteen years later, Seitz was still piqued. "'Hawk,'" the physicist complained, "was a very nasty word in those days."

Most modern presidents served twin six-year terms. Seitz managed a single, then skedaddled for Rockefeller University.

Another mismatch. It was 1968. The national mood had turned "anti-science, anti-technology." You could trace this back to "the Berkeley riots," Dr. Seitz said, and the "stir among the youth." (He was a person who called students "the youth.") This new generation—you hear a graying man's politics—"regarded education as a means of gaining access to a vantage point from which to propose radical social reforms."

He'd become, a tall man walking the Rockefeller campus, a throwback: an illustration in the textbook of something the culture had moved beyond. It happens to everyone, best case: You age past your own time into customs that are unfamiliar.

A youth hailed Dr. Seitz on the paths—"Oh, you're the President. Can I say hello?" Seitz did not return-greet. He observed coldly, "You must be new here." For Seitz, you weren't supposed to speak "in a friendly way to the President."

Rockefeller was then so stratified there were different grades of toilet paper in the lavatories, corresponding to each body's different value to the institution; the youth thumbs-downed this, too. When Seitz had been there almost a decade, he presided—according to the *Times*—over "the greatest embarrassment in the institution's 75-year history." There was a budget crisis: tenured faculty were let go. "It was the bumbling of Fred Seitz," a professor told the *Times*. Given the chance, the faculty would've dropped Dr. Seitz one tissue grade.

This was Frederick Seitz, as inherited by tobacco and denial. Not a man at the optimistic start of a career. At the grudgy close—a Cold War veteran who'd flown home to find the civilians not just ungrateful but amnesiac. In *On the Frontier*—published 1994—Seitz asks, "Are we at the end of the human quest to comprehend the stars?" Was the only available course the "cultivation of the mundane," the "triumph of the ordinary?"

This got written in the midst of the digital revolution—begun by the friend in whose Desoto Seitz had driven east, to see the moon on the Princeton buildings and the starting line. The classic mistake, from people who've once been central: if you're no longer there, there *is* no center. Because you've slowed, the world has. But to industry, he was a name and résumé that could be strip-mined for value.

>FIRST CLASS

ONCE FREDERICK SEITZ SIGNED ABOARD AS "PERMANENT
Consultant"—which sounds like corporate-ese for the staff psycholo-
gist, or an especially clear-eyed way to describe one's spouse—his contact
became an attorney named H. C. "Jack" Roemer. Roemer hailed from the
James Bond era of the industry: thin lapels and Zorro swirls of smoke above
balconies, cigarettes as the suave chaser to manly acts. Roemer grew up in
destination settings (Cuba, Paris, New York), took his degree from Harvard. Off-hours, he enjoyed high-stakes recreations like mountain climbing. He'd become general counsel to Reynolds Tobacco.

If Dr. Seitz wanted to skip ahead to the credits of the movie, watch
the sad end first, there was the career of another consultant being run by
Jack Roemer. Gary Huber, the researcher who oversaw tobacco's project
at Harvard.

This is the story of what happens when you fall into denial—and the
mixed rewards of trying to climb back out again.

Huber was a sort of emblematic figure: disadvantage. His father
never graduated from elementary school. His mother took jobs as a wall-
paperer, because you could eat the paste. The family home—Spokane,
Washington—had dirt floors. In this dry soil, Huber grew to six and a half
feet. He won admission to University of Washington on a basketball scholarship, completed medical school, did his internship at Harvard.

In 1971, Huber contacted the Tobacco Institute. He was thirty. The age at which you pick the best investment for your talents. His letter was court-ship: this is science and industry touching fingers across a tablecloth.

"For reasons I poorly understand," Gary Huber wrote the executives, "tobacco research has had more 'emotion' and less 'science' than it should, and predominantly in an anti-tobacco tone."

Institute president Earle Clements wrote back: "Being as native to Kentucky as the tobacco leaf itself, I, too, have been caught in the emotion surrounding the cigarette controversy."

Clements had been a senator and governor. You imagine suspend-ers, wide face, friendly belly: the kind of person who can write letters in drawl. "I have been alerted," Clements went on, "to the fact that the abun-dance of anti-tobacco sentiment has been nurtured in the chapels of the moralists and propagandists rather than in the laboratories of objective research investigators."

Success is quick study. Hear the notes, play the same music back.

In his next letter, Huber quoted that chapel line, and complained about "anti-smoking zealots." He offered statements that run dry on the surface, but which, to the trained eye, contain a rich underground stream of promise. "I hope we can have a truly collaborative effort with the industry." Huber used his "Harvard Medical Unit" title; this was part of his promise. "In fact, I personally believe that the magnitude of success depends a great deal on how much [the] tobacco companies are willing to contribute."

Unmistakable. The industry was overjoyed. Two unlikely develop-ments. Gary Huber had studied and rebounded his way into Harvard. And now tobacco, the incorrigible classmate—disruptive in class, cheating on exams, terrorizing the brainiacs with swirlies—had been admitted too.

"HARVARD!" writes Dan Zegart in *Civil Warriors*. "The peculiar indus-try could not believe its luck. These most notorious of companies were dining at the Harvard Faculty Club, passing carafes of wine and baskets of rolls down to their tablemates . . . at the most prestigious university in the world."

Leaf scientists and North Carolina CEOs tromped down the aisles of Huber's lab. Right from the start, Zegart writes, "the industry was planning to turn the Harvard Project into a major propaganda tool." From those dirt floors, Gary Huber had risen to become "the academic catch of a lifetime."

If you want to hook someone in a questionable enterprise, bait them with luxury. (A recruit to a later denial operation shared his travel logic with *USA Today*. "On a daily basis, my conscience was bothering me," the man said. "But you get used to being flown around first class. It tends to keep you doing what you're doing.") Hotel plaques had many stars. Air tickets were first class—the legroom must have been a blessing. Introductions to senators, industrialists, foreign dignitaries.

This is the flashy part of the movie—the champagne and limousine sequence—where even the slowest audience member knows a bill will come due.

Huber, Zegart writes, was "thoroughly seduced." (The Industry had "a lot of experience cultivating the Gary Hubers of the world.") Huber took the stage at a tobacco conference to wild applause: "I give you Gary Huber, one of the best friends the tobacco industry ever had."

His first overseer was Dave Hardy—lead hat in the posse, senior partner at Shook, Hardy and Bacon. The man who'd wrangled tobacco as a client for his firm. Philip Morris squared up to its first product liability suit in 1962. The plaintiff had lost his voice box to cancer. This left him unable to shower (throat hole) or to confidently socialize (his small talk emerged by electrolarynx). He was a dream client: his expert witness was Dr. Alton Ochsner.

It's really a story of freak legal luck. The case got filed in Missouri, which meant taking a chance on local talent: Dave Hardy. Accounts differ as to just what Hardy did to the plaintiff. One writer likes "morally degraded." Another goes with "disemboweled." And when Alton Ochsner testified, Hardy was willing to walk the extra mile. "He does not qualify as an expert," the attorney said. "What we have here is a man who is most frequently described as a zealot . . . for the fact that cigarette smoking causes lung cancer." The trial ran three weeks. The verdict was returned in less than an hour.

After that, Hardy was a star—the man who skinned Alton Ochsner. Philip Morris was dazzled. Shook, Hardy eventually represented five of the big six tobacco firms, plus the overall Tobacco Institute: It was Dave Hardy who devised their hard-shell strategy.

With Gary Huber, Hardy took an interest. He personally oversaw Harvard Project funding requests. He brought Huber to Kansas City Royals baseball games. Sat with Huber in exclusive clubs—then schooled him for use in the courtroom. ("That's the witness stand," Hardy would say, pointing at a chair. "Now you get up there.")

Dave Hardy was also a Marlboro smoker. His health was sinking. Then he was in the hospital, after heart attack number four. Hardy would phone Gary Huber—a chest specialist at Harvard—seeking midnight second opinions. This became an unburdening. Hardy offered shoptalk, the moral mechanics of representing the unpopular.

"No matter what they've done," the lawyer told Huber, "they deserve the best defense." Hardy understood what he'd cooked up: bail—possibly a mistrial—never exoneration.

"I bought them an extra twenty or thirty years," he told Huber on the phone. "But, you know, it's temporary. One of these days, we're going to have to pay up. You may live to see it. I'm glad I won't."

When Hardy died—1976, aged fifty-nine—Huber's main overseer became Jack Roemer. Together, Roemer and Huber scaled a Himalayan mountain. When Huber married, the tobacco man was by his side, tendering the ring. When the Harvard Project had lasted seven years, Huber's research on emphysema uncovered worrisome tobacco links.

That's where the funding stopped. The cigarette attorneys flew to Harvard—everybody but Jack Roemer. (When your handler doesn't show, time to make a break for it.) There was a meeting. A Philip Morris lawyer drove with Huber to the Airport Hilton. In a guest suite, more attorneys, folders, ashtrays, pacing. "You just got too close to things you weren't supposed to get into," they told Huber. That was the end of the Harvard Project.

He was the rich boy who wakes up the morning his parents lose the fortune to find that his friends have scattered. No more legroom, grand hotels, handshakes with men in charge of armies. The industry found a job for Gary Huber in its research lab at the University of Kentucky. After a year he was messily gone.

He ended up at Tyler Junior College, in Texas. Because that's what happens with denial, too. You wash ashore in places you never heard of. The boys upstairs siphon off the value, and when the process is complete they're the ones your reputation still has value for; the people who wrecked you become your only customer.

As of the mid-nineties, Huber still did work for Shook, Hardy—disputing reports by government scientists. In his office at Tyler Junior, Huber kept a black and gold Harvard chair. His wife saw video of him from the rollicking old days. Gary Huber, mounting the stage before whooping executives. He'd been tall and beautiful. Now he was stooped, gray. "What happened to you?" she asked.

Huber visited Jack Roemer one last time: found himself in North Carolina, picked up the phone, parked in the lawyer's driveway. Jack Roemer, of Paris, Harvard, and tobacco, had become a Moravian Christian. He rose at dawn to pray, wore a massive wooden cross, had pre-booked a plot in the special Moravian cemetery. Penance, obviously—Huber couldn't tell what was being repented. He never saw Roemer again.

At Tyler, Gary Huber finally turned. His daughter warned him he was being pimped. He'd been tobacco-outed by *Business Week*. And then tracked down by a team of Southern anti-tobacco lawyers. "If what this guy knows gets out," they told reporters, as in a muckraker movie, "it'll blow the lid right off this thing."

Attorneys flew the family to South Carolina. On the runway, Huber's son caught sight of the firm's private jet; the boy asked whether their flight would be first class. "This is better," Huber replied dreamily. It may be what the denial story is finally about: the short cut to a better class of passengership.

THE CHARLESTON LAWYERS staged the painful final act. They showed Gary Huber letters and memoranda, in which he was coolly appraised by tobacco, as a negotiable asset.

One of life's great shocks—encountering yourself in the unsympathetic third person. Where everyone is shorter, fatter, and more easily led than under their own narration. Tobacco turned out to have previously studied everything Huber researched at Harvard. (Huber had a great word for the useless pursuit: "What we did was artifact," he said. "Pure artifact.") This memo comes from year one—by the same attorney who'd accompanied Huber on his sad ride to the airport Hilton.

> *We had several objectives in mind in establishing the research project at Harvard. First, it was considered that the public relations value of an announcement from Harvard Medical School that it had accepted research funds from the tobacco industry would be considerable. This [helped] the Tobacco Institute in its lobbying efforts. . . . Finally, the grant put us in a position to assert that there are still important unresolved medical questions about the possible effects of smoking, else why would this eminent medical institution be interested in doing research in this area?*

His wife and kids returned from Kiawah Island Beach. Huber had spent the day under office fluorescents, sorting through pages that contained the sorriest account of his professional life. "I couldn't sleep," Huber said, "couldn't eat. It was a magnitude of shock I'd never experienced."

Huber turned. (Hearing rumors, Shook, Hardy lawyers phoned with stay-firm advice: "Keep the faith, hold the line." Also, ironically and tardily, "Don't be seduced.") Huber made the reformed man's media tour. NBC *Nightly News*: "We helped them get away with it." *Dallas Morning News*: "I've been used and people may have died." To PBS, he brought More Research warnings: "They were really, I think, funding research in different ways, the outcome of which would create confusion and disinformation. . . . Do not trust them. They cannot be trusted."

And here's the funny fact about denial. You do better if you stick. Climb aboard the missile like Slim Pickens at the end of *Dr. Strangelove*, yee-haw it all the way down. It wasn't that Gary Huber got elevated by confession. He descended to the lobby, where he was directed to public transportation. He leaves the story; then his news trail dusts over. This is what the next denial generation understood. Recant, and you declare you were never reliable in the first place. You become a long cracked face with a long unlikely story—of loss, failure, the conditions Americans are constituted above all to deny.

PHILOSOPHERS AND PRIESTS

FOR EIGHT YEARS FREDERICK SEITZ'S CIGARETTE WORK
ran stress-free. He never faced the disconcerting sight of an attorney wearing a chunky wooden cross. Aid and comfort to tobacco didn't bug him.
Which illustrates what a fixed and terrifyingly consistent thing personality
is. Same brain, plus same inputs, equals same output. Frederick Seitz and
nuclear war: "The issue of morality," he wrote, "perhaps lies in the philosopher's domain." Half a century later, Frederick Seitz and cigarettes: "I'm
not clear about this moralistic issue," he said. "I'll leave that to the philosophers and priests." Money, in the ink and void of its creation, is born without sin: "As long as it was green," he said.

He spread around about $45 million. Philanthropy with the other guy's
wallet—reviewing the bent-knee letters—it's probably a very fun time.

But it was Frederick Seitz's destiny to lend his days to a project, then
go calamitously unappreciated. Later, he claimed to have done no actual
tobacco work. (Seitz earned $585,000 for this non-service.) But this is
where denial corporations are expert. They wave you through the door:
then the value extraction begins.

Dr. Seitz disbursed their largesse: intention laundering. (Corporate
magazine *Reynolds World*: "No longer can cancer be simplistically attributed
to a single factor, such as cigarette smoking.... A scientific steering committee headed by Dr. Frederick Seitz...") He was National Academy of
Sciences arm candy at meetings and conference ballrooms. He grasped
without asking just what type of research would prove unappetizing to the

home office. "The program," Dr. Seitz wrote one luckless applicant, "is not immediately aimed at effects associated with the use of tobacco."

He hosted a strategy session re: combating "damaging claims." (A day when it would have interesting to go for a ride-along in Fredrick Seitz's brain. From safeguarding the line of Judeo-Grecian values, to filling his seat in this room.) When Camel developed a new tobacco, the physicist was on hand with marketing tips. "I doubt if the cigarette will be highly attractive to very young smokers," Seitz wrote. He advocated a test launch, "such as in Japan or Europe."

Dave Hardy—a good lawyer is an able predictor—was right about timing. He'd argued the industry an extra twenty-five years.

In 1986, the surgeon general took the dreaded next step: declared secondhand smoke (essentially, the neighbor's weeds growing in your lawn) a health risk.

THIS IS THE ISSUE that sent tobacco eyes pinwheeling, and would bring the industry to climate.

Every IOU called. Philip Morris had always been financially crazy about the arts. New York proposed tough anti-smoking measures: the company got on the horn to theatre companies, dance troupes, symphony spaces. (It's the opening subplot from *The Godfather*. First scene: "Some day, and that day may never come, I will call upon you to do a service for me." Middle: "I want you to use all your powers and all your skills.") Arts institutions reported being "so shocked they didn't know how to react." Columnist Anna Quindlen served cold tea in the *Times*. "The surprise was that so many were blindsided by the fact that the *quid* came with a *pro quo*."

Dr. Seitz got active. For his *pro quo*, the scientist would provide a vigorous surgeon general fact check. At first, the work by Seitz and his team—university researchers—looked "outstanding" to tobacco executives. Half a year later, mounting enthusiasm. "The study has progressed quite well. . . . their work is scientifically well-structured and is being carried out in an exemplary manner."

Seitz's team reached conclusion. Secondhand smoke was just as unhealthy as the surgeon general said. Perhaps even a little more so.

For any organization, a bad report presents two options: Shiatsu or torpedo, massage or sink. A Reynolds lawyer wrote angrily upstairs. The findings were "undermining," "very damaging," and would "play into the hands" of just the worst people. (The reason for this tone of affront: Reynolds had paid in advance, taken their table, and was now staring down at absolutely the wrong order.) "I fear no amount of wordsmithing can make this a helpful document." At Philip Morris, the media strategist passed along the flat Christmas news: Dr. Seitz's intended gift had become a problem. Unless "somehow countered," the report would do "substantial damage."

In the meantime, Frederick Seitz was done. Seventy-eight is not too old for a new grudge. Activist government and do-gooder science had cost Dr. Seitz his sunset job. At the end of the month, the same lawyer who'd accompanied Gary Huber on his sad trip to the airport Hilton took up the matter of Frederick Seitz. The most damaging and undermining words for a brain worker—someone whose value rests entirely in the circuit board. There was no more profit left to be had: "Dr. Seitz is quite elderly," the lawyer explained. "And not sufficiently rational to offer advice." No mercy— Frederick Seitz, exiled to Alaska.

COUNTERBLASTE

SECONDHAND SMOKE FUELED CLIMATE DENIAL; IT DREW tobacco into the game. In a way that's fitting. The issues started school around the same time—the 1950s—and grew up together. The tobacco emergency came four years before Roger Revelle's shiver observation, "Human beings are now carrying out a large scale geophysical experiment." The surgeon general's passive-smoke finding came in 1986; two years after, James Hansen told the world warming had begun. If climate science marched in tandem with the inventors, climate denial entered the three-legged race with cigarettes.

And what is global warming—your power plant exhaust is my power plant exhaust—but the largest-ever secondhand smoke story?

It would become a metaphor for the climate scientists. "It's the smoking analogy I use," President Obama's first secretary of energy, physicist Steve Chu, told a Stanford University audience in 2014. "We smoke. Our children, *their* grandchildren, suffer the consequences."

In another way it's surprising. Global warming became an alibi witness. Tobacco and fossil fuel—viewed objectively, with absolute tactical business cool—they shared a problem. Scientists. Now work backward. If you could make the scientists wrong about warming, by the transitive property they'd also be wrong about cigarettes.

FROM THE START, the question of nonsmokers got rolled up with tobacco. King James I—a first-run Shakespeare ticket holder, if you want to

get the period—disliked smoking. So much, the result was the first-ever public health treatise written by a monarch. 1604, *A Counterblaste to Tobacco*.

James had absolutely no use for tobacco. It was "vile" and "base." Also
"stinking and unsavory," "loathsome and hurtful," "venomous," "foolish,"
"hateful," and "filthy."

What was worse, strolling around London, you didn't get much choice
in the matter: you smoked whether you wanted to or not. "The public
use whereof, at all times, and in all places, has now so far prevailed," James
wrote, "as diverse men very sound both in judgment and complexion, have
been at last forced to take it also without desire." Even if you shut and
barred the door, not much choice at home. Tobacco began life as a kind of
spousal lung abuse. It was a "great iniquity and against all humanity," James
wrote, that "the husband shall not be ashamed, to reduce thereby his delicate, wholesome, and clean-complexioned wife, to that extremity, to live in
a perpetual stinking torment."

James apparently also had some run-ins with early deniers. "If a man
smoke himself to death with it (and many have done), Oh then some other
disease must bear the blame for that fault."

When cigarettes re-summoned tobacco to the public square, the *Counterblaste* came back with it. Societies formed. 1899's Anti-Cigarette League.
1915's the No-Tobacco League. The president of Stanford and the first head
of the Food and Drug Administration were founding members of the Non-
Smokers' Protective League.

Dr. Harvey Washington Wiley—the FDA man—sat down with a
reporter. "Neither I nor my compatriots say a man shall not smoke his lungs
to a frazzle," the scientist explained. "At home, or in the woods or meadows;
but not where human beings are liable to be. It is not fair."

Then war came. And, in the manner of wars, marched other topics off
the parade ground. Then the proportions shifted: nonsmoking stops being
the same kind of problem when everybody smokes.

Then the proportions shifted again. Government had barged into the
companies' lives with the 1964 surgeon general's report. In 1971, President
Nixon declared national War on Cancer. (*New York Times* query: "What
has Santa Nixon given us lately?")

Catching the government's eye, for an industry, must be like having a bear unstake the FOR SALE sign, move in next door. The bear may be a perfect neighbor. The bear—soaping his car, running his hose—might chat you up Sunday afternoons. But at night, you'll break out the field glasses and scan the bear's windows for excited guests. And in the mornings, you'll peek in his mailbox for recipe books or new culinary gear. Because it's going to come sooner or later—that bear is going to decide to eat you.

In 1972, the surgeon general's first mention; time to consider waving secondhand smoke out of the public air. Next year, Arizona became the first state to ban cigarettes in elevators, libraries, and (less foggy steering) buses. The Civil Aeronautics Board ordered nonsmoking aisles on airplanes.

In 1978, the Roper polling organization gave cigarette companies their grim physical. All the challenges up to now were "blows the cigarette industry could successfully weather." This was because "what the smoker does to himself may be his business, but what the smoker does to the non-smoker is quite a different matter." (They might as well have been quoting Harvey Washington Wiley and his Non-Smokers' Protective League.) "This we see as the most dangerous development [that] has yet occurred." Things would start slow, then gather force—the thrown stone that sends pond ripples to infinity. "As the anti-smoking forces succeed," the pollsters warned, the movement would grow "from a ripple to a tide."

Two weeks into 1981, there it was on the front page: "Cancer Study Reports Higher Risk for Wives of Smoking Husbands."

The findings came from Tokyo's National Cancer Center Institute. The conjugal unfairness that once bothered King James—"a great iniquity, against all humanity"—now typeset in the modern language of the *New York Times*: "Smoking Your Wife to Death."

An EPA scientist estimated the annual deaths somewhere between five hundred and five thousand. As of 1985, this made the other guy's cigarette—not chemical breeze, factory backwash, or toxic burp from a landfill—America's number-one airborne carcinogen.

Horror movies have it right: when pursued, you rattle every door. A Philip Morris vice chairman gathered ninety-three publishers of traditionally African American newspapers for a speech. There was, the executive warned, a "new discrimination" afoot in the land. The narrowed eyes at smokers. Then he went for it: The smokers' situation brought to mind "violence against black families moving into white neighborhoods." This was because "once the dogs of intolerance are unleashed, no one is safe— especially vulnerable minorities." Addressing the American Jewish Committee, he sketched the same villagers and pitchforks. "Today tolerance for my smoking may be under attack; tomorrow, it may be tolerance for someone else's right to pray."

In 1986, Frederick Seitz's old outfit the National Academy of Sciences went on record. Passive smoking blew lung cancer over at people who'd never even bought a pack. "Unless immediate steps are taken," the Tobacco Institute warned that year, "there will be no long run."

THE INDUSTRY HAD A very clear understanding of the place it found itself; the spot that unclogs vocabularies. "We are in deep shit," an industry lawyer announced.

This was at a secondhand smoke conference, convened on Hilton Head Island. Even when backed into a corner, executives prefer a tasteful corner.

"We are here to do something radical," Philip Morris executives announced the first morning. "To look at a problem. To achieve a solution." They added, "Nothing should be withheld." This was summer 1987—three days past the solstice—which would enter NASA's list as the warmest year then on record. I like to imagine the tobacco staff in Bermudas, smelling of aloe, soda bubbles prickling their fingers and wrists.

At Hilton Head, the tobacco men and women really did withhold nothing; not even despair. They examined charts. ("Graphs don't hold much hope for us.") They considered sales figures. ("Industry is going down.") They gave themselves room to mourn. ("We're getting killed.")

They had a preferred term. Like Frank Luntz with "climate change"— because it presented a less emotional challenge than global warming. "Passive smoke" and "secondhand smoke"—these made you picture too vividly an active hand someplace clicking a lighter. They called it ETS: "Environmental Tobacco Smoke." Just the hills and valleys enjoying a puff. "We are just at beginning of impact of ETS issue," they said. And "ETS will have a devastating effect on sales."

Knowing the solution they arrived at makes the journey a pleasure. It's like watching an organism crawl up the shore, evolving limbs and a face on the trip between water and grass.

They fiddled around with value-addeds. Whatever the short-term negatives might be for smoke, there were long-term positives associated with sex. "Condoms in cigarette packages," was one suggestion. Since no-smoking restaurants had become a problem, there was the lunk-headed but practical: "Create desirable restaurant for smokers."

Also some mustache-twirling. "Infiltrate World Health Organization." Some creative writing. Produce "best selling novel with subtle connection to evils of anti industry." (They were dead right here: when Michael Crichton wrote *State of Fear*, a hammy attack on climate research, his book proved immensely influential. President George W. Bush invited the author to a special one-on-one heart-to-heart at the White House.) "Organize 'spontaneous' protests." They really did hang scare quotes around the word "spontaneous."

Then the focus narrowed. The creature rising from the slab, tottering around the castle basement. "Change nomenclature for issue," they suggested. "Create a bigger monster." "Seek out other allies/industries." "Challenge 'bad' ETS science." "Create pan-strategic monsters." "Get together a group of respected writers and turn them loose on politics of science." "Monster theory. Making problem, we become small part." "How do we get 'bad science' into leadership consciousness?" "Need better term than 'bad science.' Weak, pseudo, uncritical."

Last day, the team presented for top brass. Philip Morris' president wanted to hear about risk. Wouldn't these tactics just make a bad situation worse? The answer came back dire. (Tobacco had become the werewolf panting by the tree, scenting the approach of the confident, well-armed hunter.) "The situation can't get any worse. ETS is the link between smokers and non-smokers and is, thus, the anti's silver bullet."

————

PHILIP MORRIS FANNED OUT. "They are spending vast sums of money"—these are the notes of a London tobacco executive—"on an international basis to keep the ETS controversy alive."

But even the seats they flew in were going rotten under their weight. In 1988, all domestic flights briefer than two hours—maximum tolerance between puffs—went no cigarette. The year's Winter Olympics banned all smoking. At the other end of the ellipse, that summer's Democratic National Convention offered a political first: smoke-free smoke-filled rooms.

Two years after Hilton Head, Philip Morris' 1989 "ETS Science Action Plan" has the sound of the restricted and harassed. No idea turned away, even under the grubbiest raincoat. The industry should "take a page from the left-wing conspiracy theorists," the strategy document counseled. Hint at cover-up. "The media loves cover-ups." Advice was thumb gouge, desert justice: "Eye for an Eye."

One path—the monster path—looked secure. If a single study couldn't be discounted, go big: discredit all science. "We need to challenge [the] popular notion that science is objective, value-free and without error," they wrote. "We are raising the scientific credibility issue each time ETS is mentioned." (Though even at Hilton Head, the Deep-Shit lawyer had cautioned: "We cannot say Environmental Tobacco Smoke is 'safe' and if we do this is a 'dangerous' statement." The difference between tobacco on health and Exxon on climate I believe is this: their records have not yet been thrown open by court order.)

Bad news lumped like a blizzard. All domestic air travel went no cigarette as of February 25, 1990. A United Airlines stewardess thumbed on the cabin PA. "So tonight's it, smokers," she said. ("If the plane starts to go down," a smoker from Queens warned a journalist, "I'm lighting one up.") The first Bush White House banned cigarettes from 1600 Pennsylvania Avenue's maintenance areas and kitchen. Never again, the crisis, the late-night snack, the squiggle of smoke rising above the respite of a presidential sandwich. When the Clintons took over the lease, smoking was exiled from every corner of the great American residential trademark.

And at the start of 1993, the Environmental Protection Agency thumped down its own report. Five research years, five hundred pages. Secondhand smoke was now officially rated a Group A Carcinogen. Stated position of the US government. Silver bullet.

And Philip Morris found its counterblast to the *Counterblaste*. Ellen Merlo, a twenty-year Philip Morris veteran, had just become vice president of corporate affairs. There was only one thing to do. She memoed the CEO: "Our overriding objective is to discredit the EPA report."

Five days later, she had the strategy. It combined parts from years of ETS action plans, stitched together to create a single Frankenstein. It was the monster strategy. Philip Morris would reach across the aisle to everyone with a grudge against science and scientists, against research when the cards were flipped over and the deal had not been in your favor. "This is the toughest challenge of all," the strategy read. "And the one that should be treated with the greatest urgency. The credibility of EPA is defeatable, but not on the basis of ETS alone. It must be part of a larger mosaic that concentrates all of the EPA's enemies against it at one time."

A CZARINA ENJOYS THE
CORPORATE CHRISTMAS PARTY

DURING THESE YEARS, THE *WASHINGTON POST* RAN A how-can-they group profile of Philip Morris management. Castmates in "perhaps the greatest morality play of late twentieth century America."

Executives bemoaned their troubles—the slight from a neighbor, the snub at a health spa. "It's these kinds of social barbs that are hurtful," they explained, "because we are people."

Back at Hilton Head, the cigarette people had stroked their chins and slapped their pockets. "How do we get 'bad science' into leadership consciousness?" Because framing the problem focuses the eye and directs the solution: it makes you the issue cinematographer. "Need better term," they fretted, "than 'bad science.' " It was a meme with dead Velcro, a slogan with no glue: it didn't *stick*.

The *Post* named Ellen Merlo ensemble standout. "No one better articulates," said the newspaper, "the philosophical worldview of the company's people." In 1993, Merlo handled these articulation services for a conference of cigarette vendors. She was in her middle fifties then—when things go right, your golden period. Full possession of your talents and an office on the power hall.

Merlo promised not to "pretty up" the situation. Not to "exaggerate it" either. Hers would be a level-headed assessment.

"Our adversaries," Merlo said, "want to destroy our industry, to eliminate it from the face of the earth. . . . The simple fact is, we are at war." An "all-out war."

The weapon they found was one syllable. Used for stuff left on the pile to the rain and seagulls. For meals that are heaven on the tongue but lava in the works. Peter Huber, an attorney with the Manhattan Institute, published a book in 1991. *Galileo's Revenge*—a big splash on the op-ed and yell-show circuit. Denial draws from a shallow pool: Huber's old sidekick was Mark Mills.

He was also a think-tank man. Huber's job was to lean back, forearms crossbowed behind his neck, and mull. Donors funded the tanks. Then came around every so often to skim off the thoughts.

In 2008, a team of academics went on a library hunt. They were chasing down the link: think tanks and antienvironmental books. They piled up 141 volumes. Eleven turned out to be natural growth, spontaneous text occurrences. The remaining 92 percent had been nurtured in the tanks. (Titles mixed the storybook with the shaken fist: *Nonsense about Nature. Haunted Housing. Eco-Imperialism: Green Power, Black Death. Panic in the Pantry.*) These were authors who could find the upbeat angle in global warming: "We are creating a world in which the winters warm and the summers do not, a world in which the nights warm and the days do not. . . . We are creating a better world."

Peter Huber's complete title was *Galileo's Revenge: Junk Science in the Courtroom.* "Bad science" did indeed lack star power. As a name it was a dud—like Archibald Leach or Thomas Mapother. It wasn't going to pose on the red carpet, star in your arguments. But "Junk Science" oozed insult charisma. It was Cary Grant and Tom Cruise. Junk Science was greasy, drive-thru: here one minute, taillights the next.

And there was the frank simplicity of need. As Harvard science historian Naomi Oreskes puts it, " 'Junk Science' was a term invented by the tobacco industry to discredit science it didn't like."

Cigarettes were not worn close to anybody's heart. (In a sense, *already* junk: sped your hours, rotted your arteries.) So the plan was to link them with "more 'politically correct' products." When Philip Morris took junk science worldwide, heading the list was "global warming."

SHELVES OF BOOKS have been written about the result. Cigarette industry watchdog Dr. Stanton Glantz was among the first to grasp the dimensions. Philip Morris had taken the next step—"gone beyond 'creating doubt' and 'controversy.'" The company was now, he writes, "attempting to change the scientific standards of proof."

In *The Republican War on Science*, Chris Mooney follows the strategy into Congress. The "movement sought to smear a wide range of scientific activities conducted by taxpayer-funded government experts, greatly undermining the public's trust in science in the process."

This seeped into climate—with, Mooney writes, "tragic and maddening" results. There might be no other issue "where a corruption of the necessary relationship between science and political decision-making has more potentially disastrous consequences." David Michaels—head of Occupational Safety and Health under President Obama—trails its entry into the bureaus. "The vilification of any research that might threaten corporate interests as 'junk science,'" he writes in *Doubt Is Their Product*, has "worked wonders."

And Naomi Oreskes' classic *Merchants of Doubt* handles personnel. "Perhaps not surprisingly," Oreskes writes, the same people who "defended tobacco now attacked the science of global warming."

Oreskes is an evolutionary biologist of debate. Tracking her argument across landscapes and decades. "In the 1950s, the tobacco industry realized they could protect their product by casting doubt on the science," Oreskes explains. Then adaptation: tougher hide, longer claws. "In the 1990s, they realized that if you could convince people that science in general was unreliable, then you didn't have to argue the merits of any particular case."

George Monbiot is the UK Naomi Oreskes: putting out the same fires on European soil. Monbiot holds the industry achievement in high regard. "The 'coalition' created by Philip Morris," Monbiot writes in *Heat*, was the "most important of the corporate-funded organizations denying that climate change is taking place. It has done more damage to the campaign to halt it than any other body."

ELLEN MERLO IS the Philip Morris executive who got maligned at that health spa. A fellow guest announced that Merlo's career path had left her "really turned off." But Merlo never went vague and mumbly about her job. "I'm very proud of it," she told the *Post*.

Merlo is the story of serving denial from the corporate side. The motivations don't only run dark-souled and mercenary. Sometimes, they can be a variety of love. And there's the surprise: exiting the canoe, parting the vines, to find Ellen Merlo at the headwaters of a science fight.

She was born in the Bronx, grew up in New Jersey; the sort of childhood that orbits a skyline. She worked part-time at a dress shop. (And stared across the Jersey cliffs to the Manhattan towers, and told herself the whole thing was a zoning error. "I'm a New Yorker.") Then two years of low-wattage college—Katherine Gibbs Secretarial—followed by entry into the nine-to-five as what used to be called a Gibbs Girl. How Merlo expresses this background is that she went to school "in the city." That is, she represents a certain kind of urban woman—skeptical and wise-eyed, hair worn dark, poofy, and short. The city woman, giving a hard look as you take the parking space, a cool stare as you arrive at the turnstile together. Now she was in corporate, bringing the same instinct, for identifying and defending turf.

She started in advertising as a secretary—Young and Rubicam—graduated to accounts work on iconic brands. Merlo smoked until she was twenty-five. Quit for what she testified under oath was "no particular reason." As it happened, she quit one year after the surgeon general's hair-raising 1964 report. She'd also quit advertising, heading west. Again, the slightly off spot: She was in Los Angeles, the movie capital, working at an auto magazine. Decades later, a lawyer failed to recognize the publication; Merlo's response grew bristles. "*Motor Trend* Car of the Year," she said, "is something that car manufacturers aspire to." From the beginning, one of her gifts was loyalty—being two-fisted and proud of the team colors she wore.

She returned to the city, joined Philip Morris, and fell in love. "I thought I had just died and gone to heaven," she said.

This was 1969, last squares on the *Mad Men* calendar. "It was the most remarkable atmosphere." A glamorous, first-names company—open doors, even the chairman's. "Around Christmastime we used to get together in the conference room," Merlo told the *Washington Post*. "And all the guys who run the company would come along with a cart and champagne and serve everybody drinks." She'd finally arrived: dream life in the towers across the river. It is a strange twist of history, that Ellen Merlo should be behind the coalition to work all the feats and wonders George Monbiot describes.

In the courtroom, a lawyer would cast it as the Cinderella story gone corporate. In a way, the beautiful story of faith, talent, and years rewarded. "You've worked your way up from secretary to be senior vice-president?" he asked. "That's correct," Merlo replied.

It must have taken iron. You can hear this in her staff nickname: the Czarina. Ellen Merlo became Philip Morris' highest-ranking woman—the only female in senior management—without a bachelor of arts. No children. Her life's abiding relationship was Philip Morris. By 1992, she was disappointed with how the Tobacco Institute, which was supposed to comb its hair and check its tie, was letting her industry go around in public.

"I just thought we could do a better job," Merlo said. "And I had a passion about the company and the people that worked there."

She put in for transfer—to corporate affairs, the image-makers. Her wish was granted three months before the EPA sent over its declaration of war.

MERLO'S RESPONSE called for filming the EPA in the low, menacing light of true-crime documentaries. Revealing it as "corrupt and guided by environmental terrorists"; bringing together all its enemies. This poor department, that had only wanted to keep rivers from catching on fire, end Thanksgiving smogs that smudged away people by the hundreds.

Victor Han was the deputy who wrote Ellen Merlo's language. A smoker, Han shared with the *Washington Post* a surprising fact: it didn't matter how hard he fought to keep cigarettes cherry-tipped in public—he took care never to light up in private, around his daughters. "As a parent," Han said, "I want to take as many risks out of their lives as I possibly can." This may be a fine statement of institutional denial: campaigning on behalf of the product you consider unfit for family use. In a way, that was Ellen Merlo's motivation too—her family, instead of everyone's.

To carry out Merlo's plan, there was the question of personnel. Merlo hired a PR firm—a startup called APCO Associates—to build and operate the Junk Science coalition. APCO declared a casting moratorium. No speaking roles for anyone from business. "Industry spokespeople," APCO informed Merlo delicately, "are not always credible or appropriate messengers."

They needed a scientist. The goal was to locate another intellectual descendant of Clarence Cook Little. And here APCO's language went on tiptoe; the publicists were "approaching" people. Three weeks after contract, APCO faxed over the résumé: they'd found Dr. S. Fred Singer.

YEARS LATER, *Scientific American* would call this the key work of the physicist's life. Dr. Singer was "best known for his denial of the dangers of secondhand smoke."

There's confusion, as there is with so many things about S. Fred Singer.

This confusion is self-made, his protective fog. Naomi Oreskes writes of receiving a nighttime call from Dr. Singer, explaining he was friends with a very famous scientist—who had himself been tight with the man at the wayback, Roger Revelle. Next morning, Naomi Oreskes telephoned that very famous scientist. He was an old European, who sighed. "Ohhhh," the scientist said. "I would not call Fred Singer a friend."

There was Dr. Singer's extra-innings denial of the ozone hole. His doughty *No* on acid rain. There were media outlets like ABC News calling him the grandfather of climate denial. Other words you encounter with S. Fred Singer are "leading," "most-quoted," and "arch-skeptic." Sir Paul Nurse, who has a Nobel and headed Britain's Royal Academy, introduced him to BBC viewers as "one of the most prominent and prolific skeptics. . . . Professor Singer's views influence skeptics all over the world."

Dr. Singer could even transform denial into a party bus. Rolling up at a European conference in 2012, *Forbes* reported that the scientist was greeted "like a rock star."

Later in life, Dr. Singer suffered an attack of tobacco cold feet. "I would not be surprised if passive smoking causes lung cancer and other diseases," he said. (Also, your neighbor's cigarette "cannot possibly be healthy.") He told a team of newshounds from the Canadian Broadcast Company he'd been unaware who he was working with—outfoxed by the anonymity of the greenback. "They don't carry a note on a dollar bill saying, 'This comes from the tobacco industry,'" Singer said.

But in March 1993, this looks iffier. Ellen Merlo wrote APCO about Dr. Singer's essay. She wanted something that really nailed "the impact Junk Science can have on the average person's life." APCO wrote back. They had "discussed with Dr. Singer Ellen's suggestion." An editorial nudge is easier to source than a dollar.

But Dr. Singer did hit her bombmaker note. The American public, he wrote, was being "repeatedly terrorized" by the Environmental Protection Agency. And he got the important thing said—getting the necessary thing said would become Dr. Singer's great skill. Secondhand smoke was "the most shocking distortion of scientific evidence yet."

He drew the connection with more politically correct issues. The EPA was responsible for other "questionable crises." Like global warming. And ozone, and acid rain. And he power-upped the title. From plain old "Junk Science" to what everybody needed—an accusation and a place: "Junk Science at the EPA."

S.

BY 2014—SUMMER, A BOLD CHOICE FOR LAS VEGAS travel—the denial movement had grown brawny enough to occupy a casino. An international conference: thirty-two co-sponsors, attendance at six-fifty. Funders and scientists had pulled off big things.

This was a modern conference: there were awards. Dr. Sherwood Idso, who'd made the old *Greening of Planet Earth* video for the coal people, received the first annual Frederick Seitz Memorial Award. Presented to him by Dr. S. Fred Singer.

And with a click—one denial grandfather named Fred celebrating the achievements of the other—climate denial closed its circle. Became one of those Möbius strip novels whose last line is really the start of its first page.

The Freds serve as main characters in Naomi Oreskes' *Merchants of Doubt*. With seething courtesy, Oreskes calls them her protagonists.

"Who could deny all that?" Oreskes marvels at one point. "The answer: Both Fred Seitz and Fred Singer."

Dr. Singer didn't leave the Strip empty-handed. He took career honors: Lifetime Achievement in Climate Science. In his testimonial, the president of the Heartland Institute (event sponsor) was not skimpy with praise. "Fred Singer is the most amazing and wonderful person participating in the global warming debate today." He looked into his audience. "If there's any person in the world responsible for the development of a skeptics movement on global warming, it's Dr. S. Fred Singer. . . . Fred is a giant. He is a hero."

He resembles a character out of Tolkien, small and twinkling. In a movie, he'd rest his chin on the grip of his cane. On TV, there are the small lively eyes, and the broad mouth—beneath a Vandyke beard—that appears constantly delighted. A figure from the old-time carnivals. It's all hucksterism, his smile says; but that's the marvelous and adult spice of life. If my science is flimflam, could be the other fellow's science is also flimflam—maybe all science is flimflam. And now perhaps we can return to business? ABC News, 2008, asking Dr. Singer's opinion of warming research. "All bunk," he replied. But those other scientists; the learned societies; NASA. With mounting happiness: "What can I say? They're wrong!" All of them? With his soft accent, and greater joy: "Yes!"

He was born a Jew in 1924 Vienna. And if life is joy, it's a willed joy. Because the facts are heavy, and sometimes they're sad. Fred Singer fled to England the day German tanks rolled across Czechoslovakia. Burned a year in damp Northumberland. He was fourteen, and worked for an optician. (Which may have been the wrong job, message-wise. Presenting life to the young S. Fred Singer as a question of lenses. What you hear in this climate versus denial remark is the bell at the optometrist's door. "Nobody tells an untruth," Dr. Singer told the *Washington Post*. "But nobody tells the whole truth either. It all depends on the ideological outlook.") The Singer family reunited in the States. And not a Vienna substitute: Ohio. It was a life that keeps tugging you from the spot you aimed at—you have to adjust what you'll accept—the rocket launched a few degrees off course.

His birth name was Siegfried. Flight across Europe had scraped most of this away, like the sticker on an old steamer trunk, until only the *S.* was left. In America, he shook hands as a man capable of denying his own name. He took his PhD (physics) from Princeton.

He served. Everybody did. If you had S. Fred Singer's background, it was probably sweeter. After the war Dr. Singer stayed on—working with the navy's High Altitude Research Group. Rocket stuff. Balloons lift the missile above the boat, for clearance. Or you arrive at the equator. Warm wind, water slapping the bow. The rocket ignites. Flash, hiss, a dot against the horizon. One time, the captain didn't want any part of it—a sloshy deck isn't exactly the ideal platform to launch from. The rocket and the chins went up. One crewman kept watching. The dot got bigger. Then it got way bigger. The klaxon sounds, the rocket smashes down a few hundred feet from the boat. The captain announces, "I've had enough of this nonsense," and steams back to port. Singer—working with the scientist James Van Allen—was measuring cosmic rays, ionosphere, the sky.

He just missed being part of the team that discovered the Van Allen Belt. Dr. Van Allen had invited him to the University of Iowa. The navy pitched him science liaison officer in England. London versus the cornbelt: how is that a fair fight?

It was the moment of Fred Singer's life. "I had a choice basically between completing this work with Van Allen," he later said, or "traveling to Europe." Dr. Singer decided on accents, and fountains, and flexing his passport. It was the kind of mistake whose size you don't recognize until after, when you can see all the things it obstructed you from.

James Van Allen would become climate pioneer Jim Hansen's mentor: and Hansen would rise and rise. The denial side of climate change is the story of people who just missed. The people who *weren't* picked. The lucky team heads off . . . and you're after all still there, with a life to fill. Discord is the Greek god closest to denial. She's left off the invites to an Olympian wedding. In revenge, she starts the Trojan war. It's understandable: excluded from the guest list, you trash the party.

It's now—London and after—that Dr. Singer's career lost altitude. "They never met with great success, Singer's endeavors," a fellow researcher recalled to the Center for History of Physics. You'd work all week with Singer, putting together the rocket payloads. "And then he'd go off somewhere and shoot them and come back without any results." He was developing a reputation—for big words, little show. Another physicist boiled it down to a seven-word life: "You know Fred Singer. It didn't work." (This was before denial, the family squabble that hurls sharp words across the table. This cast Dr. Singer in a new light. "Speaking of idiots," is a third physicist.)

Part of it was a matter of style. Singer graduated college precociously young—then wrote about this as "the tender age of 18." It could be the pitfall of near brilliance. Not enough to be confident—the brink level that must be certified by others. He told *Time* magazine this was the point of science: "To be able to say 'I told you so.'" Which is probably no more fun to hear beside a particle accelerator than anywhere else.

He loved the sudden mirror, the curves and angles of his name reflected from a news story. He was famous for it, as another might be for exploits, or a hat.

"You're Fred and I'm me," one physicist explained to an interviewer. "And a reporter comes along, obviously someone from the press. . . . Why, suddenly your attention is going to shift, and it is going to go to talking to that fellow." Singer was the wingman who deserts you at the bar for the better offer. "That's Fred. . . . The guy is a burr under the saddle, you can hardly stand him sometimes." Life finds your flaw: People began speaking of him as a self-promoter, a loose cannon.

So this is the story of how Siegfried Singer became S. Fred Singer: the most prominent denier, a rocket scientist who couldn't secure a contract from NASA. In the fifties, the community decided to put a satellite in orbit. National prestige, plus maybe some spying.

Dr. Singer talked up his own model: articles, speeches, interviews. He would not pipe down. Singer got carried away, began speaking as if the whole satellite concept had been hatched by him. (Jim Van Allen later said the idea was like pollen: "in the air, all over the place," everywhere.) This caused, a space historian writes, in the gentle language of the dentist's chair, "discomfort."

Homer Newell—one of the heroic bureaucrats of space exploration— turns biting about Singer. Newell ran the military's project: his scientists, working navy contracts, had voices on lockdown. And there was Singer, spouting off, his mouth out of uniform.

This comes from Newell's history of NASA. And you can feel irritation lick the edges of his words—this is the genre of writing meant to be read and recovered from by its subject:

> First, Singer's manner gave the impression that the idea for such a satellite was original with him, whereas behind the scenes many had already had the idea. . . . Muzzled by classification restrictions, they could not engage Singer in debate.

Second, being unable to speak out, those who had dug into the subject in much greater depth [Singer had performed only some "fairly simple calculations"] could not point out that Singer's estimates overshot the mark somewhat, and that his suggested approach was not as workable as others that couldn't be mentioned.

A weird thing: Dr. Singer actually did the project good. He had a gift for attracting attention. But he was the hypeman turned away at the door to his own show.

When the satellite launched, no Singer experiment was aboard. Dr. Van Allen's was; it discovered the radiation belt. (His "I told you so" orbits the planet.) Fifteen years later, Dr. Van Allen would send Jim Hansen along to the space agency. Homer Newell ascended to NASA's third most powerful desk. Science has a long memory. "I never got a NASA contract," Dr. Singer told an interviewer. "No, never."

A bunch of rocket scientists (including, Singer grumbled, Van Allen) later got inducted into the National Academy. Dr. Singer was not in their number. It was this exclusion, Singer said, that "handicapped me most in my own personal relations."

His mistake had been "taking a visible role," Dr. Singer said. "I would think that was much resented." And why? "I think probably just jealousy." It challenges our notions of human perfectibility. You can have the experience, even spot the lesson-bearing zone. And miss the moral.

EMPEROR OF THE UNIVERSE

THESE ARE THE MIDDLE ACTS OF A DENIAL LIFE: BUILD-
ing the résumé, testing your value, locating the sweet wells where you can
draw a salary.

In the late sixties, Dr. Singer had a Washington office. One of those
double-barrel titles, a managerial fraction of a fraction: S. Fred Singer was
deputy assistant secretary at the Department of the Interior.

His beat was water quality. A descent like one of his rockets—as the
denial grandfather put it, "from the stars to the sewers."

Environmentalism had become the national music, the song playing in
every room. Dr. Singer's version had a folky sound. Listeners on blankets,
grass stains at the knee, the speech between sets: "This is our last chance."
Fred Singer was the alert type of conservationist whose words are to be
found on the *Time* magazine letters page: "Human activities—whether by
neglect or by accident or by intent," he wrote, "are constantly damaging the
environment." He was firmly anti-business. This too was in the songbook. It
was "up to industry to furnish its own funds to clean up industrial wastes."
Dr. Singer's first book was called *Global Effects of Environmental Pollution*.

When the Environmental Protection Agency launched, the conservationist was rewarded. Another deep-breath title: deputy assistant administrator of policy. He exited soon after—angled for promotion, got shot
down. After that, environmentalism became another Siegfried. A sticker
with the letters faded and gone.

At the end of the decade, he'd reformatted himself—pulled on the opposing team's jersey. Signs you see from the service road: Exxon, ARCO, Sun Oil, Shell. Dr. Singer hired out his expertise to the oil companies.

He also began writing about oil in Senate waiting room magazines—the *New Republic, Foreign Affairs*. His credentials wandered all over the road. "Fred Singer is an economist," read his *New Republic* byline. A journalist described him as a "political scientist." The *Boston Globe* called him a "maverick" economist—exceptionally maverick, Dr. Singer had no degree in economics. (Once he took up warming, all non-science flags would be permanently retired.)

These were the oil shock years: lines at the pumps, fistfights on the lines, talk of taming the oil companies. Dr. Singer dwelt on the industry positives. Things a CEO might yearn to say, if they weren't so shy. "The oil crisis is largely a media event," Dr. Singer wrote. A problem that would repair itself, "if we don't interfere too much with the natural processes of the market."

His forecasts could be thrilling and reassuring. No need for panic—by the nineties, Singer wrote in the *Times*, worldwide oil consumption would fall to twenty million barrels a day.

You know Fred Singer; it didn't work. By the mid-nineties, the world was chugging through seventy million daily barrels.

When it comes to work, everyone is a recycler, a conservationist of effort. Dr. Singer bundled together these pieces, mailed them to Reagan White House Human Resources. Job hunt. Dr. Singer was on tap, he wrote, for leadership of the EPA or NASA. Those seats were unfortunately occupied by other gentlemen. He got staffed onto an acid rain panel.

To the 1970 environmentalist Fred Singer, this was a simple question of physics. "Sulfur, which occurs as an impurity in fossil fuels, is among the most troublesome of the air pollutants," Dr. Singer had written in *Scientific American*. Sulfur raised "the acidity of the rainfall. . . . As a result small lakes and rivers have begun to show increased acidity that endangers the stability of their ecosystems."

It was all a question of the ideological outlook. Now you needed "further research," Dr. Singer wrote in *Science*, "to remove the uncertainties linking emissions, acid deposition, and ecological effects."

He was named chief scientist at the Department of Transportation. Runways and highways—but at least not the sewers, and nobody's deputy. He stayed aboard till 1989. When in short order he retired and reported back to the oil companies for more consulting work.

In a deposition, Dr. Singer approached issues of timeline with the delicacy of the bomb squad. "I, of course, received no money until I left the federal government," he said. He added, "I don't know if this is a trick question." So, no oil money in 1989? "No, I wouldn't say that. I left the government in April of 1989." So after that? "Yes." Did any of this work involve climate change? "Yes."

He was restored to the hot seat in 2006, this time by the Canadian Broadcast Company. (Their documentary was called "The Denial Machine.") "As with tobacco," the host explained, "Dr. Singer has denied taking any oil-industry money." The former rocket man looked very dapper on screen. Suave tie, neat beard. "Have they paid me?" Dr. Singer asked. "I don't remember. I don't think so."

The host promptly furnished proof: two oil contributions, both 1998. The mists of time parted.

"I hadn't heard of the $65,000," Dr. Singer said. "I got $10,000—*once*." Merry smile; human-to-human. "Came in over the transom. I cashed the check. Wouldn't you?" As it happened, he'd received this same $10,000 transom surprise twice. And his official website detailed an additional Exxon gift, made annually, for "the past three years." With an internet thank-you: "We are happy to get it." The past is a tangle. Joy resides in the sun-washed present.

HEADING AWAY FROM the Department of Transportation, Dr. Singer climbed into a strange van. At the wheel was the Reverend Sun Myung Moon.

There's that great term in science for the unforeseeable jump: Nonlinear. (It's the elevator that climbs to two, then five, then all the way to the 107th floor.) The alliance with Reverend Moon was nonlinear. Dr. Singer's first organized denial group wasn't climate: it was religion.

At first, they were like the mismatch in the movie you fall asleep to on the airplane.

Dr. Singer was a US physicist with an economics sideline. The reverend was a Korean billionaire with a Westchester mansion (it featured an indoor waterfall, bowling alley, and six pizza ovens; the full greasy weekend) who claimed to be the messiah. Dr. Singer alienated potential mentors. Reverend Moon had been on visiting terms with God, Moses, and Jesus since adolescence. The former chief scientist was without children. Reverend and Mrs. Moon were the True Parents of All Humanity—so, a houseful of kids. (When the reverend went on fishing trips, his tackle box said TRUE FATHER'S.) Dr. Singer was a bachelor. The Unification Church was so couple-friendly that paradise was designated off-limits to singles. (The reverend smoothed this over for Jesus personally, wedding his spirit in absentia to a Korean lady he knew.) The doctrine would later bring on a crisis in the church.

You'd think this would tax all Dr. Singer's powers of alibi.

"In fairness to members of the Unification Church," Dr. Singer explains on his website, "their beliefs and religious rituals—to an outsider—appear no more odd than those of Catholicism, Mormonism, Christian Science." Like smoking and warming, it was all a question of the ideological outlook.

And I feel you could make the opposite case. Sex is often an X-factor with new religions. Shame or appetite—too much, not at all, or inventive configurations that can be seen all the way from Heaven. In the wedding ceremony, bride and groom struck each other with a bat called the Indemnity Stick. Then newlyweds embarked on the Three Day Ceremony: sex beneath photographs of Reverend and Mrs. Moon. ("After the act of love, both spouses should wipe their sexual areas with the Holy Handkerchief. . . . [It] should never be laundered.") Now purified, a genital might be ridden to the stars. "I wish that you could center on the absolute sex organ," the reverend encouraged parishioners, the "unchanging and eternal sexual organ, and use this as your foundation to pursue God."

There were his plans for America—big and strange. "History will make the position of Reverend Moon clear," Moon promised, in his 1987 New Year's sermon. Like many oversize personalities, the reverend experienced his own journey in the third person. "And his enemies, the American population and government, will bow down to him. That is father's tactic,"

Reverend Moon explained. "The natural subjugation of the American government and population."

He had a thing about crowns. In a tasteful ceremony attended by friends and wellwishers, he arranged for himself and Mrs. Moon to be declared Emperor and Empress of the Universe. And at the Dirksen Senate Office Building, before a startled banquet crowd, he summoned a crown on a velvet pillow. It was brought by a congressman—the tall, knotty type you see in crusader movies. Reverend Moon then revealed he was King of Peace and "none other than humanity's Savior, Messiah, Returning Lord." (Voice-over artists tend to be the unshockable pros of the entertainment industry. But on the church-produced video, you can hear a clenched, what-about-my-soul? pause before the words, "God's painful heart was eased.")

And there was spillover. The worldwide True Family; the domestic one being raised with on-site bowling in the Manhattan suburbs. In 1984, Reverend Moon's teenaged son Heung Jin died in a car wreck. A high-ranking church official rated the loss as of "far greater importance" than the death of Jesus. It also activated the no-singles-in-heaven clause. Reverend Moon's solution was posthumous marriage. The bride, a ballerina, was the daughter of Moon's closest aide; she walked the aisle veiled and gowned, hugging a photo of Heung Jin. (The *Washington Post* had some grisly fun: this was "a unique church ceremony.") The sight, a family member recalled, could "only be described as bizarre."

This brought on the faith crisis. As the son of the messiah, specially powered, Heung Jin's voice began sounding in parishioners' heads. A word, encouragement, a condemnation of homosexuality. Reverend Moon explained his son was leading a splendid afterlife. Jesus had "bowed down" to him, Heung Jin had been crowned youth-king of heaven.

Some messages have the sound of beta tests, for Instagram or Snapchat. "Dear Brothers and Sisters of the Bay Area: Hi! This is the team of Heung Jin Nim and Jesus here. We need to establish a foothold among you." He had the frisky, improvisational style of the adolescent: no telling where or when. "I received my own (and only) revelation from Heung Jin," reported one church elder, "during an airing of the popular song 'Love is a battlefield,' by Pat Benatar."

A Zimbabwean convert caught the signal with special fidelity; then he received the full download. Reverend Moon's son was now a former hotel waiter, medium height, wearing aviator-frame glasses. A church theologian got on a plane and confirmed this miracle. "Now we can speak with Heung Jin Nim directly in a physical body!" Known as the Black Heung Jin, he headed out on a worldwide tour.

In New York there were welcome-home banners. The Moons embraced him—though he had no former-life memories and spoke no Korean—as their restored son. There was the ticklish question of marriage in absentia. As the new Heung Jin addressed the membership, his bride slipped quietly into the hall. There's a photo; the young ballerina looking apprehensive and miserable. The new Heung Jin explained that as this was a case of spirit possession, ordinary customs—the Indemnity Stick, the Three-Day Ceremony—could go unobserved.

This was the big church news, in the year before Dr. S. Fred Singer found his way to Reverend Moon's organization, and discovered a faith like any other. The new Heung Jin brought a renewed vigor to counseling. He was especially troubled by romantic expression. Certain kinds. "Heung Jin Nim was very angry about oral sex," a church member recalled. He slapped; he beat parishioners with his fists; with a bat; with rope. He became the Indemnity Stick. He handcuffed them to radiator pipes and went after the True Mother's housemaid with a chair. (Mrs. Moon provided an explanation. Heung Jin, a teenager when he died, remained immature about sex.) In a memoir, the reverend's daughter-in-law describes the home office reaction—leaving us a portrait of the True Father in a light mood. "He would laugh raucously if someone out of favor had been dealt an especially hard blow." Heung Jin counseled the father of his postmortem bride so thoroughly he required open-brain surgery.

Then the whole thing got to be a bother. Reverend Moon announced the spirit had passed out of the body and sent the Zimbabwean home. (The official church history is that the new Heung Jin had picked up, like a flu, an additional, evil spirit.) But he couldn't keep the story out the papers: It made the *Washington Post* front page in 1988, while Dr. Singer was wrapping up at the Department of Transportation. THEOLOGICAL UPROAR

IN UNIFICATION CHURCH. Then papers all around the country. "This is quite in line," a spokesman told *Newsweek* mildly, "with what we've always believed."

It doesn't sound especially like St. Patrick's or the *Christian Science Monitor*. But it is helpful. It gives us the opportunity to test a determination reached by S. Fred Singer: what he was willing to have us believe. Deceased bridegroom marriage, handcuffs, national subjugation (with a bow to Reverend Moon). As with climate, either Dr. Singer hadn't researched, and so shouldn't have spoken. Or he had, and hoped others were in the dark.

WHO DIGESTED THE SCIENTISTS?

THIS IS A STORY OF DENIAL ON THE SPIRITUAL SIDE.
Ozone, recycling, and cults were features of the American seventies. Reverend Moon arrived in 1965 with a humble dream: "A place where you can come to pray and not be bothered by Satan." Nine years later, he was giving private counsel at the White House to President Nixon. In 1976, he gathered two hundred thousand followers for a mongo religious observance on the DC Mall. (Followers collected two thousand blood samples in a vat beside the Washington monument, then set the blend on fire. I still have no idea what this meant.) By the decade's end he was famous. But loathed. In a Gallup poll, the only figures to share his murky, unloved depths were the dictator of Cuba and the leader of the Soviet Union's campaign of smiles.

Part of this was the mass marriages. The reverend handled spouse selection, then conducted the vows in sports arenas. Don DeLillo opens his novel *Mao II* with a Moon wedding, because it's among the creepiest of modern visuals. Choice and personality (what marriage is, basically) excluded. Just row after row of identically dressed couples, like the assembly line at a wedding cake factory.

Why join? The seventies were touched by the fresh wind of disorganization: the field trip with the substitute teacher, the moment before taking seats in the school auditorium, extended ten years. New arrangements seemed possible because they felt necessary.

"In a country whose young tripped out on radical politics or drugs in the sixties," reported the *Times*, "religious cults seem to be the opiate of the seventies."

There was Scientology, Hare Krishna (shaved heads, orange robes), EST (three-day conference, no bathroom break, squinched insights). In a spiritual energy crisis, people turning over every rock. There was the UFO cult, whose leaders, Bo and Peep, construed *Star Wars* as a personal message— ideas beamed by aliens into the "minds of creative humans"—and who later committed team suicide under the name Heaven's Gate. There were the Children of God, whose leader had received a Red Sea vision of the West Coast crumbling into the Pacific, so anything went. (Their most famous ritual was called Flirty Fishing—divine prostitution, for converts and alms. Church mimeograph: "God is interested in *details* and *specifics* . . . Q: If I make love three times with a fish on one occasion, do I count it as three times or just as one time? A: Credit to whom credit is due! If you loved him or her three times, you *deserve* to count it as such. God bless you!") The Peoples Temple, mass worshipper suicide by beverage, leaving behind the ice-cold analogy about Kool-Aid. (For years, Kraft Foods has maintained a delicate, losing fact check: pointing out that what the Temple really served was the budget alternative, Flavor Aid. "We all try to protect the value of our brands," a spokesman explained. "This one just kind of got away from us.") A member of your family, a resident on the dorm hall, met a new person, had a thrilling conversation. Next day, gone.

WHAT UNNERVED PEOPLE most were the eyes. You find it in piece after piece. Young people whose gaze didn't quite reflect local stimuli: "Blissed out," "spaced out," "the thousand-mile stare." (John Updike, who in this sad era wrote two novels with cults, favored "starry-eyed." Don DeLillo takes a few different stabs before settling on "the transfixed gaze.") The Moonies were probably just tired. They sold roses and snacks from dawn till dusk, stealing catnaps in the van—all proceeds kicked up to the church. If you tried to leave, brothers and sisters warned the tug had come from Satan and you'd lose your soul forever.

"I've seen kids take years to recover," a former church leader said. Then continued, as if deliberately refuting S. Fred Singer, "They want you to think of them as just another church, but that doesn't happen with mainstream churches—that sort of massive residual psychological damage."

A public relations hurdle, for sure. Reverend Moon had been excommunicated by the Presbyterian Church of Korea. (Followers dumped a bucket of urine and feces over a Seoul University religion professor. This one I understand.) Denied membership in the New York Council of Churches. Blackballed by the National Council (Moon was "incompatible with Christian teaching and belief"), rejected by the World Council. An international faith outlaw. He'd later be banned from setting foot in the entire nation of England—as Marsellus Wallace would say, he'd lose his London privileges. All this interfered with the simple objective. "If we can manipulate seven nations at least," Moon said, "we can get hold of the whole world."

So what do you do? You could chalk it on the board like a B-school problem. The reverend was in a position similar to the fuel companies, the cigarette folk. He had an attractive product—enlightenment. With a bum side effect—you'd joined a cult.

He took up investment in denial: spending yourself clean. The reverend expressed this principle very crisply in a sermon. "All of a sudden they didn't see Moonies anymore, just the money."

The church purchased the New Yorker Hotel, where Nikola Tesla had aged away his life on the thirty-third floor. They later acquired United Press International, one of the wire services made possible by the luckless Samuel Morse, via the machine that was the source of all Thomas Edison's luck. They bought the waterfall mansion from the owners of the Maidenform bra company. (The estate had a creepy, bra-sounding name: Exquisite Acres.) Property tells a history. First the inventors. Then the manufacturers. Then the mood artists, people who fiddle with the mists inside your head.

The church founded newspapers, magazines, a publishing house, a pharmaceutical firm, hospitals, a company that turned out rifle and grenade parts (in which line the reverend probably didn't face much competition from fellow messiahs), a bank, snapped up farms, restaurants, day care, a video production house, a concert hall and recording studio (NBC's *America's Got Talent* taped there), a whole cable network, an entire Northeastern campus, the University of Bridgeport. As the *New Republic* wrote, "Sun Myung Moon is to cults what Henry Ford was to cars."

Like Keep America Beautiful with their Indian, fossil fuel and the ICE campaign, Moon's people weren't above clapping on a wig and fake mustache. You might slam coffee, lace up Nikes, go for a run with the DC Striders Track Club—then discover you were doing laps for Reverend Moon. As the magazines put it, "a dizzying array of front groups." To beguile reporters and politicians, there were international organizations with bombastic goodwill names. The International Cultural Foundation, the World Family Movement, the Professors World Peace Academy—and a young person's dance troupe, the Little Angels. This was deployed by the reverend for missions of an especially sinister and adorable nature. "Our young Little Angels . . . enabled us to influence political figures in Japan, and now we will influence the congressmen of other Asiatic countries." This is Reverend Moon in a sermon. "We must approach from every angle of life; otherwise, we cannot absorb the whole population of the world," he said. "We must besiege them."

The church built the raw-seafood empire True World. Still America's largest supplier to sushi restaurants. If you dipped hamachi in soy last night, odds are you are among the besieged. And far below, at the base of all the turtles, were the kids with starry and transfixed eyes—who, wandering a city, could take in thousands per day. Half of church revenue generated by people who, once the fever had passed, the rescue vehicle arrived, would publish tell-alls like *I Was a Robot for Sun Myung Moon*. The "organization we're setting up wants to be utilized as an instrument," said the reverend's second-in-command in an interview. This is the man later sent to the hospital by his daughter's husband, the reincarnation of Reverend Moon's dead son. "The instrument to be used by God."

The central heavenly instrument was the *Washington Times*. Ultra conservative, it would become reliably vicious on climate change. A place for Dr. S. Fred Singer to write columns with nothing-to-see-here, cigarette-scare titles. "Chilling Out on Warming," "No Proof Man Causes Global Warming," and the immortal "Climate Claims Wither under Luminous Lights of Science."

The reverend could start it, in 1981, because news jockeys are hard-bitten people: a byline is a byline. "I've worked for a lot of publishers who thought they were God," one editor shrugged. It succeeded because the reverend was fine with burning fifty million dollars a year. Also because politicians are equally hard-bitten: support is support. It became President Reagan's favorite paper, his go-to morning read. ("Without knowing it," the reverend boasted, "even President Reagan is being guided by Father.") Former President Nixon got on board. The first President Bush used a word you don't immediately associate with Reverend Moon: the newspaper brought "sanity to Washington."

It caught the eye of the cigarette folk. Opinion maintenance is a daily grind: facts stay fact, but opinion comes unglued overnight.

Philip Morris noted the reverend's sweet advance in a planning document: "Consider acquiring a major media vehicle," they wrote, like the "Moonies [and] *Washington Times*." Never again, having to bargain or cringe with some East Coast jerk reporter. Conservative activists would relearn this lesson decades later, with fake news items planted on Facebook. The "grassroots readership," Philip Morris wrote, "cuts the legs from under effete criticism."

What the reverend was denying overall was that his group was a cult. Theological denial works the same way as climate. Moon began circulating funds in great gusts up the right-wing side of the Capitol. Numbers large and powerful as a storm front—figures at which money *becomes* weather. Two hundred million. Three hundred. "The Unification Church is trying to buy its way into the conservative movement," a right-wing lobbyist blabbed to the *Washington Post* in 1984. "It's frightening."

Still, it worked. "The Church has established a network of affiliated organizations and connections," *U.S. News & World Report* was explaining, five years later. "Almost all conservative organizations in Washington have some ties to the Church." Political money is a sort of miracle cream: it heightens, brightens, sharpens, erases. The conservative *American Spectator* saw the outcome in the loss of a key word. A "creeping reluctance to call a cult a cult." Instead, "among Washington conservatives, the Moonies are legit." Political groups saw money, not the Moonies.

It is among the era's most shameful episodes. Moon, in this hurricane of dollars, did a thirteen-month stretch at Danbury Federal for tax fraud. The acceptance came *after*. The publisher of *National Review* supplied a character reference: "The Unification Church has settled into the landscape of Washington as a good influence."

There was another outpost to besiege. The most important. As Frank Luntz would later advise on climate, "people are more willing to trust scientists."

Moon, new to America, grasped this as a nutritionist does a deficiency in the shopping cart. "More than anything else we need scholars in the scientific fields," the reverend said, in a 1973 sermon with the boot-marching title Our Future Path of Advancement. He envisioned conference halls packed with academic guests. Then, "back in their home countries," Moon said, "these scholars will influence their own national policies in a joint effort." By this method, "We will surely influence the policies of the whole world in the near future."

The deck seemed stacked against him. As of 1978, Moon had been investigated by Congress. The reverend's aim, they determined, was global theocracy—in "which the separation of church and state would be abolished, and which would be governed by Moon and his followers." Not a party lots of scientists would drop labwork to attend.

He graced his annual meeting with another goodwill name. The International Conference on the Unity of the Sciences. Staged it in aspirational locales, the church picking up airfare and hotel. If you spoke or handled leadership tasks, there were honoraria. (So unlike most vacations, you headed home with a fatter wallet.) Who came? Not Roger Revelle, Dave Keeling, or Jim Hansen. And attendance was fraught, a matter of wrestling and pinning your conscience. Because what the church wanted was obvious. "The presence of distinguished academics at church-sponsored gatherings," the *Washington Post* reported, "gives Moon the aura of power and influence he seeks."

There were scientist-on-scientist accusations, of being "bad for the reputation of scholarship." Parliament advised the entire British scientific roster to sit on its hands.

But Frederick Seitz served on the advisory board for eight Moon conventions. He's right there in the program: as chairman, vice chairman. And S. Fred Singer—who would sacrifice a lot for a good pre-paid air ticket—hit up ten church conferences. As vice chairman, as speaker. It remains among the enduring Reverend Moon achievements. Matchmaking: bringing the two men together. The first time the denial grandfathers appeared side-by-side in a magazine story.

Nature reported the Freds as "luminaries on the organizing committee" at the 1983 event. That's the start, for these two names that would enter history together. It's a wonderful thing. Frederick Seitz had always wanted the spiritual element deleted from his hard drive. In the fifties he'd complained, "the issue of morality perhaps lies in the philosopher's domain," when it came to world-cruncher weapons. So "long as it was green," was Dr. Seitz's policy on tobacco cash. As for morals, "I'll leave that to the philosophers and priests." Well, here the moment was at last. An actual priest, in an inexcusable domain—and Fred Seitz did what he did. He took the money.

"All these people should know better," sighed the leader of a group for lonely mothers with children in the cult. "The names are used inside, too," a former church official explained, "to keep kids in the Church. They say, Look, these people are with us. If we weren't on the up-and-up, why would these people be helping?" Another glum mother, helming another sad parents group: "These academics are selling themselves." She explained to a reporter, "All these conferences are taped and those materials are used in recruiting programs all over the world." She could just as easily have been speaking about any denial cause; celebrated faces advocating for smoke, for heat. "It adds credibility to their organization."

Frederick Seitz and S. Fred Singer went above and beyond. Dr. Seitz interrupted the 1988 conference action. To stand and deliver a short, sympathetic speech on the reverend's tax woes. "I hope," Dr. Seitz said, "that this kind of persecution against Reverend Moon will soon come to an end."

The second Fred sat for an official goodwill photo with the reverend, matching flowers in their buttonholes, and provided a testimonial. "The conferences," Dr. Singer said, "have produced a tremendous intellectual output." His words were reprinted in church literature. "The academic participants have carried the message back to the classrooms." It's what Reverend Moon had called for, back in the seventies. And it's where the rocket had landed Dr. Singer.

Any victory—especially one dependent on human weakness—tickled the reverend. "Who digested whom?" Moon asked his followers. "Did Reverend Moon digest the scientists? Or did the scientists digest Reverend Moon? . . . The umpire has declared a winner. Some think they can still stop Reverend Moon, but it is impossible."

Digestion was a big topic with the reverend: what goes in, what comes out. "Once you digest my teachings, your accomplishments are incredible," Moon explained. "But the problem is that often you don't digest [and] you begin to suffer from spiritual diarrhea." He was among the most lavatory-minded of religious leaders. "Can we say that the anus of a holy man is a holy anus?" is a riddle he left readers of the *New Yorker*. "Father never flushes," he explained in sermons. "And toilet paper! Father does not use lots. . . . Love manifests especially beautifully in the bathroom. . . . When there is true love, you will open the door where your spouse is using the toilet and feel you are smelling a perfume." Odor habituation was the critical tool. "In the end," Moon asked followers, "who will become the winner—the Chinese who don't mind bad smells, or the rootless Americans who can't tolerate the odor of human waste?" The reverend was crazy. But what comes through his words is joy, the gratified surprise of one soul at the gentle reward system of the world.

And sometimes, when parishioners giggled, or expected something along more standard theological lines, the reverend sank into musing self-pity. "No one understands Father," he'd say. "Not even Mother and his children."

The Reagan time ended, Democrats came back to the White House. And that was the reverend's end, influence-wise. They had taken his money but never the man—he'd been, in the political phrase of the era, *rolled*. (Tobacco filled the vacuum. By 1995, Frank Rich wrote in the *New York Times*, the Republican Party was "all but choking on tobacco loot.") He grew plumper, older; lines score every face, even a savior's. His daughter-in-law published a memoir. It killed the movement. Her husband, Moon's heir apparent, had been a pornography buff and a cocaine enthusiast. (Often

simultaneously: "One evening, after he had spent the entire week snort-
ing cocaine and watching pornographic videos . . .") Also a gun nut and a
domestic sadist: he beat his wife bloody during pregnancy. "He kept saying,
'I'm going to kill the baby!'" the elegant Mrs. Moon shared with *60 Minutes*. Then he "licked his hand and said, 'It tastes good. This is fun.'" Her
former husband employed a one-time-only domestic threat. "You think
I'm afraid of the police? Me? The son of the Messiah?"

You might walk away from a media crash like this, but you don't walk
the same. Moon expired, on the cross of pneumonia, at the age of ninety-
two. His following had dwindled from the millions to the thousands. In
the spirit world, he presumably took over the management reins from the
team of Jesus and Heung Jin. His family gave up on America in 2012—after
four decades, complained his widow, with "such little results." America
remained unsubjugated. As with ozone, acid rain, and cigarettes, the Freds
received their names back, with minimal wear and staining.

There was a spot of the type of postmortem action the reverend
enjoyed. "It will be interesting to see how many respectable figures in the
conservative firmament find nice things to say about him now that he's
gone," the *Washington Monthly* wrote. And how many would imitate the
reverend's dishonesty, "by denying they ever had a thing to do with him."

MILLIONS OF GUINEA PIGS

S. FRED SINGER DIDN'T WAIT FOR A BODY IN THE GROUND.
He was never one to linger with a denial.

In the mid-nineties, the physicist assured the *New Republic* that he had
received no Moon financial support. Explained by website that the connection was an "old rumor," a "dead horse." Dr. Singer had to account for the
whole mess on a 1994 episode of ABC News *Nightline*. Where he admitted
that the reverend had been an ideal landlord: free office space for a year.

It's a perfect crisis-publicity model. A nibble of truth, until the beast
forgets it's hungry. (Cornered, he always gave up just enough data to call off
the search.) Dr. Singer was a Euclid of portion size.

Actually, once the reverend digested you, you stayed digested. You
traveled all the way down the alimentary canal. Singer served on the editorial board of Reverend Moon's magazine *The World & I*—the most humble
and grandiose entry in the *Readers Guide to Periodicals*. He became a columnist for Reverend Moon's *Washington Times*: hundreds of appearances,
right up to yesterday. He never stopped working with Reverend Moon. The
church published three of his books. He joined the reverend's think tank,
the Washington Institute for Values in Public Policy. In his semi-retirement,
Dr. Singer received a monthly stipend: $5,000, from the denial think tank
the Heartland Institute, for career services rendered. So expenses were covered between occasions when Dr. Singer was rolled out to declare climate
change "all bunk." Regular science gave him attitude at the door. But denial

showed Dr. Singer to his table and knew his drink. It would have taken a titanic act of willpower—of self-denial—to refuse.

WASHINGTON INSTITUTE UPKEEP cost the reverend $1.5 million per annum. It was spring 1989 when Dr. Singer pushed through government doors for the last time—into the powdery, expectant air of the capital, where every street and car ride might end in the rooms where decisions are made—and autumn when he walked the few blocks over to Reverend Moon's place.

Why join? You're nothing in DC without letterhead. Just another schnook, one ballot under the sky. Dr. Singer lent the reverend his name, and received the dignity of a title. He became director of the Science and Environmental Policy Project at Reverend Moon's Washington Institute for Values in Public Policy.

By autumn, that's how he was signing himself. In the *New York Times,* the *Wall Street Journal*; in *Science.* "S. Fred Singer, Washington Institute." The reason to greet his opinions with especially perked ears.

And you know Fred Singer—it wasn't just a year. In October 1991, he was still at it. Slagging climate change in the *Post.* "I'd be more worried about future cooling than about greenhouse warming...S. Fred Singer, Washington Institute." (Denier predictions have remained a kind of anti-marvel. Here's the skeptic Patrick Michaels. "Twenty years from now, we will more fully appreciate the effects of modest warming and CO_2 enhancement," Michaels wrote in 1992. "Longer growing seasons, summer temperatures that do not change much." 2012 was the then-warmest year on record for the United States. "It's like farming in hell," a plant biologist told *Business Week*.) Testifying before Congress, Dr. Singer rustled papers and cleared his throat as the Institute's man on the Hill. Addressing the federal Arctic Research panel, same deal. (Climate change; Dr. Singer's recommendation was "additional research.")

Fruitful years. In this marriage-crazy environment—where even the lonesome dead attract spouses—Dr. Singer took a wife. The organization was good to the newlyweds. Let Dr. Singer hook up his bride with a Reverend Moon job. Mrs. Singer became "a project editorial director with the Washington Institute for Values in Public Policy."

Even after the relationship soured, Mrs. Singer could still provide value. In 2005, the UK climate writer George Monbiot went hunting a phantom statistic. One of those juicy, counterintuitive Mark Mills numbers.

Monbiot had been startled to discover it in the magazine *New Scientist*. "555 of all the 625 glaciers under observation by the World Glacier Monitoring Service in Zurich, Switzerland," Monbiot read, "have been growing since 1980." Just the sort of number that diminishes faith in science, that melts and impedes belief in climate change.

What piqued George Monbiot's interest was its ubiquity; the number being "all over the Internet."

So Monbiot phoned Zurich. The glacier service called the stat fiction. (Their actual metaphor was digestive: "This is complete bullshit." Which might have brought happiness to the reverend.) The vast majority of the world's glaciers were relaxing into water, just as you'd expect.

The "555" part turned out to be a typo—keyboard trouble. A British denier told Monbiot he'd meant to enter "55," then the percent symbol, then missed. That's the whole thing: the sensitivity of the information system. How an incorrect fact can wriggle its way into national thinking. But the 55 percent number was excremental, too.

Monbiot took a breath, then tracked the number upstream. He splashed through big-name and eccentric newspapers, waded across think tanks and conservative websites . . . and when he looked up, Monbiot was standing at Dr. Singer's door. The 55 percent figure had issued from the Science and Environmental Policy Project that Dr. Singer started for Reverend Moon.

At first, the scientist was indignant. "Monbiot is wrong," Singer fumed. "Or simply lying. . . . Obviously, he's been smoking something or other."

Then he relented. The glacier number, Singer owned up, was his. It "appears to be incorrect," he said, "and has been updated." Not that we should think any less of Dr. Singer. He put the blame squarely on his "former associate Candace Crandall." He did not add that she was also his "former wife Candace Crandall." She'd been Dr. Singer's executive vice president for a decade. Like many people temporarily enlisted and enriched by denial, she had drifted away. Into TV production—she provided voice-over to the convivial PBS series *Wine, Food & Friends*—and then the world arranged a noncombatant happy ending. She now shares her home, according to the bio, with a new husband and many companion animals: "They live with three dogs and a cat in suburban Washington, D.C."

When George Monbiot checked back almost a year later, the glacier figure was still emitting its pale, incorrect glow on Dr. Singer's website.

In these months, the real-life economist Paul Krugman offered climate people some play-mean halftime advice. "You're going to have to get tougher," Krugman wrote in the *New York Times*. "Because the other side doesn't play by any known rules."

NO-RULE GAMEPLAY suited Dr. Singer. It's what it meant to be a columnist at the *Washington Times*. Every quality that once caused career bruising—the big mouth, the passion for journalists, the lightly researched certainty—became a bonus. It's how it feels to find your métier: every-thing *fits*.

In 1992, three years after joining the reverend's Washington Institute for Values in Public Policy, Dr. Singer incorporated the Science and Environmental Policy Project outside the confines of the church. He named fellow Moon luminary Dr. Frederick Seitz chairman of his board. Apart from the job title, free office, free office computer, editorial gigs, writing assignments, three books, congressional visits, the job for his wife, conference podiums, conference chairmanships, his column, Dr. Singer had not been supported in any way by his association with Reverend Moon.

THERE'S A LINE IN Joan Didion. About a religious fraud—the former bishop of California—who completed his earthly adventures by driving a rental car into the Israeli desert. His body was later winched out of a canyon. The bishop's journey had taken him, Didion writes, through "every charlatanic thicket in American life."

That's Dr. Singer's trail map, too. Here's how he came to the attention of tobacco.

There once was a great schism in the consumer protection movement. In 1933, a man named F. J. Schlink wrote a book: it was called *100,000,000 Guinea Pigs*.

This sounds like a Disney project with a casting headache. Really it starred the whole population. All of us—how lax oversight made every shopper a lab rat, every purchase a spin of the wheel. "These people violated no law," the book relates darkly. "In the eyes of the law we are all guinea pigs, and any scoundrel who takes it into his head to enter the food or drug business can experiment on us."

The book, among other things, helped cement a metaphor. "Guinea Pig means they are being used experimentally," the *Times* had to explain, sort of touchingly, in its review. The way "real guinea pigs are in scientific laboratories."

Schlink and his cowriter examined products like Radithor, a male virility treatment that claimed its radioactive elements would bring about "a great improvement in the sex organs." Also Koremlu, a depilatory for women, supported by big, dainty ads in *Vogue*: "The achievement of a noted French scientist—not only removes the hair *for all time*, but actually is most beneficial."

Well, that French researcher must have been bending the elbow with pest control. Rat poison was the active ingredient. Some consumers who trusted *Vogue* and Koremlu became entirely hairless; others went blind, paralyzed.

And Radithor really *did* contain radium. There were deaths. Including a fifty-one-year-old socialite—a former national amateur golf champion—who after three years on the stuff had lost his teeth, his jaw, had holes in his skull; he died weighing 92 pounds. (Ideal hospital couple: blind Koremlu woman, withered Radithor man.) The writers' warning, as of 1933, summed up the next eight decades of fights. It's like a video where every screen grab recapitulates the entire plot. "The average manufacturer will resist to the end any interference with his business," explain the writers. "Even when it has been proved beyond doubt that his product is a menace to health and life." Just what Louis McCabe relearned two decades later in Raymond

Chandler Los Angeles. What Sherwood Rowland was taught by DuPont. As if every generation must hire its own detectives, to bring back evidence of the same ongoing crime.

The *Times* called *100,000,000 Guinea Pigs* "astonishing and devastating." It remained on bestsellers lists for years, brought in a slew of federal laws. F. J. Schlink is the reason you can open your Amazon package without safety goggles.

Schlink had started the first product-testing magazine: *Consumers' Research*. There was a strike—the vicious kind, with smashed windows and police presence. It cracked something in Schlink.

His coauthor broke away with a radical consumer-protection splinter group. This faction founded a much better product-testing magazine: *Consumer Reports*. One of the great publishing success stories. Safer water, less germy hospitals, explosion-free driving, microwaves that leave brain and hand unbroiled: that's the legacy of *Consumer Reports*. You've never heard of Schlink's magazine.

He was a frustrated-looking man with a small mustache and mournful eyes; a fellow from the provinces. Who for one moment had stood at the center—felt the rush of having done the necessary thing. Maybe touching success once is more deranging than never coming close; leaving with a single memory of the shouts and gladness of the hall.

Schlink turned. Developed a sweaty interest in communism—became the sort of publisher who mutters about Reds and plots. Schlink drifted into the stormier precincts of the Right. And his magazine became a place where the businessmen he'd previously called "scoundrels" could go for a sympathetic hearing. "In a final irony," writes one historian, "Schlink found a lasting home among his former enemies."

But the similarity left room for confusion. *Consumers' Research, Consumer Reports*—it really was just one syllable.

If you were a corporation, and arranged a supportive tidbit in *Consumers' Research*, then quoted it in ads or speeches, some portion of your audience was bound to believe they'd heard the reliable name.

It's the way of deniers: They are feasters on the off-brand, snouts deep in the soundalike. Years later, with his twinkle, Fred Singer would cook up something called the *NIPCC Report*—the Non-Governmental International Panel on Climate Change. If you didn't squint extra hard at the label, it sounded just like the United Nations' Intergovernmental Panel: the name brand you trusted.

In a documentary, Dr. Singer grins and snickers behind a two-book comparison. "The actual IPCC report is about *so* thick," Dr. Singer says. Identical girths. "We use strictly the same thickness." The head of the debunking Skeptical Society appears onscreen a few moments later, with other comparisons. "If you're a Congressman, how do you tell the difference between this, and *this*?" he asks. "And *that's* the point of this deception."

It was another charlatanic thicket Dr. Singer walked happily through. Of course he and his associate Candace Crandall published stuff—"Global Warming Science: Fact versus Fiction," "Drastic Remedies Are Not Needed"—in *Consumers' Research*. (Though a series on the wickedness of abortion, including "The Fetus Beat Us," was a project that Mrs. Singer tackled solo.)

On a gray spring day in 1991, with the cover of *Time* reporting 125,000 cyclone deaths in Bangladesh, Dr. Singer climbed the steps to a *Consumers' Research* symposium.

This was held at the Dirksen Senate Office Building. (Poor, martyred Dirksen. It's where Reverend Moon received his Crown of Peace.) One expert spoke about the perils of fuel efficiency—improved gas mileage was "an absolute disaster." Another cautioned against the thrill ride at the organic grocery. "There are lots of carcinogens in natural foods," he explained. "If you take the time to look."

Not the fellowship you want to end up in. Dr. Singer probably did not know or appreciate the full irony: being hosted by a magazine that had traveled the same path as him. Promise, setback—and then the chance to excel at something grubbier.

Dr. Singer spoke that day about climate and temperatures. "They will not show a discernible [upward] trend in the 1990s," he said.

You know Fred Singer—it didn't work. "At the time, the 1980s was the hottest decade on record," the National Oceanic and Atmospheric Administration later reported. "In the 1990s, every year was warmer than the average of the previous decade. The 2000s were warmer still." As number-crunchers from that organization calculated, if you were born after February 1985, you have never experienced a cooler-than-average month on this planet.

So I wonder how it felt to stand in that room. This was the final installment on an air ticket to London instead of Iowa City; the cost of non-admission to the National Academy of Sciences. I wonder if Dr. Singer regretted it. Probably not. Any self has a survival instinct, wants to get on with the sweaty business of being itself. In a way, that was Reverend Moon's insight. We become comfortable with our own stink.

Dr. Singer was about to bring his special gifts—thirst for press ink, a greater-than-average tolerance for noisy public error—to cigarettes and warming. "The reason I bring all this up," Dr. Singer told listeners that day, "is because the tendency to misuse science, or to ignore it, is very strong." Maybe there's a dial inside everybody that's permanently set to *confession*. He had delivered a one-sentence summary of his entire career.

A Tobacco Institute consultant also spoke that day. Gary Huber—the long-legged Harvard Project doctor who'd been kicked down the map to Texas. That's how small-scale the denial operation is. Always the same faces, cycling endlessly through the chowline. Huber told his audience the push against tobacco wasn't really science, but a social movement. ("I can't turn back the clock," Huber told NBC News. "I wish I'd been smarter or wiser.")

Cigarette publicists sent the Tobacco Institute a *Consumers' Research* piece about Huber and the symposium. You always wonder about the feeling tone of such people: Do they lament? Crack jokes? Gnash teeth? The publicists wrote, "I know you'll get a kick out of the ad on page 22." This was a cartoon tombstone, above notice from the American Heart Association that cigarettes caused death by cardiac event.

And S. Fred Singer's name appeared for the first time on a list of potentially useful experts in Philip Morris files. He was sixty-seven. On that gray morning, it probably would have surprised him to know that his best days were still ahead.

DINOSAURS

WITH THE CLINTON PRESIDENCY, WARMING ACHIEVES
narrative perfection. Best intentions plus the worst possible timing.

Bill Clinton and Al Gore decorated their White House workstations
in January 1993. They were the first Pennsylvania Avenue Democrats for
twelve years. And when your party has been out of power, it's like delayed
Christmas—voters expect a ton of gifts. Al Gore had written one of the
first climate bestsellers. A title you picture in oozy letters on the splash
page of a comic book: *Earth in the Balance*. (Captain America and the Red
Skull, wrestling over a single parachute as they plummet.) Climate advocates could pop out a thought balloon of the West Wing, and know it contained a member of their tribe.

Gore played the sidekick. Second hat on the eight-year ride, the time-
passing voice in the president's ear. A few weeks after their Washington
arrival, he showed up for Friday lunch with a prop. A giant folding graph.
The sea level where carbon dioxide started, the sharp peak of its ascent.
"Just in case you missed it in my book," was the vice president's joke.

Gore had studied with Roger Revelle at Harvard: the Revelle-Dave
Keeling climate tutorial had made it to the world's most exclusive classroom.

From the first bell, Clinton was a reluctant pupil. The kind who is
respectful at the desk but cracks wise on the school bus—the customer
who glances away during the embarrassment of the sales pitch. His biggest
impression was Al Gore. "I was convinced that *he* was convinced," the president said.

By the end of Clinton's first year, green lobbyists felt mussed—romanced for votes, abandoned at sunrise. "What started out like a love affair," the head of the National Wildlife Federation told *Newsweek*, had come to feel like "like date rape."

White House greens felt a reciprocal irritation toward their own supporters. Who didn't understand the ground game, the slow and steady pickups, the gridiron of Washington. "Environmentalists," they complained, "are like a football team with only one play: a 95-yard touchdown pass."

Gore kept at it. President and vice president sat down to a weekly lunch—the one unbudgeable item on the Clinton schedule. (*Inviolate* is Bob Woodward's word in *The Choice*. "Clinton had come to rely on Gore as his indispensable chief adviser.") Those Fridays, White House silver ticking against White House china, Gore delivered his climate brief. By spring 1995, he felt he'd briefed enough. Clinton later joked he could "now pass Al Gore's climate test." For the first time, an American president was on board for "specific targets and timetables" at worldwide climate negotiations.

BUT BY THEN it was too late. In November 1994, Newt Gingrich and the Republicans had tusked and stampeded through Washington—taking both houses of Congress for the first time in forty years.

Newt Gingrich became House Speaker. He had the look of a faded infant prodigy. Cranky toddler features, a wise man's head of snowy hair above. That head was crammed with peculiar notions. He thought America should just go ahead and build *Jurassic Park* already—"Wouldn't that be one of the most spectacular accomplishments?" Which meant he'd either misunderstood the movie or hadn't stayed around for the crunchy second half. He believed "honeymoons in space" would become routine by 2020. ("Imagine weightlessness and its effects," Gingrich explained, as to an after-hours meeting of the Science Club. "And you'll understand some of the attractions.") He fully expected to become president one fine November, and to that end expected aides to record every speech. "He thought there'd probably be a museum someday," a staffer explained "Where you could go and check one of those tapes out."

For his first 1995 speech on the environment, Gingrich chose from the argument-starting section of his wardrobe. A red tie, with a baby Triceratops. He flapped this tie at a glum audience of environmentalists. "They're extinct," Gingrich said. "Take that as a warning. These things happen. Life is like that."

In the House, polarities swapped, as opinions out of power rose to the podium. The head of the Energy and Environment Subcommittee— it had global warming—was now Representative Dana Rohrabacher. Who declared federal climate research programs "scientific nonsense." (Rohrabacher would later propose his own hypothesis for the source of warming: dinosaur farts.)

The new Republican majority did not "operate on the assumption that global warming is a proven phenomenon," Rohrabacher said. "In fact, it is assumed at best to be unproven, and at worst to be liberal claptrap, trendy but soon to be out of style in our Newt Congress." Warming had been shoved outside the chamber, cooling its heels. It's the kind of frustrating travel experience climate has had on its long trip through Washington.

International climate negotiations were set for December 1997, in Kyoto, Japan—a milepost at the dark edge of the year.

Clinton and Gore talked up the issue to whoever would listen. That summer, they invited Nobel laureates to kick around the science at the White House. In October, they bused in a hundred weathermen—hairdos who could spread the gospel at five and eleven. After the meeting, the news teams set up on the White House lawn and forecast the weather. A rare meteorological event; for an hour, the weather was solid weathermen.

Five days later, the two men hosted a warming conference at Georgetown University. No euphemism. This White House conference on climate change was called "The White House Conference on Climate Change." The gifts were being delivered at last.

From Georgetown's main hall, the president addressed the cameras. He talked data and predictions, simple and fair. "The great majority of the world's climate scientists have concluded," Clinton said. "If we don't cut our emission of greenhouse gases, temperatures will rise and will disrupt the global climate."

He talked perspective. How it had felt for the first astronauts to ride above the planet. "Looking down on Earth from a vantage point that revealed no political boundaries," Clinton said, these astronauts all formed the same terrified and loving impression.

"They were simply awestruck by how tiny and fragile our planet is," the president said. "Every astronaut since has experienced the same insight, and they've even given it a name: the Overview Effect." He quoted one of the first Apollo astronauts—a traveler's moment, the surprise of seeing the world reduced to blue memory through the capsule's porthole, home becoming tiny nostalgia at the push of a button. "You realize that on that little blue-and-white thing, there is everything that means anything to you, all the history and music and poetry and art and death and birth and love." The astronaut had added, "All of it on that little spot out there you can cover with your thumb."

The president, a Georgetown graduate, was also visiting a former campus—halls he'd blurred through as a student three decades earlier. It put him in the alumnus' reflective mood. The person he'd been, the one he'd become. He was amazed, he said, to find his presidency more than half over.

"What seems at the beginning of your life a very long time," Clinton said, "seems to have passed in the flash of an eye." This, the slim gap between arrival and departure—between plans and seeing if they work—was on his mind too.

"Popular democracies are far more well-organized to take advantage of opportunities," the president said, "or deal with immediate crises, than they are to do the responsible thing. Which is to take a moderate but disciplined approach far enough in advance of a train coming down the track to avoid leaving our children, or grandchildren, with a catastrophe."

He looked into the cameras. "So I ask you to think about that. We do not want the young people who sat on the steps today—for whom thirty-three years will also pass in the flash of an eye—to have to be burdened, or to burden their children, with our failure to act."

It had been nine summers since Jim Hansen declared the warming in effect; this was the world still moving on a reasonable schedule. Kyoto would be the fight. So industry and the deniers were reaching out wherever they could, too. This is where Dr. Frederick Seitz made his greatest impact.

COMMITTEE ON THE
CARE OF CHILDREN

TOBACCO HAD SHOWN HIM ITS BACK. FREDERICK SEITZ, Philip Morris wrote in 1989, was "quite elderly and not sufficiently rational to offer advice." That is, the scientist had proven no more gifted at obstructing the passage of time than anybody else in history, and had allowed himself to become advanced in years.

His rehabilitation involved a single sentence. And a DC pathfinder—one of those pressure and influence experts with a map to every corridor and keys to every forgotten door—and it also involved aspirin.

Who would've thought? That in the distance waited tobacco congratulations? And a Las Vegas skyline where men in suits would applaud the Frederick Seitz Memorial Climate Change Award? That "remarkably," as *Vanity Fair*'s Mark Hertsgaard writes, climate denial would rely on "the same tactics and strategies—even the same scientists—that the tobacco industry had previously used." Which gives both Freds the appeal of the damp soap bar in the service station washroom. "No man did more to assist them than Frederick W. Seitz."

HERE'S THE ASPIRIN PART. In 1980, the Federal Centers for Disease Control alerted the pediatric community. They'd discovered a link between the pain reliever and the childhood ailment known as Reye's syndrome.

Reye's was especially cruel. In speed, in the stunned and culpable what-ifs it left parents. "The hell of self-questioning reproach," writes Rudyard Kipling, "which is reserved for those who have lost a child."

Reye's struck mostly between the ages of five and fifteen. Your child ran a fever—flu or chicken pox, Reye's happened only with viruses. After consulting the children's aspirin label (*for symptoms of flu*) you shook out the tablets. Then vomiting. Followed by convulsion, coma, brain damage, death. You stood by, helpless and implicated, having become the executioner of your child.

ABC *World News Tonight* reported that Reye's was killing or crippling five hundred children per year. Dick Van Dyke earned his first good notices in the Elvis musical *Bye Bye Birdie*. Two generations later, his granddaughter was cast in a middle school revival. Nice story, musical theater continuity. Jessica Van Dyke came down with chicken pox, accepted aspirin from her parents, died before curtain.

"We had no conception," her family said. "We never knew about Reye's syndrome." The Scaggs family fed their twenty-month-old the pain reliever—"I thought that was what you were supposed to do when children ran fevers," his mother told ABC. Then shaking, emesis, their son lost before the nurses and fluorescents of the emergency room.

Four separate studies demonstrated the link. In 1982, the government called for a warning label on aspirin.

The rule landed on the desk of a quiet, savagely efficient lawmaker named Jim Tozzi. Handsome, even a little forbidding, and locally famous for his practice of the interoffice martial arts. The Aspirin Foundation of America lobbied Tozzi a little bit. He picked up the phone. (And here you have an instance of just the guinea-pigging that once bothered F. J. Schlink.) Tozzi called the FDA. "You have not made your case," he said.

Warning labels got blocked. The capsule version of the climate fight. So compressed you can see the whole sweep.

By now, you could write it yourself, and I could include it in a story about America's climate. The fight ate through four years. The tiny lives didn't weigh on anyone in particular.

The industry fielded a denial and publicity organization. The name it bears is the chilliest in this book: the Committee on the Care of Children. The one where, after the meeting broke, the executives must've dusted off their hands and grinned. "Well boys, pack your bags, read your Dante—we're going to hell."

The committee leaned on broadcasters: don't run the government's spot about Reye's syndrome. They intimidated knock-kneed and valiant pediatricians. (The head of the Aspirin Foundation spoke to reporters, straight-faced. "Someday, somebody might find out the cause of Reye's syndrome," he said. "Now that's going to be an interesting day.") When the FDA tried to go wide with a half million pamphlets—for distribution at supermarkets to families like the Scaggs and Van Dykes—the Committee on the Care of Children argued data and crisis out of the language. Then printed up literature of their own: "All medical experts agreed that aspirin does not cause Reye's syndrome."

And through it all, even on the children's bottle, the old labels stayed put. *For general relief of symptoms of the flu.* "Basically telling parents," a doctor marveled, "to do something that could kill their kids."

In 1985, one last study, from the National Academy of Sciences, propped open the door, and labels squeaked through. The *New England Journal of Medicine* called it "a public health triumph."

Annual cases instantly plunged. From 658 before government intervention, to 36, then 2. But in the years consumed, epidemiologist David Michaels noted, "an untold number of children died or were disabled." You could tally that figure in many slightly quieter homes—where there'd once been children, now "buried long ago," a journalist wrote in our century, "forgotten by all except the families."

COMMUNITY LIFE IS made bearable by the human habit of reinterpretation. At first, Jim Tozzi took credit for the label blockade. Without his own desk between the FDA and the shopping carts, Tozzi said, "The rule would have gone out."

In 1985, he bragged about it—to the *Boston Globe*—as a career highlight: "Procedurally," Jim Tozzi said, "that was the best decision I ever made."

Post-labels, Tozzi held out. Reye's gave birth to the alternative pain remedy—one reason for the Advil, Tylenol, and Aleve that crowd supermarket shelves. ("If you ever wanted a case of an industry shooting itself in the foot," an FDA official said, "that's the aspirin industry.") Tozzi meanwhile was tracking a whole new risk. "We could end up with a whole generation of kids hooked on Tylenol," he warned. As the figures rolled in—the estimate is 1,470 childhood deaths—he took to claiming he'd had no part in the aspirin fight at all.

CORPORATE DENIAL IS a flashlight in the woods at midnight; it draws interesting fauna. Roger Ailes, sway-bellied and outspoken founder of Fox News, worked the trade for Philip Morris during these ashtray brawl years. So did Karl Rove, chief campaign mastermind for George W. Bush.

Jim Tozzi became a—really *the*—Philip Morris political consultant. Built trim, around a plutonium core of confidence. Tozzi is the invisible guy. Not the public leader, out on a balcony somewhere waving his arms, who yells the trains into running on time. The quiet guy who knows a guy, and drops by the printing office, and has the timetable adjusted without fuss.

Jim Tozzi was Frederick Seitz's smoke angel—leading the cast-out scientist back inside the gates.

He was born in 1938, in the woodsy nowhere of Waynesburg, Ohio. (One vowel away from Winesburg, and literature.) His father ran a grocery. Tozzi heard broadcasts of the party being thrown elsewhere, and grew up wanting to play jazz trumpet. He enrolled at Tulane, failed to become a jazz trumpeter, took an advanced degree, sat behind a desk in the army.

He turned out to have an eerie gift for forms. Where others saw a sleepy afternoon—a page to be initialed, then forgotten—Tozzi heard music, saw dotted lines leading to the X of treasure. It's the player who absorbs the rules who understands they are a guide to freedom.

He grew a reputation. "That nerd over there in the Pentagon who likes to review rules." In 1970, President Nixon signed the Environmental Protection Act—a pearl-clutching moment for his chief of staff. "What did we let out of the box?" he asked.

So they hired Jim Tozzi: to leash the agency, keep it next to the original packaging. Soon the journal *Environmental Forum* was ranking Tozzi as "the single most influential person in the United States" for shaping environmental policy.

In a *Washington Post* profile, he's just another commuter—a shave in a suit—but trailing an aura of staff lists and outcomes. "Every weekday evening," the paper reported in 1981, "Jim J. Tozzi rides a Metrobus home towards Tysons Corner with part of the federal government's future in his lap."

The grocer's son, the former jazz nerd, now gave all federal changes the thumbs up or thumbs down. He had become "the career official responsible for reviewing every one of the 3,500 regulations issued by the government in a year." (Like his environmental work, this followed a strict code of *omerta*. No bragging until after. Why complicate the job by tipping witnesses to your effort? "Environmentalists said I operated in the back room," he told an interviewer years later. "And listened to lobbyists who told me to water down environmental regulations. I sure as hell did.") Because he could fly at a regulation with no radar signature, just the boom and smoke curl after he passed, Tozzi's nickname was military—"Stealth." His own motto he pinched from the police locker: "I don't want to leave fingerprints," he said.

Something else got left behind. Journalists would call Tozzi "flamboyant," an "eccentric." He'd mix them drinks from hooch stashed in his office globe. Show off an indulgence of city living—the expensive brass telescope rig he aimed at neighbors' windows.

"Things get really interesting at night," Tozzi would tell them. Or, "I'm a dirty old man. I love it."

Whatever human switches must be deactivated to do the work stay down after-hours. You develop a deafness to the sound of your own life. After Clinton and Gore's election, Tozzi, a conservative, donated thousands. A fellow lobbyist was surprised to discover him at a black-tie Clinton event.

"Tozzi," the lobbyist said. "You're such a whore."

"So was Mary Magdalene," Tozzi answered. "And it didn't hurt her billings."

But when journalists discussed the work self, this commanded absolute respect. Tozzi was "almost legendary," "the premier tactician"—"the ultimate insider" and "master of the chessboard." The best person ever to do the thing he did.

He'd entered private practice by then. Tangling up regulations in the calendar, in the black hole of rules, for industry. His company was called

Multinational Business Services. Because why broadcast the objective to every yokel on the ground? (The equivalent would be Superman operating under the name Metropolis Flight & X-Ray. The Hulk going around as Green Redesign.)

He spun off his own nonprofit, a foundation called Federal Focus. As of 1993, the year S. Fred Singer began running cigarette errands, Philip Morris was describing Jim Tozzi in espionage terms: as an asset. They paid their asset's company $40,000 a month. Next year, in the manner of an NBA franchise, they capped this annual salary, at $610,000. Jim Tozzi wore tailored suits, ranged the capital by limo—from that Ohio grocery, he'd become a man to whom the Philip Morris senior vice president could send a $200,000 check, with the sentiments of a principal at graduation: "Best wishes for your continuing success." It was then that Jim Tozzi rehabilitated Frederick Seitz.

DR. SEITZ HAD STARTED a foundation of his own: the George Marshall Institute. During climate hearings years later, Senator John McCain came across some colorfully incorrect data—a product of the Marshall Institute. The group had taken its name from one of the Second World War's most beloved generals. This momentarily puzzled Senator McCain.

"George Marshall was a great American," he announced to the hearing room. "I think he might be very embarrassed to know that his name is being used in this disgraceful fashion."

A moment later, picturing the razor and the foggy mirror, the senator asked, "I wonder how these people shave." Thirty minutes later, still bugged: "George Marshall must be turning over in his *grave*."

In June 1993, Jim Tozzi wrote Philip Morris. "Initiated discussion with Dr. Seitz of Rockefeller University to support [our] position" on secondhand smoke.

It took six months. Tozzi wrote again in September: he was busy arranging "a meeting between Dr. Seitz and Philip Morris officials." Seitz began work. And in the slushy, deliciously vacant days between Christmas and New Year's, Jim Tozzi received an advance copy of Dr. Seitz's report.

To a master of the chessboard, it must have come as a shock; like suddenly losing your queen. Seitz's paper wasn't about cigarettes at all. It was about denying climate change.

"But," Tozzi told Philip Morris brightly, "it mentions Environmental Tobacco Smoke."

That's all it took. A single mention—fifteen words in thirty-four pages. To fire up a machine so big it could kick you out everywhere.

Jim Tozzi got busy. He organized a federal symposium: officials from the EPA, Defense, Energy, executives from the private sector. They'd come to hear climate change wasn't true, and while there passively inhale the smoke.

He ballyhooed Frederick Seitz in his Federal Focus newsletter. He recommended Philip Morris donate to Dr. Seitz's Marshall Institute. But please—for appearances' sake, for all good reasons of safe financial conduct, route it through Jim Tozzi.

This is why deniers can often say they are "not supported by" or "receive no funds from" industry. The money heads out on a little transalpine adventure, like a deserter in Hemingway. Exits the corporations, swaps passports and cools its heels at a mountain lodge like Jim Tozzi's Federal Focus, before hiking down into the denier's bank account. It was by this route, Tozzi told Philip Morris, that denial groups like the Marshall Institute retained their "considerable credibility."

Carried away, Tozzi started referring to it as "the Federal Focus/Seitz Report." The paper—Dr. Seitz's longest individual testament on climate change—was a co-release of Jim Tozzi's Federal Focus and the Marshall Institute. No one in DC seemed to find any of this odd, scientist and sharpie together. Just the way business got done, in a city becoming blind to reflections of itself.

As *Vanity Fair*'s Mark Hertsgaard wrote, Dr. Seitz gave both denials—climate and tobacco—"scientific respectability" and "instant credibility."

And Philip Morris saw to it the report wouldn't just touch America. The Seitz paper got dispatched to bureaus worldwide: The official plan was, "utilize Seitz piece to stimulate science debate." And everywhere debate got stimulated, climate took a shellacking.

And now time's ravages fell away. In Philip Morris memos, strategies, and press releases, the physicist was rehabilitated from "Dr. Seitz is quite elderly" to "Dr. Seitz is a world-renowned scientist." Made a full recovery from "not sufficiently rational" to "is one of the most distinguished scientists in the United States." Because to enhance your reputation in denial, you have only to deny.

He was installed on the seven-member Advisory Board of Philip Morris' Junk Science group. As well as serving as chairman for Dr. S. Fred Singer's outfit. He accepted an offer to travel Europe on behalf of Philip Morris. Scientists later worried that S. Fred Singer had corrupted the older man. "The community of established, highly respected senior scientists," a physicist explained, "can hardly stand Fred. And they think that Frederick Seitz, former chief of the National Academy, has just been taken in." But if they'd checked chain of title, Frederick Seitz was Reynolds Tobacco, to Reverend Moon, to Jim Tozzi, to Philip Morris. Any moral oxidation had taken place long ago.

With a sad outcome. Whatever else he achieved, here is what headlined the scientist's 2003 *Times* obituary. "FREDERICK SEITZ, PHYSICIST WHO LED SKEPTICS OF GLOBAL WARMING. . . . In the 1990s, as consensus about global warming was building, Dr. Seitz's contrarian views became a spark for debate."

There was a happier outcome for Jim Tozzi. His Philip Morris salary cap was lifted to $780,000. His brass telescope became a Dupont Circle voyeurism landmark. He became part-owner of a high-end capital restaurant and a winery, Villa Tozzi. And he'd made good on the original dream: He sponsored his own music combo, the Federal Focus Jazz Band. For student musicians—children like himself, who grew up loving jazz.

AN EXCEPTIONAL CASE

IN HIS GLUM FINAL BOOK, GLOBAL WARMING ANTECES-
sor Svante Arrhenius explained, "Statesmen are only in exceptional cases
interested in nature and science."

Al Gore had enrolled in Roger Revelle's climate class at Harvard—
where the professor, clad in size-fifteen shoes, had escorted students down
the hallways of charts and figures. "The implications were startling," Gore
later wrote. "Professor Revelle's study taught me that nature is not immune
to our presence, and that we could actually change the makeup of the entire
earth's atmosphere in a fundamental way."

Gore graduated, served in Vietnam, got himself elected to Congress—
and twelve years later, former student called former teacher as lead witness
at the first-ever Washington hearings on climate change.

That is, here was a best-case story: education to action. Gore imagined
Revelle's testimony would motivate fellow lawmakers. ("I assumed that if he
just laid out the facts as clearly as he had back in that college class," recalls
Gore in *Earth in the Balance*, "my colleagues and everybody else in the hearing room would be just as shocked as I had been.") It did not. Gore was so
committed he'd flown in 1988 to the bottom of the world, taken an eight-hour
hop from New Zealand, hiked up the side of a Transantarctic mountain, and
stood at shivering attention while a scientist showed him the ice cores; he'd
stepped into Roger Revelle's world. He voyaged by submarine to the opposite
pole. "I slept under the midnight sun," Gore writes, "in a small tent pitched on
a twelve-foot-thick slab of ice floating in the Arctic Ocean." Here was another

irreplaceable zone of sharp peaks and vistas. "And then I was standing in an eerily beautiful snowscape, windswept and sparkling white."

Gore's *Earth in the Balance* became the best-selling environmental work since 1962, when Rachel Carson's *Silent Spring* kicked off the whole please-don't-poison-us movement. That is, here was an entirely honorable work, from an entirely surprising source. With an astonishing outcome: a global-warming author in the White House. Gore was Arrhenius' exceptional case.

So early on, people couldn't fathom why he'd taken the trouble. It was "widely considered to be Mr. Gore's midlife crisis book," reported the *Times*. And of course, it *was* a midlife crisis book. But for the book's subject, rather than its author.

And in December 1997, Gore's wish was being granted. The world's nations were gathering in Kyoto to limit greenhouse gases.

THAT FALL, industry raised a stink and a shout. Thirteen million dollars' worth of anti-Kyoto advertising. The TV weather felt unseasonable— political spots in a nonelection year. NPR put the campaign's effectiveness in Godzilla territory: It was "a monster."

The Global Climate Coalition worked the international angle, Frank Luntz's emotional home run. TV ads scored for two voices, the Ralph-and-Alice you might have over dinner, or with a spouse in your head, getting more and more steamed.

FEMALE VOICE: "What do you know about the United Nations proposed climate treaty?... Countries like China, India, and Mexico are exempt." MALE VOICE: "We pay the price and *they're* exempt?" FEMALE VOICE: "It's not global, and it won't work." One environmental leader heard the Godzilla steps on her own staircase. "I knew we were in trouble," she said, "when my 13-year-old son was swayed by a commercial."

Click over to globalwarming.org—skeptics got their hands on the domain names first—and you'd find tips for educating the young. "Explain to them that at the time dinosaurs lived, the atmosphere had carbon dioxide levels that were at least five times greater than what we now have," advised the site. "These high levels of CO_2 contributed to the era's rich vegetation." By winter, the *Times* reported, lobbying had reached a "fevered pitch."

It's a savage pleasure—it is science hate reading—to cast your eyes back on the Global Climate Coalition. While the name suggests a group working the environmental side of the street, the coalition was really a manufacturer's clubhouse. You'd shake hands with the Aluminum Association (sponsor of Iron Eyes Cody and Keep America Beautiful), wave to the National Coal Association and Western Fuels (coal had created the ICE group), grunt at Ford and General Motors (the showrooms), share a drink with Chevron, Exxon, and Shell (the pumps).

Years later, the *New York Times* ran it as a page-one story: the coalition's own researchers had never disagreed with mainstream science. They'd looked into denier explanations—of course they had. Which didn't hold up. (Their lead scientist recalled trying "unsuccessfully" to make business leaders accept "scientific reality.") The coalition researchers' position, as of 1995, was just what you'd find in the science journals: "The Greenhouse Effect and the potential impact of emissions of greenhouse gases such as CO_2 on climate is well-established and cannot be denied."

Kyoto negotiations ran the first ten days of December. One hundred sixty nations, attendees in the thousands. The conference center had the angled, racy, many-decked look of a cruise ship; a boat sailing above talk.

Inside, it felt like a corporate retreat—faces you knew from the elevator bank, the coffee dispenser, refreshed by the novel setting. "It is as if a large chunk of the lobbying community," Senator Joe Lieberman told a journalist, "has been transported from Washington to Kyoto for two weeks."

Geologist Jeremy Leggett attended as a reporter. He saw high stakes: the "defining battle." If the deniers lost and "a protocol—any protocol—with legally binding cuts was negotiated," writes Dr. Leggett in *The Carbon Wars*, "they would probably always be on the defensive." Denial's future would be a sloggy retreat, "one trench after another." Environmentalists were finally throwing their 95-yard touchdown pass.

First week, talks crested, then stalled. Lists elongated, pages crumpled.

In the final days, Vice President Gore flew to Kyoto. He'd been waiting since 1965 and Roger Revelle. He phoned the Japanese prime minister at

2 a.m. He gave the American delegation new instructions: do what you must, return with a treaty.

The conference blew through its final day. It was the all-nighter—with the home flights booked, the paper due, the chairman gaveling in article after article. Delegates sacked out on tabletops, on the floor.

"We are about to blow any chance of an agreement," the chairman announced at 4 a.m., "in all six languages." The pace sped. At dawn, half the treaty remained. After nine, one last treaty point. The chairman announced, "I will now forward this protocol for adoption." Delegates embraced and applauded; some wept.

It had taken 103 Decembers—from the Christmas Svante Arrhenius sat down to his mood-cracking, pencil-and-paper calculations. Statesmen, he'd said, cared about science only in exceptional cases. Here *was* the exceptional case, and politicians had just set the first international limits on the blanket, on his overcoat, on the hothouse.

THERE ARE TWO kinds of magic. The Vegas kind is stage magic—special effects in huge venues. Then there's the close-up kind, where a performer works person-to-person, balanced on a high wire of pure skill.

Presidential campaigns are Vegas magic. Working the Senate for votes is close-up stuff, and rewards an extraordinary performance.

President Clinton's job would now be close-up magic. Rolling his sleeves, wading in, winning Senate approval vote by vote. Five months earlier, senators had passed a nonbinding resolution 95–0 against any treaty. Conservative leader Chuck Hagel made a gruff, Western-sheriff promise while still in Japan: "We will kill this bill." Back in Washington, an outraged Newt Gingrich declared the treaty "an outrage."

But all the lobbying hadn't taken. Newspapers ran polls: 65 percent of Americans believed the United States ought to make emissions cuts "regardless of what other countries do." Which, for all the sighs and nearlys in these 103 years, is an occasion for pride.

Kyoto had acted in the capacity of a heavyweight promoter. Scheduling the capital's next twelve rounds of argument.

"Now the administration has a larger fight on its hands," the *Times* wrote that December. One that would "help define not only budget negotiations next year, but the 1998 Congressional elections and even the next Presidential race." Action, the *Times* foresaw, would be "brutal."

The president felt confident. "But," Clinton said, in a special Oval Office interview, "it will require a very disciplined, organized, coordinated effort." What the *Times* called "an extraordinary level of presidential energy."

Ten days later, just before Christmas, the paper drew cheer from national opinion polls. "President Clinton's biggest challenge in the new year," the *Times* wrote, could simply be "leading the Senate where the public wants it to go."

Stay a moment here. This is the place where one kind of future becomes another.

Three weeks later, an abrupt climate change hit Washington. Something nonlinear; it posted on the internet at half past nine, January 17. "Blockbuster Report: 23-Year-Old, Former White House Intern, Sex Relationship with President."

It really is impossible to imagine a political issue with worse timing than global warming. 1998 represented an unexpected outlet for presidential energy.

Instead of *Kyoto, emissions, ratification*, the year's words were *Monica Lewinsky, blue dress, impeachment*. And, I suppose, still *emissions*. Warming at the beginning couldn't hold a candle to the Arrhenius theory of panspermia; what resources did it have against a blue dress? (The garment had been spared from dry-cleaning oblivion on the advice of disgraced O. J. Simpson detective Mark Fuhrman. Life is connections: the detective and a friend of Lewinsky's shared a literary agent.) 1998 also turned out to be—on a channel nobody was still watching—the warmest year in recorded history.

"ARTHUR ROBINSON IS A GOOD SCIENTIST"—ARTHUR ROBINSON

IT IS THE GREAT IRONY OF FREDERICK SEITZ'S LIFE. HE deplored PC—"what we did to the Indians and so forth," by his definition—but the scientist helped create something with true multicultural appeal. Fist-waving progressives could yell the stat at former Vice President Gore. This happened at a 2009 Chicago book signing. Gore shielded by a hedge of black-suited security men. "This is a eugenics operation (don't touch me). Look it up! 31,000 scientists are speaking out against your global government agenda! (You do not need to assault me, sir.)"

And it warmed the cockles of the Right. That same year, complaint journalist James Delingpole sat in on the BBC audience-and-argument show *Any Questions*. Delingpole has the look of a malicious science experiment: as if someone took a high-strung Oxford undergraduate and, for whatever cruel reasons, crossbred him with a river eel.

One audience member ventured an opinion. "I would like to say that the media finds a need to polarize this discussion," he said. "But the science is quite unequivocal."

Delingpole cut in. It is a strangely passionate voice—enthusiastically stricken. "I can't accept that. Because the science is *not* settled. 30,000 scientists have signed a document, saying that we do not *believe* in this anthropogenic global warming."

"Thirty thousand scientists?" confirmed the host.

"From around the *world*," Delingpole insisted.

What force on Earth could find the intersection between James Delingpole and We Are Change Chicago? Between the gripe Right and the fist Left?

I was watching Sean Hannity, in 2007, harry an unfortunate guest who'd come on the broadcast to support science.

Hannity did the Fox News lacemaker thing: Started plain, then embroidered as he grew more riled. "There are a lot of credible scientists that disagree with you that are on the record," Hannity said.

Moments later, and with no time for additional field work, those résumés had strengthened. "I have a stack this thick," Hannity said. "Our leading scientists have decided—climatologists, they're all saying it's bogus.... Scientist after scientist like Ted Ball saying it's bogus.... How do you respond to that?"

The science advocate did not waver.

"I've never heard of him," she said. "No credible scientist—"

Hannity consulted a fact sheet on Dr. Ted Ball. "He's the *first* Canadian PhD in climatology. He has his doctorate of science, the University of London. Was a climatology professor at the University of Winnipeg." Throwing to commercial, Hannity added, "Checkmate."

Dr. Ball was so impressive, a few weeks later Hannity invited him on the show. Where his name turned out to be Timothy Ball, not Ted. "Consensus is not a scientific fact," explained Dr. Ball—then backed it up with the Oregon Petition number. "I could get you to a site that Fred Singer has got up, where he's got the signatures of 19,000 scientists."

It wasn't just the Timothy versus Ted part. Dr. Ball was not, as he liked to say, Canada's first climatology PhD. There were plenty before his doctorate; Canada is a big, science-loving country. Which was in geography anyway. At the University of Winnipeg, he'd been a professor of geography.

Ball eventually sued the *Calgary Herald* for printing the facts above. This entitled the paper to introduce all kinds of pulverizing material into evidence. To note that Tim Ball never published on carbon dioxide in the science press. (Places where he did publish were called things like *History and Social Science Teacher* and the powerhouse *Manitoba Social Science Teachers Journal*.) And that Dr. Ball was "viewed as a paid promoter of the agenda of the oil and gas industry rather than as a practicing scientist." You can't sell that kind of publicity—and can't pick that kind of fight, when your opponent's best response is showing you a mirror. As the physics professor John Farley wrote, "his reputation in ruins, Ball dropped his lawsuit."

The signatures bewitched industrialists—drew a teasing hand down their faces. Exxon Chairman Lee Raymond uncorked them on an unruly shareholder's meeting as room Xanax.

"This is a petition signed by seventeen thousand scientists," Raymond said, a slide projecting the Oregon document that Frederick Seitz had helped arrange. "So contrary to the assertion that's just been made," the CEO continued, "there is a substantial difference of opinion as to what exactly is going on."

The same ink offered automakers a smooth, high-octane ride. Here is General Motors Vice Chairman Bob Lutz on the *Colbert Report*. "Like many noted scientists," Lutz began, "I do not believe in the CO_2 theory."

Colbert deadpanned that the heat had to be coming from somewhere— perhaps the national stockpile of toaster ovens. The GM executive gave a curt, boardroom nod. "In the opinion of 32,000 of the world's leading scientists," he said, "yes."

Dr. S. Fred Singer marched the signatures through the Capitol. (The *leadershipness* of an S. Fred Singer press release: "REPRESENTATIVE OF 17,000 SKEPTICS TESTIFIES ON CLIMATE CHANGE TO THE U.S. SENATE." In a movie, you'd shoot from a low angle, as Dr. Singer mounts the dreadful steps.) Senator Chuck Hagel—later secretary of defense—cited the document as grounds for opposing the Kyoto treaty. "Nearly all of these 15,000 scientists," the senator said, "have technical training suitable for evaluating climate research data." This was nowhere near true—in fact, almost none of them did, and the majority weren't scientists at all.

For reporters around the world, it became the heavyweight datum you sent into the ring—flattening opponents with a single blow. It's "been cited," writes England's George Monbiot, "by almost every journalist who claims climate change is a myth." For the environmental writer Jeff Goodell, the effect was more structural—"one of the pillars of global warming skepticism." You heard tell of it on NPR, in the *Wall Street Journal*, the *Washington Post*. And of course on the internet, where it was an everywhere fact, bulking out a conservative argument, or just keeping you company during the long nights: "Global Warming—It's Not Just Me Anymore."

So you'd expect the second Bush White House to show the signatures love. When the Oregon Petition was a gray-muzzled ten years old—Dr. Seitz had released it in early 1998, to chase away the Kyoto Accords—there was a morning press conference in the Brady Briefing Room.

"More than 31,000 U.S. scientists," a reporter began, "have now signed a petition rejecting global warming. My question: What is the White House reaction to these 31,000 scientists?" And here the signatures' appeal went cold. "I would say everyone is entitled to their opinion," answered Press Secretary Dana Perino. ("That's all?" asked the reporter. Answer: "That's all I'm going to say.") Basic veterinary science. The White House heard woofing, and recognized a dog.

RACHEL MADDOW WELCOMED Arthur Robinson to her broadcast in October 2010. "He's a man who runs something he calls the Oregon Institute of Science and Medicine," the host explained.

"I say it's 'something he calls' that," Maddow continued. "Because despite the lofty name, of the eight people listed as faculty of this Institute, two are dead, and two are Mr. Robinson's sons. Plus there's *him*."

The Oregon Institute is the raggedy far edge of climate denial; it's the place you find when you travel all the way. Jeff Goodell once interviewed Dr. Arthur Robinson. He was a biochemist on bad terms with evolution— which he described as "a hypothesis that is unproven." (Just interview good manners. Alone and unsupervised at his keyboard, Dr. Robinson calls natural selection "pseudo-science" and "pornography.") Here was another cautionary tale. About the people who have slipped in, reached down, and redirected our conversation on warming.

His institute did not feature tolling bells, teeming footpaths, or any student. It was a barnlike structure on the researcher's lawn. Goodell called Robinson "a unique breed of scientist." Someone you'd trust half an hour, then every minute a little less.

That's the thing about the proofs of denial. They're put together with just enough paint and nails to withstand one look; they collapse on look two.

I once spent an afternoon on the New York lot at Warner studios. Four lovingly recreated blocks—Manhattan simulated and amplified. (You've seen the streets on *Seinfeld* and *Friends*.) And as long as I never tried a door, rapped a lamppost, or walked down any subway steps, it was impossible to convince my body it wasn't back home, that I was in fact in the middle of a desert.

This is the gate where denial lands: our brains are designed to fill in. Catch a scrap, complete the picture. See four streets and populate a city. Hear *petition*, *climate*, *thousands*, and understand a world in the grip of science debate.

So I was excited to see Arthur Robinson on the Maddow show. I'd read about the Oregon signatures for years. He wore an open Carhartt jacket, and had something of Clint Eastwood's craggy magnetism—a person with the confidence of having been looked at their whole life with pleasure. Another figure like S. Fred Singer. Often, with deniers, what you see is the wreck, the residue of what had once sailed out of the harbor as promising. Rachel Maddow quoted odd things Robinson had written—shutter the public schools, AIDS is a myth—and Robinson shot back that they were smears.

For eighteen minutes it was riveting TV. Twice Robinson told Maddow, "I'm a scientist. And a good one." The compliment economy, with the sadness of its trade imbalance. What you don't get from outside, you must manufacture yourself. Also one of the sudden TV intimacies. When the simmering internal conversation—everybody's I-am-undervalued monologue—boils over into public view. I was seeing the spot through which all those TV fights and articles and speeches had entered the world. Arthur Robinson is the man who prepared the Oregon Petition.

THIS IS A HORROR STORY. It's also about Linus Pauling, a scientist for whom the waves broke just right. The only person in history to win two unshared Nobel Prizes.

By the time the men partnered up, Pauling had reached a strange elevation in the cultural sky: he'd become a "walking god." (These aren't the words of a TED groupie, or some labcoat sniffer; it was a director with the National Cancer Institute.) Between acceptance speeches, Pauling taught. Arthur Robinson enrolled in his 1959 Caltech chem class. "I don't know which is more fun," a friend told Arthur. "Watching Pauling lecture, or watching you watch Pauling."

A tight grader, Pauling gave Robinson A-minus. He later called him, "My principal and most valued collaborator."

Robinson completed his doctorate at San Diego. So brilliantly, he was awarded a position on the faculty; a future stacked up safe and orderly ahead of him. The soft September to May of the academic career.

That's when Linus Pauling sharked back into his life. Together they founded the Linus Pauling Institute of Science and Medicine. Later, when everything had ended, both men would be left to wander the sadness and rubble of a legal action. Robinson naming that thing on his lawn the Oregon Institute of Science and Medicine was defiance: the comeback when the offense has slipped from every mind but yours.

Here's how dashing a figure Linus Pauling cut then. He once spent an afternoon picketing the White House. No more nuclear weapons! Then buttoned into evening clothes, breezed through his own pickets, dined with President and Mrs. Kennedy under chandeliers. (Jacqueline Kennedy joked with the scientist: whenever their daughter saw protestors, she asked, "What has daddy done now?") Pauling contacted fellow laureates; they agreed to sit on his board. Arthur Robinson must have experienced one of those spells of lucky career vertigo—still in his early thirties, but traveling among the stars. It's in the index to a Pauling biography: *Arthur Robinson, surrogate son.*

LINUS PAULING IS the spirit at the end of the CVS aisle: the phantom grinning as our carts fill with Airborne, Spectrum, and Emergen-C. Pauling published his best-selling *Vitamin C and the Common Cold* in 1970. One chalky pill could put an end to all that winter sogginess.

It's like the montage from the movie: the doohickey gets invented, then rolls off the line, past customers digging in their pockets, then flies off the shelves.

Vitamin makers called it the Pauling Effect. "Sales doubled, tripled, and quadrupled," writes Thomas Hager in *Linus Pauling and the Chemistry of Life.* "Manufacturers began building new factories to keep up with the demand.

The vitamin industry had never seen anything like it." By the mid-seventies, nearly a quarter of Americans were staving off colds with vitamin C.

These were battlefront years, when President Nixon had declared a national War on Cancer. And maybe Linus Pauling and Arthur Robinson could be the troops leading us to victory—tipping back their helmets, having rolled that big vitamin C tank all the way to the White House lawn.

Results weren't quite there. So Linus Pauling nudged them. He discovered in himself an inner Fred Singer. He grew strong in publicity, granting interviews everywhere. To the *National Enquirer*, the *Mother Earth News*, even a sub-*Enquirer* with the werewolf name *Midnight*.

The laureate made great copy. Linus and his beloved wife Ava Helen—married half a century—swallowed ten daily grams of vitamin C. And, to be frank, he bent the truth a little. The Paulings still caught bugs. ("Both Mama and I have suffered from colds during the last few weeks," Linus wrote his children. "Never bad, but dragging on.") With reporters, however, he walked tall and dry-nosed. "I always stop the cold," Pauling would say. "In fact, I no longer talk much about the common cold. It's moot."

These early days play like a science buddy picture. The two researchers meeting fate shoulder to shoulder, paving over shortfalls with their wallets. Arthur became facility president, and toward the end of the decade helped land the institute millions in federal funding.

That's when the whole deal went to hell three ways. First, Arthur Robinson turned out not to be the sort of person you leave in charge of anything. There's the listening and resolving kind of boss—Tina Fey on *30 Rock*, any of the good ears on *The West Wing*. And there's the Robinson kind—who yells and then fires you.

Second strike was politics. Pauling was a liberal. (A tribute to Ava, the couple's progressive wing. Pauling embraced certain causes, he said, "in order to retain the respect of my wife.") In 1978, Arthur came out as conservative: delivering the first political speech of his life at the libertarian Cato Institute. Capitalism, property rights, the no-lifeguard swim of laissez-faire—these were the fiscal policies of the Lord. The speech, one biography shudders, represented "everything that Linus and Ava Pauling had always detested."

Last and worst, Arthur was conducting experiments on cancer and vitamin C.

Linus came to believe—who can say why?—that fatalities could be reduced 75 percent with vitamin C alone. The road to science cuts through the pet cemetery. Arthur took mice; blitzed them with radiation—a nightmare, ozone-layer dose. Now they had skin cancer. Then Arthur fed his tiny patients the rodent equivalent of ten grams of vitamin C.

Worst case. It gave them *more* cancer. (Arthur offered an explanation years later. Vitamin C nourishes all systems, tumors included. "Unfortunately," he told PBS, "you're feeding the cancer along with the animal.")

Linus' beloved wife Ava had been diagnosed with stomach cancer. She was dying. The couple turned their back on traditional remedies. Ava would now be living proof—rescued by vitamin C alone. "More than scientific," Pauling's biographer writes, the vitamin had become "a personal cause."

And here was Arthur, in the clench of the family dynamic. The adopted son telling the surrogate father that his treatment of the mother was poison.

A few weeks after the God is Business speech, Linus Pauling burst into Arthur's office. Everything was over; he was the police chief demanding gun and badge. Arthur must turn over his building pass, surrender all research material, exit the premises. Pauling himself would be taking charge of the mouse project. (The older scientist did take charge. The animals, with whatever they verified, were all destroyed.)

When Arthur resisted, Linus pointed to a magazine cover. The Nobelist was waving around the loaded weapon of fame. "Magazines will write bad articles about you, you will never work in science again," Linus said. "I will see to it. I will see that you never work in science again."

Pauling was the name on the door. Arthur's work was impounded: Lab notes, shelves of books stretching all the way back to grad school. Arthur's past was carted up and rolled away. In *Forces of Nature*, Thomas Hager gives this the feel of a vocational murder. At one stroke, Pauling was "ruining his academic and professional career."

The self-defense of last resort is the gavel. Arthur sued. (There were tricky emotional crosswinds. A trustee told Arthur: "He views you as an ungrateful son.") It became the sort of battle that combusts everything—where victory consumes the prize.

Pauling sacrificed his disciple to the press. Arthur Robinson was incompetent, an amateur, a bad scientist, a quack. The suit also did, Thomas Hager writes, "incalculable damage to the Institute's reputation." Attracting reporters, scaring away grant committees. After five years, thicker smoke hung over the Robinson side. Linus had "callously demolished" the young scientist's career. Arthur accepted $575,000, the price for having one sort of life ended.

He rolled the settlement into his Oregon Institute. Arthur told the journal *Nature* it was running on "three employees and hope."

ARTHUR AND THE WORLD

AVA PAULING DIED IN 1981. FOR A LONG TIME AFTER, Linus wept at the mention of her name. He carried on one-sided husband-and-wife conversations, spoon tinkling as he stirred vitamin C into his morning juice.

He claimed his vitamin would enhance intelligence. Treat herpes. When AIDS came around, you can imagine the Linus Pauling recommendation.

He was eighty. Then he was ninety. Science has a stone heart. By the end, he was no longer the charismatic double Nobel with White House dinner plans. He was the oddball whose coat pocket rattled with supplements. "From being a public figure of high stature with an idealistic philosophy," Oxford's R. J. P. Williams wrote in *Nature*, "to being viewed as a lonely crank, is indeed as great a fall as in any classical tragedy."

ARTHUR AND HIS WIFE established the Oregon Institute on 350 acres of fences and long sky views, under the smell of fir trees.

Lauralee Robinson had daily commitments: she homeschooled their children. But Arthur was at loose ends. He was the man who has experienced the wickedness of the city, then retreats to the wilderness with scars and prohibitions. All other places became the same place: "the World." Where he was well-known, but the wrong way. Fame precedes you through a door and softens up the room. Notoriety sticks the hinges. What do you do, when you can't go back?

He arrived at a solution. Take the words *Linus Pauling*. Add a minus sign. He became, as another biographer wrote, the "vocal, perennial adversary."

Linus had campaigned against nuclear weapons—the occasion for Nobel number two. Arthur's institute published a self-help guide called *Nuclear War Survival Skills*: mushroom cloud dos-and-don'ts.

Pauling was a proud public atheist. Robinson burrowed in the reverse direction: "I do not know of any verified scientific facts," Arthur said, "that are inconsistent with the literal truth of every aspect of the Bible." He wrote of sharing this belief with Sir Isaac Newton, whose access to verified science fact had been discontinued in 1627 by natural causes.

Pauling was the outspoken kind of Democrat. Arthur drafted the civil defense plank for the 1988 Republican Party.

He found his way there via a book he'd written. *Fighting Chance— Ten Feet to Survival*. Ostensibly, this was the blueprint for a nation of fall-out shelters.

The actual book is lots weirder. His collaborator, Gary North, has enriched the public record with another unique career. Son of an FBI man, former staffer to the Tea Party's Ron Paul, husband to the daughter of Rousas John Rushdoony, bearded patriarch of the homeschool movement. North had previously written *None Dare Call It Witchcraft*. This revealed the threat to America posed by acupuncturists and the horoscope chart. North also expressed nostalgia for some Bible-age legal traditions. "That modern Christians never consider the possibility of the re-introduction of stoning," he wrote, "indicates how thoroughly humanistic concepts of punishment have influenced the thinking."

Half a million copies of their unusual volume got distributed. Since any Armageddon will in a flash eliminate all smirks and foes, you don't have to read very far to recognize such works as wish fulfillment. "Heaven," the Robinson and North study argues, "is a nicer place than anywhere else on this sin-cursed Earth, too. It's even nicer than a three-bedroom house with air conditioning. Yes, even if the mortgage is paid off!"

By its dark final pages, the book has left flashlights and calamity shoveling far behind. "Our problem is not lack of hardware," the men advise.

"Our problem is a lack of faith in God, as revealed by our lack of obedience to God's laws. We will learn how costly such rebellion is." The World, Arthur believed, had it coming.

In November 1988, Lauralee Robinson took to bed with a stomach complaint. Next morning, the children couldn't wake her. A single-day fatal illness: acute hemorrhagic pancreatitis. At forty-seven, Arthur had become, like Linus Pauling, a widower. With a windblown institute, a family classroom, and six motherless students to educate.

IF HE HAD NEVER DRIFTED into climate, this would be the source of Arthur's renown. Educating six Robinsons. He became a columnist with the movement publication *Practical Homeschooling.* Public schools were another of the World's shortcomings. "Moral sewers," Arthur explained, "socialist institutions," providing at best "institutionalized child abuse."

It's proof of the Magellan nature of opinion. Arthur had become so conservative, he'd sailed all the way round and washed ashore on the island of the hippies. He declared a candy moratorium: "Sugar diminishes brain function and increases irritability." He even drew the blinds on the World's window, television. "I know one man who took his out on the back porch and shot it. This is extreme," Arthur wrote. But "destruction of the thing is the best option."

The aim was a big HIGHWAY CLOSED sign. No traffic between his family and the World, where everything that happened to him had happened. "Home schooling is no more than a tool," Arthur explained on his teaching website, "to keep a child out of the World."

Home became a stalled place, a world of frozen clocks. Christmas decorations twinkled half the year. "Traditions have become somewhat unusual," Arthur admitted on his website, which records the results of a particular experiment. A voyage deep into Robinsonness: what happens when a highly intelligent personality is cut off from the herd.

"Have your daughter at your side 24 hours per day, seven days per week," Arthur advised fellow homeschoolers. Life would be "immeasurably enriched." Same went for sons. Arthur and his youngest, the baby, had become "virtually inseparable" in the years since his wife's death. "I even moved him into bed with me, so that I could watch at night to see that he was okay," Arthur wrote a decade later. "Well, he is still there. ... Add to this unusually close relations with all six children and I have had more happiness in the past ten years than most people dare to even dream about."

College in this environment was a chronological menace, a scheduled heartache. "Eventually," he promised readers, "this threat will be removed by homeschool universities."

When the oldest boy did matriculate—Oregon State University—there were bitter complaints. Campus tradition was thoroughly non-Robinson. "I'm tired of wading knee-deep through homosexuals," his son said, "just to get to the library."

This lifestyle, which contains aspects of a tragedy, had become something to market. Arthur boxed up lesson plans, texts, method, turned his Oregon farm into product.

The Robinson Curriculum. Open the lid, out popped your classroom. Public domain reference works, beyond-copyright geniuses like Kipling and Shakespeare, chestnuts on the order of an 1892 Life of George Washington. ("He was kind to his slaves ... but quick to see that they did not shirk their work.") Also a personal copy of *Nuclear War Survival Skills* (because you never know) and the King James Bible. ("It is the foundational book of the curriculum.") It retailed for $195. Twenty thousand kits were sold the first four years. It became the nation's third most popular homeschool—hailed by *Practical Homeschooling* as "the famous Robinson curriculum."

And you can imagine how these words must have solaced the harried Christian parent: "The Robinson Curriculum is a product of the Oregon Institute of Science and Medicine."

I've seen video from this period and I was wrong: Arthur was not always handsome. He looks crabbed and depleted, as attractive as anyone else on a desert island. Only denial restored him.

A couple years later, the famous curriculum got supplemented; the complete works of Victorian novelist G. A. Henty. In the books, school-age heroes lit out for history's trouble spots, then subjected them to a British tidying. After the First World War, Henty was junked, in the manner of a disproved theorem. Arthur loved these stories. "Taken together," he said, "they constitute a superb course in world history and an education in some of the highest aspects of human behavior."

Also one of the lowest: racism. They were the distant relative who opens their mouth and silences the holiday dinner. I don't like quoting the texts, so here's just one. Henty on post-Civil War America: "The freed slaves soon discovered that their lot was a far harder one than it had been before, and that freedom so suddenly given was a curse rather than a blessing." (For additional flavor, Henty on people of color: "I did not think that the earth contained such monsters.")

A decade later—a perk of climate service—the Republican Party would nominate Arthur as a candidate for Congress. Ran five times, got skunked five times. The curriculum assisted his sinking. Arthur couldn't make sense of modern objections. He'd stood so long outside the World, his attitudes had got frozen, too.

ARTHUR HAD TRAVELED. From the free range of science to the snug fit of the Bible. From the Pauling Institute to the one outside his door. I wonder if he ever regretted Linus coming a cropper with vitamin C and those mice. But liking the self you are now—repairing it, retailing it—is the healthy, in-house version of patriotism.

Writing about science, Arthur's mood darkened. Linus had not treated him to the sort of experience that breeds faith in our institutions.

Now, when Arthur gazed past his fence, he saw enemies and intentions. The World reaching with thick soiled fingers for Oregon. Multiple-choice entrance exams were a conspiracy. "The SAT tests themselves," Arthur said, "are being used as tools for social engineering." Multiculturalism was another conspiracy. "This word has been brought to us by the dead hand of government and by political agitators." TV and print news, both conspiracy puddles—"Shallow sources of propaganda."

Conspiracy was Arthur's vitamin C. His cure-all and explain-all. Driving east one January—a field trip with the entire Robinson student body—Arthur was approached by two policemen and a social worker. "The state of New Jersey made a grab for my children," is how Arthur understood this. They had been notified of five eerily silent kids, sitting alone in a pickup truck, in the dead of winter. But Arthur recognized conspiracy. "When a child is seized," he explained, "these programs provide over $100,000 per child." His family thus represented "at least $500,000 to the social services industry of New Jersey."

Big science had a fishy look. Françoise Barré-Sinoussi and Luc Montagnier eventually shared a Nobel for identifying the HIV virus. Arthur had previously diagnosed conspiracy.

"AIDS may be little more," he explained, "than a general classification of deaths resulting from exposure to homosexual behavior." This was 1994: the year Tom Hanks won an Oscar for the AIDS drama *Philadelphia,* and the virus had become the leading cause of death for all Americans aged twenty-five to forty-four. "Only government reclassification of more and more disease types as AIDS," Arthur wrote, "has kept the number of victims at politically necessary levels."

Life is a repertory company: you age your way up and play all the roles. When Rachel Maddow quoted these remarks back to Arthur, he became Linus Pauling, on that long-ago day in his office: red-faced and sputtering, confronted with results he didn't like.

"You are trying to smear me," Arthur said. What does it mean, when smearing and quoting you have become the same thing?

The ozone hole: another Nobel, another conspiracy. "International bureaucrats," Arthur wrote, "have even seized upon the 'ozone crisis' as a new reason for world government."

"There was a guy like Art Robinson, I'm guessing, in your town somewhere," Rachel Maddow told viewers. "Maybe on the outskirts of town.... There are a lot of unusual people in this country."

When Arthur took up climate, he uncovered the greatest conspiracy of all. A twilight struggle, against shadowy forces with "evil and dishonest plans."

For a man who'd lost his career because a boss would accept only one answer, this made sense. It's also what you get from S. Fred Singer: The world of science, of any closed system, as experienced by somebody not present at the meeting—when the backs of the more fortunate are what you see. "For those of us who understand our accountability to our creator," Arthur wrote, the myth of global warming raised a much larger issue. Americans already permitted "the slaughter of millions of children each year by abortion.... Will we now allow an even greater act of technological genocide?" he asked. "How much longer do we think God will stay His hand?" Because personality is a terrifyingly consistent thing. (It's the product of a long-range conspiracy inside the self.) Because since the eighties, Arthur had been waiting for God to make some time and straighten a few things out.

That is, Arthur wasn't necessarily the first person you'd turn to for a problem that affected the whole world. In a sense, he was the last such person: Arthur hated the World.

If this next is true, it's one of this story's richest ironies. The report Arthur wrote about warming—it accompanied his Oregon Petition—has become, he tells us, "the most-read scientific article on this subject." During the 2016 election, his petition became the most-shared climate data on Facebook. All the core-drillers, heat-measurers, and air-samplers—all the National Academy, the NASA—and the honor of greatest educator goes to the man with his own homeschool and institute. And the Oregon Petition was arrived at *by* a conspiracy.

JASON BOURNE'S
CRESTFALLEN ITINERARY

DAY TO DAY, S. FRED SINGER HAD LESS FUN THAN FREDER-
ick Seitz.

His was the grubbier portion. Singer pulled engine-room detail, deep
in denial's clanking and oily cellar. Seitz at least got to work alongside Jim
Tozzi, an elegant titan in his field. Singer's company was fellow shovelers.

Men like Kent Jeffreys, an Alexis de Tocqueville Institution adjunct
scholar. Who, in his non-Singer hours, could produce a style of free-market
nature beat poetry. "Special-interest bees," Jeffreys wrote, "will always find
the political honey, and the average citizen will usually get stung." And,
"Constantly claiming that the sky is falling, the environmentalist Chicken
Littles have become the geese that lay golden eggs." He could even wring
this music from the humble implements of the kitchen. "The result is a
layer cake of half-baked state policies, made according to a market socialist recipe."

In spring 1993, Dr. Singer co-hosted a two-day Junk Science conference. Secondhand smoke, ozone, and, of course, climate change. (Journalists from the *Times*, *Newsweek*, and *Washington Post* in attendance.) Philip
Morris' director of scientific affairs cruised the ballroom—"encouraging
speakers," he reported back, to raise cigarettes "as an example of bad science." This event was taped, then sifted and analyzed by Philip Morris
experts for clips to "educate policy makers on the use of bad science."

A year later, Fred Singer was issuing hot denials to host Ted Koppel on

ABC *Nightline*. "No one has ever accused me of being a mouthpiece for industry, or shill."

Well, maybe he just wasn't letting people get to know him very well. A week after airdate, the executive director at the free-market Alexis de Tocqueville Institution was acting as a science sports agent. Negotiating Dr. Singer's fee, as a mouthpiece for industry.

The Tobacco Institute needed a pro-smoking report. The executive director sent over the scientist's CV.

"Worth the $20K we discussed," an executive determined coolly. So Fred Singer and Kent Jeffreys wrote up a report, spreading the good word. Secondhand smoke plus a host of other pollutants were fine. Their study was unveiled via conference at the Rayburn Congressional Office Building. A nice ROI on that initial $20K. Per the World Health Organization's Derek Yach, the paper was "widely used by the tobacco industry."

And still Dr. Singer, in his public role, maintained the crusade for fairness and objectivity. Lamenting to the *Wall Street Journal* just after New Year's 1996, "Scientific critics of the climate doomsday scenario risk being attacked as lackeys of industry."

A few weeks earlier, he'd been prevailed upon by Brown & Williamson Tobacco (Viceroy, Pall Mall, and Kool, the misspelled green menthol) to do a little something extra on behalf of industry.

The public relations firm Shandwick International brought Singer their request. Dr. Singer needed persuading. He was becoming prominent, and reluctant about cigarettes. Shandwick's bright idea was to sneak the true work inside the luggage of something public-spirited. The publicists "recommended the concept of creating a 'myths list.'" Cigarettes would remain the "focal point"; Dr. Singer would "package" them with "four other issues."

I don't know if any Shandwick personnel ever served in uniform. But it's the classic soldier's strategy. Avoid the volunteering front row—too visible. And ducking and lurking at the back is too blatant. Your best hiding place is the middle.

When Dr. Singer published the "Top Five Environmental Policy Myths of 1995," the Second Hand Smoke Scare was packaged and accounted for at central number three. Shandwick put together an "aggressive media inter-view schedule." Local and national TV and radio. Dr. Singer especially enjoyed the call-in segments. The top slot on the myths list—flying out over all that aggressive media—went to Global Warming.

A year later, Dr. Singer panned a book. This is a piece he must have liked: it ran in four separate Reverend Moon publications. Names descend-ing from aboveground normal to the full-on shadows-and-illusions crazy of the basement. The *Washington Times, Insight on the News, The World and I, The International Journal on World Peace.*

The work, *The Betrayal of Science and Reason,* was in effect a denial yearbook. A record of just how Dr. Singer, Dr. Seitz, and company had been spending their semesters. It addressed the way scientific misinfor-mation was "disseminated by public relations firms and right-wing think tanks." (Like, from Dr. Singer's own recent personal experience, Shandwick International and the Alexis de Tocqueville Institution.) Such efforts were "successfully sowing seeds of doubt among journalists, policy-makers, and the public." It mentioned Dr. Singer by name.

The scientist's response was canny. "They cannot prove a conspiracy exists, and indeed there is none." I am a stone admirer of Dr. Singer, for the part of the sentence he decided should run first.

When he repeated the Top-Five stunt a year later, *The Betrayal of Sci-ence and Reason* weighed in at number five. (Both lists airing in Reverend Moon's *Washington Times.*) And how could a *book* be a myth? Top slot again went to global warming.

Meanwhile, Dr. Singer, and his subject, was becoming a star. In sum-mer 1996, Tom Brokaw threw to the physicist on *NBC Nightly News.* "Of course, there is a spirited debate in the scientific community," the anchor-man said. "With another view tonight, Dr. Fred Singer."

And there the former Siegfried was, bringing his message of hope to the climatically anxious. "What would happen if the climate *did* warm?" Dr. Singer asked. "Well . . . the climate has warmed many times in human history, and every time it warmed it has benefited the population." But no celebrating just yet. "We now have data of nearly twenty years. And actu-ally, they show a slight cooling."

You know Fred Singer. It didn't work. The *New York Times* had run it months earlier on the front page: HOTTEST YEAR ON RECORD.

THERE'S A REASON Dr. Singer shared this vital data through columns, broadcast, and the safe, nonjudgmental space of the press release.

Science journals are peer-reviewed. The back-and-forth can swallow years. Studies are road-tested, in the manner of a promising medication or new model car.

Imagine this as movies. Shoot a big-budget tentpole, and if the multiplex were peer-reviewed, your first cut would be screened for Kevin Feige or Jim Cameron. To catch botches you made, climaxes you whiffed. Gangster pictures would head to Martin Scorsese. Stiletto-heeled workplace drama to Shonda Rhimes. (Your anthem about getting baked would pass to Wiz Khalifa.) "After all, any one can *say* anything," Harvard's Naomi Oreskes points out. "But not anyone can get research results published in a refereed journal . . . . They must be supported by sufficient evidence to convince others who know the subject well."

The deniers pretended their distaste for science journals was in fact a rejection of clubbiness. Mainstream researchers were hogging the tables of contents. An Australian scientist held this to be a mistake—refereed science would have welcomed them.

"If you came up with an alternative hypothesis that disproved global warming," Dr. Andrew Pitman told BBC. "You would win the Nobel prize." For the classic science reason. "You'd win because you had replaced a century-old theory with something new." By not submitting research, deniers spared themselves the distraction and inconvenience of a trip to Sweden.

Skeptics determined their campaign must be conducted via journalism—on the opinion pages and talk shows, where data flows at a speed not subject to peer review. This is why Dr. Seitz self-published his Federal Focus paper. Why Linus Pauling gave interviews to the *National Enquirer* and *Midnight*. Why, when Fred Singer published in a journal like *Nature*, it took the form favored by *Titanic*'s Jack Dawson: a strongly worded letter to the editor.

In 2004, Naomi Oreskes examined 928 refereed papers published during the decade of peak Dr. Singer and peak Dr. Seitz. She could not find a single peer-reviewed study to dispute that current climate change was man-made. The piece (published in the refereed journal *Science*) made her name. "Politicians, economists, journalists, and others may have the impression of confusion, disagreement, or discord among climate scientists," Oreskes wrote. "That impression is incorrect."

There was a 2010 follow-up, in the (refereed) *Proceedings of the National Academy of Science*. Among active-duty climate researchers, acceptance of man-made change ran at 97 percent.

This is what Ralph Cicerone, head of the National Academy of Sciences, was getting at with his testimony to Congress: "I think we understand the mechanisms of carbon dioxide and climate," the scientist said, "better than we do what causes lung cancer." The seabed and the cryosphere; the oceans and desert; the present, the future, the incredibly distant and ferny past—"Global warming," Dr. Cicerone said, "may be the most carefully and fully studied topic in human history."

Science is never 100 percent. Science is the statue holding a magnifying glass over its own chisel. For a reference point, the Pew Research Center surveyed an unrelated proposition in 2009. The same ninety-seven to three figure matches scientists who believe and don't in evolution.

And still Dr. Singer could crouch, aim, take potshots. "The scientific basis for climate warming has come under attack," he told the London *Times* in 1995, "in the world's leading scientific journals."

This was Dr. Singer's edge. He never let himself get hemmed in: By what he said yesterday, by anxiety over what might be required tomorrow. Scientists had once worried about global cooling. Dr. Singer made light of this—promising the *Miami Herald*, "If we have a real cold winter, you'll see all those guys (new Ice Age prophets) coming back. I'm not kidding."

And you know Fred Singer: but in this case, he was proved right. One scientist *did* come back. And his name was Dr. S. Fred Singer. "There is a sure threat to human existence," Dr. Singer wrote, after the cold winter of 1995, "and that is the near certainty of a coming ice age. . . . Perhaps greenhouse warming can save us from an icy fate."

This work required yoga-level displays of flexibility. On sea level: "The best thinking might be that it is a lowering of the land surface." In 2000, Dr. Singer assured the *Wall Street Journal*, "The polar regions have not warmed appreciably in recent decades." Hearing that the melt had in fact opened Arctic shipping lanes, he made a bright observation to the *Washington Post*. "We spent five hundred years looking for a Northwest Passage," Dr. Singer said. "And now we've got one."

This is why the fun lay all on the denier side. The scientists were bound by rules, starched by convention—peer review, observable reality. Skeptics said whatever. The climate scientists had to build roadworthy reports— vehicles that could endure the highway through splashes and years. Deniers made one day smash-'em-ups. And the track was cleared and the stands emptied before the next race.

LET'S SAY you've shipped out as a denier.

You're in it for the action, the dollars, the travel, the fun. (Jim Tozzi, post-government: "I loved it. It was bloody. It was the action.") And you shade your eyes, glance up at those tall numbers. Your enemy is that 97 percent.

This is what pollster Frank Luntz understood. "Voters believe that there is *no consensus* about global warming," Luntz wrote, in his famous battle memo. "Should the public come to believe that the scientific issues are settled, their views about global warming will change accordingly."

And this is what Fred Singer and Frederick Seitz also understood. A word—a concept, a percentage—was your enemy. And every six years the IPCC, the international climate science body, would stamp along on its five thousand legs and drop down another big dose of *consensus*. Plant it in the headlines of every newspaper. Here was the spot on the tree to carve your "X." As you spit in your palms and lifted the axe.

Dr. Singer tried twice. The anti-consensus petitions have names: The Leipzig Declaration, The Heidelberg Appeal. They sound like spy movies: lovelorn and crestfallen thrillers starring a tongue-tied Jason Bourne, about the cities where he tried to make his feelings known.

The appeal came first, in 1992. (A style that perhaps descended from

Reverend Moon. You could crew on a church project without ever knowing you'd joined. Stand-up comedian and future pariah Bill Cosby once
accepted $150,000 to perform at something called the Family Federation
for World Peace. Discovering this meant Reverend Moon—also that his
contract was airtight—Cosby delivered exactly sixteen minutes of unsmil-
ing comedy. Then, according to a dispirited organizer, he "just got off the
stage and ran out of the building.") Dr. Singer and an associate helped
arrange a conference in Heidelberg, Germany. Scientists were invited to
sign a petition.

At first, Dr. Singer called it a "statement." Time passed, coasts cleared.
And he was like a man alone at the breakfast bar, filling his plate. Dr. Singer
called it "strongly worded." Said the appeal "expressed skepticism on the
urgency for global action to restrict greenhouse-gas emissions." That it
"urged statesmen to go slow on climate-change policies."

As it happens, the Heidelberg Appeal never once mentions global
warming. It's very pro-science. It's just not at all anti-climate science.

But it was a list of science names and got weaponized anyway. When
denial Senator James Inhofe quoted the petition in Congress, this is how
the message ran. "The Heidelberg Appeal, which says that no compelling
evidence exists to justify controls of anthropogenic greenhouse gas emissions. They agree it is a hoax." Two possibilities: either the senator had
never read the appeal, or he hoped you hadn't.

Dr. Singer took a firmer hand on the next go-round. New and
improved—now with global warming.

This was 1995. Earlier that year, Dr. Singer had sent a fossil fuel company his prospectus. For a very reasonable $95,000, the scientist promised
to help "stem the tide towards ever more onerous controls on energy use."

His hook was ozone. The spray cans that had been phased out, Dr.
Singer explained, "all on the basis of quite insubstantial science."

So if funds were provided "without delay," Dr. Singer could deliver: an
event, a panel, and a round number—"a Statement of Support by a hundred or more climate scientists." With the Singer specialty: "This Statement could then be quoted or reprinted in newspapers."

I don't know whether Dr. Singer ever secured his funding. But that

November, a panel did convene: in Leipzig, Germany. And one year later, his Statement did appear: the Leipzig Declaration. With the promised one hundred signatures.

The names crinkled brows. (Harvard's John Holdren, later science advisor to the Obama White House, wrote of them as a mirage or the dream you reconstruct over breakfast: the list "dissolves under scrutiny.") Sleuths from Danish Broadcasting attempted to track down the thirty-three European signers. Four could not be located. Twelve denied signing or even knowing about any Leipzig Declaration. Three were offended to hear their names were associated with it. The Statement had also been signed by dentists, lab techs, engineers, and one off-course entomologist who landed briefly on the page.

But the Leipzig Declaration packed its bags and coast-to-coasted anyway—from the *Wall Street Journal* to the *Orange County Register*, migrating also to Canada, London, Scotland, Australia, New Zealand. "It is widely cited by conservative voices," write journalists Sheldon Rampton and John Stauber. "And is regarded in some circles as the gold standard of scientific expertise on the issue."

Dr. Singer identified a hardy, *Band of Brothers* spirit among his "one hundred climate scientists." As he explained in the *Wall Street Journal*, "It takes a certain amount of courage to do this."

What it didn't necessarily take was a degree in science. Florida's *Saint Petersburg Times* ran their Leipzig story on the front page. Because (a) Florida, sea level. And (b), one signer was a local, the weather guy over at Tampa Bay's WTVT. Who lacked "a Ph.D. in any scientific field," the paper noted. "Or, for that matter, a bachelor's."

Dr. Singer had met his quota by reaching out to these sportscasters of the air. Twenty-five weathermen signed in, a big klatch from the state of Ohio. This included Richard Groeber, owner and operator of Dick's Weather Service: you dialed his phone number and he told you the weather.

The Petersburg newshound dialed. Was Dick Groeber, he asked, really a scientist?

"I sort of consider myself so," Groeber replied. "I had two or three years of training in the scientific area, and thirty or forty years of self-study."

The reporter brought his concerns to the keeper of the signatures, Dr. Singer. The scientist's answer is a testament to the virtue of persistence, of keeping an eye fixed always on the prize. What was truly important, Dr. Singer said, was "the fact that we can demonstrate that 100 or so scientists would put their names down."

And I wonder if it bothered Dr. Singer. If it's the story of his outranked life. That for the Oregon Petition—the signature list that did go over the top—the push came from the bigger, better honored, more consequential Fred.

AN UNEXPECTED GIFT

I DON'T KNOW WHO TOOK CARE OF THE INTRODUCTIONS. I do know S. Fred Singer sent Arthur Robinson material to beef up the research paper that accompanied the Oregon Petition. And I know that Dr. Seitz's Marshall Institute dispatched two specialists—climate Sherpas, to lug and guide Arthur along the trickier science crevasses.

One of them was later exposed on the front page of the *New York Times*. Dr. Willie Soon had been the beneficiary of $1.2 million in fossil fuel largesse. The last of his dinosaur generation to find their way into the tar pits.

"In correspondence with his corporate funders," the *Times* reported in 2015, Dr. Soon "described many of his scientific papers as 'deliverables' that he completed in exchange for their money."

And then a beautiful single-sentence short story: capturing the whole project and spirit of denial. "Though often described on conservative news programs as a 'Harvard astrophysicist,' Dr. Soon is not an astrophysicist and has never been employed by Harvard."

Arthur cowrote his paper with the two Dr. Seitz specialists, and a fellow member of the Oregon Institute faculty. His twenty-one-year-old son, Zachary—the Robinson for whom gay classmates had once posed a unique library access challenge.

This father-son teamwork produced something strange. First, their paper said climate change would not occur. Then, somewhat unexpectedly, it reversed field and explained that the change was already in progress and accomplishing marvels. Their concluding sentences drop the effort of science entirely. (It's like a marathon hiker wriggling off her pack; you feel the relief. Two National Academy experts later gave their work a Kleenex valuation: "a tissue of mistakes and distortions.") The language pans across streams and meadows—takes in a drowsy summer morning, with the sound of bees. "We are living in an increasingly lush environment of plants and animals," the Robinsons write, a little dreamily, "as a result of the CO_2 increase. Our children will enjoy an Earth with far more plant and animal life than that with which we are now blessed. This is a wonderful and unexpected gift of the Industrial Revolution."

Arthur's paper had never been published or peer-reviewed. It was entirely homeschool.

And here's where you can appreciate the great, freewheeling advantage of having fun. Arthur Robinson and Frederick Seitz collaborated on a tremendous prank.

Arthur had his report professionally printed. Now this home-cooked meal, this sloppy Joe, resembled an entrée at the end of a Food Network episode. The National Academy of Sciences produces one of the world's most distinguished journals. Garnishing with font and layout, Robinson labored until his blessing looked, in the words of the journal *Nature*, "exactly like a paper from the *Proceedings of the National Academy of Sciences.*"

Everybody has the one résumé line they lean on. It's whispered before you sweep over to shake hands; it will lead the obituaries when you step away forever. Frederick Seitz was the former National Academy president—publishers of the *Proceedings* journal whose format Arthur had copied.

Dr. Seitz wrote the letter that accompanied the Oregon Petition.

> *The United States is very close to adopting an international agreement that would ration the use of energy. . . . This treaty is, in our opinion, based upon flawed ideas. . . . We urge you to sign and return the petition card.*

Dr. Seitz signed with his résumé line: *Past President, The National Academy of Sciences.*

A cover letter from an Academy president. A paper formatted to look exactly as if it had been published in the Academy magazine. (Plus the plural *we urge*, the institutional *in our opinion*—the speaking voice of an organization.) Arthur and Seitz had pulled off the greatest soundalike in denial history.

The package was then sent all across America—as one researcher wrote, to "virtually every scientist in every field." And how could recipients fail to believe, tearing open their envelopes, that the Academy was reaching out to them, at an hour of scientific need?

IN 1996, *Nature* had written about the "dwindling band of skeptics." You picture palm fronds and breakers, the shoreline from *Lord of the Flies*: a rocky atoll among rising seas.

This line vexed deniers. It so bugged S. Fred Singer he ascribed it, for ease of attack, to Al Gore. (The scientist loved to attack the vice president. He also knocked Gore for a tendency to "accuse people who disagree with him" of being "liars, lunatics, or tools of industry." Which, to be fair, would make a pretty good *Jeopardy!* question. With the answer being, "Who are the denial scientists?") So the other aim of the petition: to grow the movement, at least in the eyes of key readerships in the Washington metro area.

It really was their weakness: Demographics. Max Planck once made an ice-eyed observation about scientific change. It doesn't result from fresh evidence, or the Kevlar argument. Positions get too dug in for that. It steals on gradually, in calendars and gravesides. "A new scientific truth does not triumph by convincing its opponents and making them see the light," the physicist wrote. "But rather because its opponents eventually die, and a new generation grows up that is familiar with it."

The plain truth was the deniers weren't getting any younger. Actual science was drawing the young PhDs. (When S. Fred Singer addressed a roomful of such climatologists in the spring of the Oregon Petition, the reception was not hostile. It was charity. His audience "politely pointed to datasets and to scientific research," wrote science journalist Myanna Lahsen, "none of which Dr. Singer appeared to be familiar with.") It's why the great denial work was brought off by Frederick Seitz, eighty-six, and

S. Fred Singer, seventy-eight; and by Arthur Robinson, aged fifty-six, whose footsteps Linus Pauling had long ago banished from institutional hallways.

"What will happen is clear," Arthur told supporters, in a sort of pre-invasion essay, as his envelopes mustered at the post office. "The warmers will be deprived of the central pillar that underlies their entire campaign."

This was that tall, shade-throwing word: *consensus*. "Remove their facade of scientific consensus, and they will likely lose—*if* it is removed in time."

AND IT WORKED. In the House and Senate, lawmakers said the petition proved climate change was "bogus"—a non-issue for "the vast majority" of scientists. (They needed something like it to be true. So they went ahead and believed it into truth.) It worked because it's a big library, and we're all busy people. And, as the bibliothecary Jorge Luis Borges once observed, "The person does not exist who, outside their own specialty, is not credulous."

"Happy Earth Day, Al Gore!" Fred Singer wrote in his *Washington Times* column. "Your much-touted 'scientific consensus' on global warming has just been exposed as phony." They'd finally found a way to bring down the tree.

AND IT WORKED because the National Academy didn't get wind of the business until April 1998. Frantic calls from officials and scientists. So the Academy took what the *Times* called an "extraordinary step": public denunciation. (One of the fresh experiences denial was making available to everybody: authority theft. Having to explain that the thing going around with your ID and typeface wasn't really you.) "The NAS Council would like to make it clear that this petition has nothing to do with the National Academy," went the press release. "The manuscript was not published in the *Proceedings of the National Academy of Sciences* or in any other peer-reviewed journal. The petition does not reflect the conclusions of expert reports of the Academy."

But Frederick Seitz and Arthur Robinson's Oregon Petition had been hunting and gathering through the mails since February. It had already bagged its fifteen thousand signatures.

There's the famous honesty and footwear observation that's usually attributed to Mark Twain. How a lie can travel halfway around the world while the truth is still futzing with its shoelaces. (Jonathan Swift was the first pessimist to post results of this sad honesty time trial. In a 1710 essay on the Art of Political Lying: "Falsehood flies, and the Truth comes limping after it.") This was the truth moseying onto the sun porch with a coffee mug. While the lie—clothes unpacked, souvenirs distributed, snapshots liked on Instagram—is home from the airport. Arthur was quite right to call his petition "the most effective single action" ever taken against the climate scientists.

A man like James Mountain Inhofe (actual middle name) is a big feature on the landscape. He headed the Senate committee responsible for climate. Years later, Inhofe told Sean Hannity how the Oregon Petition had cleared his vision and set him on the denial path.

"When I became chairman of the Senate Committee on Environment and Public Works, I was a *believer*," Senator Inhofe said. "All I heard was that global warming is due to human activity. . . . Then I thought, 'I better look at the science.' Well, one of the first things I looked at was the Oregon Petition." From then on, the senator knew. Climate change was "the greatest hoax ever perpetrated on the American people."

Two decades later, the head of a denial think tank would sweep his eyes across a denial ballroom, take in the many denial faces at the denial tables that made up his denial audience. "Art Robinson is the reason many of us are in this room," he said. No Arthur Robinson, no petition, no resistance. "If it wasn't for him, we wouldn't be here."

IN 2001, *Scientific American* went through his signature books. Present on Arthur's list were names submitted in a spirit of substitute-teacher abuse. (Arthur told the Associated Press that he had "no way of filtering out a fake.") There was Shirl E. Cook and Richard Cool and Dr. House, and the presumably dependable Knight and the presumably less steady Dr. Red Wine, also the accommodating Betty Will, the in-terrible-distress

W. C. Lust. Also someone who gave their name only as Looney. Plus a dash of celebrity like Michael J. Fox and John Grisham and the *dramatis personae* of the medical series *M*A*S*H*. Even some businesses, like R. C. Kannan & Associates, and Glenn Springs Holdings, Inc., had found a way to lift the pen and get involved. Dick Groeber—Dick's Weather Service—had once again elected to lend the effort the weight of his endorsement. All these names appeared on Arthur's petition as it was cited in Congress.

Arthur claimed only one false name was ever found to soil his list. (Some jokester had snuck on Dr. Geri Halliwell—Ginger Spice, of the empowerment band Spice Girls.) But post-media, all these names were quietly withdrawn. W. C. Lust and Betty Will and Glenn Springs Holdings, Inc., and Dick's Weather Service, scrubbed from history.

The names *Scientific American* examined were real. Barrier to entry was not high. If you claimed a bachelor's in math, science, or engineering, to Arthur's way of thinking you were a climate scientist. (Even so, Dick Groeber had no real business being on this list.) Your kid's math teacher could sign. So could her shop teacher, and the veterinarian.

These names were Styrofoam peanuts, packaging, and brushed aside. *Scientific American* took "a random sample of thirty of the 1,400 signatories claiming to hold a Ph.D. in a climate-related science."

Of the twenty-six names they could identify through the databases, "eleven said they still agreed with the petition." The magazine went on, "One was an active climate researcher, two others had relevant expertise, and eight signed based on an informal evaluation. Six said they would not sign the petition today, three did not remember any such petition, one had died, and five did not answer repeated messages." The magazine estimated that Arthur had managed about two hundred climate researchers—"a small fraction of the climatological community." Remove number from box, shake off the packaging: What Arthur Robinson and Frederick Seitz had delivered was a sweaty means of *confirming* the consensus.

And still there were international headlines ("NO SCIENTIFIC CON-SENSUS ON GLOBAL WARMING"). And still Frederick Seitz and S. Fred Singer could make their use of the data.

Dr. Seitz told reporters the petition represented "the silent majority of the scientific community." (Which meant at least 51 laconic percent.) And Dr. Singer called it "the largest group of scientists ever," as if the petition combined a Caltech homecoming weekend with an especially congested Burning Man.

Arthur kept up the petition drive. Yet among supporters, he couldn't quite bring himself to call the signers colleagues. The tongue values what it values.

"We've got now about 17,000 scien—"Arthur caught himself. "People with *degrees* in science." As of 2008, he'd nearly doubled his figure.

S. Fred Singer experienced the same performance trouble. In 2012 he was *still* quoting it. Because it was the only thing—Arthur had given the movement the strongest evidence it ever had. But even the famously reliable Singer tongue went rogue. "There's hundreds of us—thousands," he said on PBS. "Look, 31,000 scientists and engineers signed a statement." Then the scientist went a bit green. "Look, they're not specialists in climate."

But in 1998, when the ground was fresh, Dr. Singer told Congress that signers were "specialists in fields related to global warming." He told readers, while the issue was being contested, they were "experts in the pertinent scientific fields."

Arthur's website gives his patriotic side of the figure. "31,487 American scientists," he writes. "Including 9,029 Ph.D.s." You needed a data point, a comparison.

So, for the doctoral number: America is home to half a million science and engineering PhDs. Arthur netted 1.8 percent. His yield was small. And for the bachelor's number: We've awarded ten million first degrees in science and engineering. Here Arthur's petition was an absolute crash: 0.3 percent.

Arthur again sounded the Academy horn in a press release. "More than 40 signatories are members of the prestigious National Academy of Sciences." But Arthur had withheld the comparison. The Academy's got 2,200 members. His yield was eerily consistent: 1.8 percent. The generally accepted number for climate scientists and warming is 97 percent to 3 percent. Arthur's fate was to spend twenty-five years as superintendent of a consensus he loathed.

———

THE NUMBER IS still there, at the Oregon Institute of Science and Medicine website. With the Robinson Family Curriculum, it is the institute's most widely adopted product. Spreading the truth about climate is what that barnlike structure on his lawn turned out to be for.

And stepping into success, Arthur left Frederick Seitz and S. Fred Singer out in the cold. (Dr. Singer's innovations with the petition form, Dr. Seitz's experts, his Academy letter, all forgotten.) Arthur was speaking with Public Broadcasting. The signature concept, he said, dated all the way back to his former partner—to the mentor and surrogate parent, Linus Pauling. "He had two Nobels. The second was the Peace Prize," Arthur explained. "Which he received for circulating a petition against nuclear weapons. So maybe I got the idea from Linus." One more shot across the old man's bow.

And whether or not vitamins really do, fame enhances vitality. (Think of all those Academy Awards attendees, the Thalberg winners, out-good-living everybody.) The Republicans ran Arthur five times for Congress. He became chairman of the Oregon Republican Party. He appeared on television, accepted honors, gave speeches. In 2016, his petition was Facebook's most shared item about climate change. After Donald Trump's election, he was proposed as a candidate for White House science advisor. Denial restored Arthur Robinson to the World.

EDITING TURNS THE MILD
INTO WEATHER GODS

UNDER GEORGE W. BUSH, WARMING BECAME A REBOOT. The same people, data, arguments—a cycle from the past fifty years. But with the speed cranked way up, so everything appeared and then reappeared fast again like background in a cartoon. Questions that had been settled reopened. A face you thought behind you waited around the next turn.

Climate changed the president into a White House version of Wile E. Coyote. Devices that looked perfectly reliable in the Acme catalog detonated at the worst possible moment. Jet skates blowtorched the presidential hindquarters. A rubber band drilled him snout-first into a mesa. President Bush announced America's complete withdrawal from the Kyoto treaty in March 2001. (To Van Helsing applause from the deniers. "The Kyoto Protocol is like a vampire," S. Fred Singer told the *Times*, advocating some Transylvanian police work. "You need to drive a stake through its heart.")

The president gave as his reason the energy shortages then browning-out California. A few years later, it emerged that the crisis had been deliberately manufactured by Enron—coincidentally, the president's largest corporate supporter. (During the campaign, George Bush had bummed rides on the Enron jet. In November, staffers boarded Air Enron to oversee the Florida recount that gave their boss the presidency.) CBS News got their hands on audiotape. Private conversations between Enron traders. In a corporate satire—something to make business seem foolish and sinister—you wouldn't believe the dialogue.

"All that money you guys stole from those poor grandmothers in California?" asks Enron Trader 1.

"Yeah," replies Enron Trader 2. "Grandma Millie, man."

"Yeah. Now she wants her fucking money back for all the power you've charged," says Enron Trader 1. "Jammed right up her ass for fucking $250 a megawatt hour."

In May 2001 the Bush White House charged the National Academy with a full review of climate science. The twentieth second opinion? The two-hundredth?

The response was delivered quickly and sooted the president's face. (The administration had asked by mail. White House stationery is not encumbered by street data or zip code. It states, wonderfully and airily, the most intimidating return address in the world: THE WHITE HOUSE. WASHINGTON.) Their first sentence removed the matter from doubt. "Greenhouse gases are accumulating in Earth's atmosphere as a result of human activities, causing surface air temperatures and subsurface ocean temperatures to rise. Temperatures are, in fact, rising."

The White House had to send National Security Advisor Condoleezza Rice creeping onstage with soft-soled pronouncements. "This is a President who takes extremely seriously what we do know about climate change," Rice said. "Which is essentially that there is warming taking place." Which didn't fool anybody. Newspapers: "The report thus all but eliminates one reason the administration has been using to forestall any action on global warming."

It was just another reboot. "They asked a string of questions that might have been appropriate in 1990," one National Academy scientist complained to the *Times*. "Hello? Where have you been for the last decade?"

It had an aversive effect on the White House: ash-faced president. The coyote never orders the same product from the Acme catalog twice. "The President did not ask the Academy for advice about global warming again," writes Jim Hansen, "during the remainder of his eight years in power."

THE PLOT HAD TURNED so hyper-circular that different actors kept delivering the same speeches, their episodes separated by years. It's what happens when a TV series hangs around half a century. Every plot combination has been tried, every character has gotten enmeshed with every other character.

In 1989, Al Gore looked at the environment, and searched himself for a metaphor. You crunched on the slivery evidence everywhere you walked. The world was looking, Gore said in the *New York Times*, at "an ecological Kristallnacht." Eighteen years later, Jim Hansen searched himself for a metaphor: The evidence crunched under every shoestep. "Can these crashing glaciers serve as a Krystal Nacht?" Hansen asked, in 2007, again in the *New York Times*.

In November 1990, English and American troops were on the verge of invading Iraq to overcome Saddam Hussein; the British prime minister rose to the podium.

"The threat to our world comes not only from tyrants and their tanks," Margaret Thatcher said. "The danger of global warming is as yet unseen but real enough to make changes and sacrifices."

In March 2003, English and American troops were on the verge of invading Iraq to overcome Saddam Hussein; the UN chief weapons inspector rose to the podium.

"To me, the question of the environment is more ominous than that of peace and war," Hans Blix said. "I'm more worried about global warming than I am of any major military conflict."

Blix took a media shelling on his position. "Who is this guy? And why are we still listening to him?" asked Joe Scarborough on MSNBC. "There's still snow on the ground in New York City, one of the coldest years on record." 2003 entered the record as history's fourth-warmest year: in the blistering European summer heat wave, fifty thousand people lost their lives.

In 2004, the UK's chief scientist, Sir David King, was still, without realizing it, doing the Thatcher material. "Climate change is the most severe problem we are facing today," Sir David wrote in *Science*. "More serious even than the threat of terrorism." The chief scientist sent a nudge across the Atlantic: "We and the rest of the world are now looking to the U.S.A. to play its leading part."

TALL ORDER. The US—where, as one EPA official told the *Times*, there was now "a complete paranoia about anything on climate."

The second President Bush had taken a page from the first Bush's playbook. You can't stop the reports scientists kept sending the American public. And you couldn't block the inboxes where voters and journalists opened them. But you could take control of the email server.

At the climate agencies, political appointees became a kind of universal mistranslator. Scientists said *yes*—and what came out was *maybe*, or *no*.

The style was sly. Not a lie, exactly. Just enough extra words to crowd out the truth. "There's a bear outside the door," might become: "A wide variety of animal life is often thought to reside in the vicinity of human dwelling places." Research veterans had never seen anything like it. When the story broke, in 2004, the *Times* headline was BUSH VS. THE LAUREATES.

Science and the White House were "at war. . . . Nowhere has the clash been more intense or sustained than in the area of climate change."

The National Oceanic and Atmospheric Association might try to provide a warming update: "NOAA reports record and near-record July heat in the West, cooler than average in the East, global temperature much warmer than average." This came out in the placid voice of drive-time radio. "NOAA reports cooler, wetter than average in the East, hot in the West."

NASA's watch-this-space anxiety—"Cool Antarctica May Warm Rapidly This Century"—was smoothed into a bland and forward-looking item from the school newspaper: "Study Shows Potential for Antarctic Climate Change." Officials at the Department of the Interior would publicize research only after the words "climate change," "warming climate," and "global warming" were purged from the press release. As the *Washington Post* reported, there was now a full-on "battle over climate change."

One NOAA scientist put the situation in terms of value for investment. "American taxpayers are paying the bill," he said. "And they have a right to know what we're doing."

IF YOU WERE A COYOTE, there was the long-running problem of Dr. Jim Hansen. In 2003, the scientist was ordered by NASA to maintain radio silence: no more public talk about the human influence on climate.

Jim Hansen went public anyway. (A decision he struggled with. Family again beat job. "I did not want my grandchildren, someday in the future, to look back and say, 'Opa understood what was happening, but he did not make it clear.'") In a November speech, Hansen said government inaction represented a "colossal risk." At a meeting of the American Geophysical Union, Hansen warned that warming would result in "a different planet."

NASA now upgraded its threat: continue, and there would be "dire consequences."

So the White House put a minder on him. George Deutsch III, a twenty-four-year-old Texas A&M graduate with spiky hair and widely spaced eyes who'd interned on the Bush reelection effort. Deutsch looked like someone about to mount an erratic run for homecoming king: now this unlikely person had become the conduit between James Hansen— climate's most influential scientist—and the world.

Deutsch kept Jim Hansen off NPR—"the most liberal news outlet," he said, "in the country." He told coworkers serenely that his job was to "make the President look good."

Then it came out that George Deutsch III was ladling uncertainty over fairly certain things like Big Bang, insisting that room be made available for the Creator. (At-risk youth might turn to the agency in a moment of spiritual crisis. "This is more than a science issue, it is a religious issue," Deutsch explained. "And I would hate to think that young people would be getting only half of this debate from NASA.") Then it came out that the uncertainty extended even to his degree status. Deutsch had exaggerated his résumé; he'd never graduated from Texas A&M.

More singed fur on the president's snout. Deutsch resigned—and Jim Hansen ran free again.

PHILIP COONEY HAS the look of Washington middle-management—soft body sealed inside dark suit, tight Republican skullcap of hair. (Democratic hair is more catch-as-catch-can.) An air of cultivated invisibility.

Cooney graduated from the University of Richmond in 1981, headed north to Villanova Law, then signed on with the American Petroleum Institute. There followed fifteen years of lawyering and lobbying on behalf of oil. Junior attorney; senior attorney; then leader of the Petroleum Institute's Climate Team.

Here is the Petroleum Institute on warming's importance. From a 1999 strategy document, when Phil Cooney ran the team: "Policies limiting carbon emissions reduce petroleum product use. That is why it is API's highest priority issue."

Then Phil Cooney, with his air of not especially wanting to be noticed—when you are stepping many times a day into the vault filled with power and money, two things people limitlessly crave, it's best to look as though you have no drives at all—went into government.

He arrived in summer 2001, a mid-level hire. Here's the *Wall Street Journal*, their first sentence a wry smile. "Climate-change treaty foe Philip Cooney is the new chief of staff for the White House Council on Environmental Quality.... At the American Petroleum Institute, he helped develop the oil lobby's opposition to the Kyoto Protocol." The *Times* described Cooney as a member of Washington's phantom powerful. The toadstool successors of Jim Tozzi—workers who emerge in the moonlight, handle the necessary, are gone with the dew. Officials who "hold less visible jobs in the administration and agencies that allow them to make far-reaching policy and regulatory changes."

Phil Cooney is the man who rewrote the climate reports. He'd get a paper from the EPA, from the United States Climate Change Science Program, then set to work dismantling it. He was the dreamy humanities professor who is able to pick apart a poem, see through its surface sense, mine the rich seam of ambiguity below. Under Cooney's pen, *yes* was replaced by *maybe*, *potentially* nudged aside *will*.

From the government scientists, Cooney would receive "Many scientific observations indicate that the Earth is undergoing a period of relatively rapid change." And with a sprinkle of wording, a snip of edit, Philip Cooney changed the weather. "Many scientific observations *point to the conclusion* that the Earth *may be* undergoing," he'd write.

Incoming would be, "Reduce the causes or effects of human-induced climate change." And outgoing would be much the same—but with *human-induced* deleted. Presto. In a sense, problem solved. People no longer changed their climate.

Some stuff he just cut. "Energy production contributes to warming"— gone. "Use of fossil fuels"—gone. He was, loyally, the former leader of oil's Climate Team. Other stuff he'd slip in. "The uncertainties remain so great as to preclude meaningfully informed decision making." Left alone, this mild man grew the bravery of a conquistador.

Sometimes his changes were so bald-faced as to be fun. "Delete 'disaster reduction,'" he'd write, "replace with 'opportunities.' . . . Delete 'famine,' replace with 'impacts.' . . . Delete 'human actions.' . . . Delete 'knowledge.'"

In 2002, Phil Cooney's former boss at the Petroleum Institute faxed over a backslap: "You are doing a great job." In a call with the Exxon-sponsored Competitive Enterprise Institute, Cooney revealed the mysteries of his process. "We'd take the text from the EPA," Cooney explained. "And then we'd add a sentence like, 'We don't really know if this is really happening.'"

With any art, proficiency leads to ambition. Cooney had the warming chapter from an EPA pollution report stripped. As the House Committee on Oversight later marveled, Cooney's Council on Environmental Quality "went beyond editing and simply vetoed the entire climate change section." To another report—with the briskly cheerful title "Our Changing Planet"—Cooney made 110 edits. Scientists protested; Cooney scribbled the budge-free word *No* in the margin. One scientist later told the *Times* the changes had "a chilling effect."

Another study was due: the EPA's overall "Report on the Environment." Cooney yanked the sentence "Climate change has global consequences for human health and the environment." (Though without this, what was a report's point?) The agency quoted the killer first line from the National Academy: "Greenhouse gases are accumulating in the atmosphere as the result of human activities." Cooney replaced this with an every-which-way of his own devising: "Some activities emit greenhouse gases and other substances that directly or indirectly may affect . . ." Phil Cooney now outranked the National Academy of Sciences. He'd become—with scissor and inserts—the most powerful literary editor in the world.

EPA debated the merits of ditching the climate material entirely. "Uncertainty is inserted (with 'potentially' or 'may')," an official complained, "where there is essentially none." So that the study "no longer accurately represents scientific consensus on climate change." EPA head Christine Whitman felt Cooney's changes would denigrate the report as a whole—so "Report on the Environment" was simply never published. Presto. Problem solved. Four days later, Christine Whitman recognized a similar exit door for herself, and resigned from government.

On a report detailing the nation's entire climate science program, Cooney and his staff made 294 edits. To, by the House of Representatives' count, "exaggerate or emphasize scientific uncertainties" and "deemphasize or diminish the importance of the human role in global warming."

Scientists were now leaking documents. Phil Cooney was "someone who had no scientific training," explained the *Times* climate writer Andrew Revkin. "And also with, essentially, a vested interest for years in not accepting the science on global warming. That he was doing that kind of revision was horrifying to scientists within the government. And that's why they came to me with the documents."

INTIMIDATION IS A numbers game with a fixed outcome. Eventually, you run across the person willing, at the cost of salary and career, to end it.

That person was named Rick Piltz. He'd spent ten years with the US Climate Change Science Program; he resigned in early 2005 over the Cooney edits.

A documentary crew was shooting a piece about warming and politics. Piltz, who now had time on his hands, led them on an old haunts tour: Good nine-to-five footage, the pleasures of the day when not at the bidding of the clock. He toured them past the Council on Environmental Quality, a plum-colored townhouse on Jackson Place—Phil Cooney's office.

"I sort of felt when he was there," Piltz said, "that Exxon-Mobil was in the room," Piltz said.

It is an illustration of the ancient wisdom against speaking of the devil. At that moment, Philip Cooney stepped through the townhouse door.

It was spring and brisk. Piltz wearing a Saturday jacket, carried a Starbucks coffee in its stiff brown paper cummerbund. Cooney was sealed within his suit-and-tie armor. The camera records a fascinating confrontation. What you see is a whistleblower, at the precise moment of bringing whistle to lips.

You can hear his anxiety. The oil man comes down the steps. "There's Phil Cooney right *there*," Piltz whispers.

This was perhaps the only video of Phil Cooney then available—until he was called to testify by the House Investigation on Political Interference with Climate Change Science. (Because of Piltz's actions, the committee would find, "The Bush Administration has engaged in a systematic effort to manipulate climate change science, and mislead policymakers and the public about the dangers of global warming.") Never acknowledging the film crew, Cooney veers coolly left. His face registers as a pink blur. "He doesn't like publicity," Piltz whispers.

The producer can't hear. "He does *not* like publicity," Piltz repeats. "If you Google on him, you will not find much. He's a behind-the-scenes guy..."

At the end of the block, Cooney fishes in his pocket. Out comes the cell phone.

"Did he see you?" the producer asks.

On *Sixty Minutes* a year later, Piltz would explain, "I mean, even to raise issues internally is immediately career-limiting. That's why you will not find too many people in federal agencies who will speak freely."

Piltz laughs to himself in shock. "So now he knows that someone is sticking a microphone in my face and I'm looking at his office."

And you see the whole thing register on his face. Piltz takes a sharp step back from the film crew, as you would from a barking dog or hot stove.

Piltz had preserved Cooney's edits. Each time Phil Cooney noodled out a "shown that," a "confirmed that," and replaced them with data pasta,

information linguine. Piltz turned this material over to the Government
Accounting Project.

Seventy-two hours later, Phillip Cooney was on the covers of the newspapers. Forty-eight hours after that, the lawyer resigned. "Phil Cooney did
a great job," said a White House spokesperson. "We wish him well."

And then Phil Cooney's life took a turn. The publicity-shy attorney
had to suffer the sink-or-swim worst case for those with a media phobia,
and hear his name batted around on CNN, MSNBC, NBC, CBS, NPR,
on *Meet the Press*. If he'd had the stomach, he could have read about himself
in England, France (*Agence France Presse*: "Philip Cooney . . . often would
subtly alter documents to create an air of doubt about findings few scientists dispute"), Germany, Glasgow ("Philip Cooney . . . is not even a scientist"), as far away as Qatar, and Australia's *Northern Territory News* ("White
House Man Bites Dust"). Or he could catch it in the *USA Today* on the
rug outside the hotel room door. "Oil Industry Lobbyist Philip Cooney . . .
talk about the modern-day equivalent of the Flat-Earth Brigade." The piece
pointed an accusing finger up Jackson Place toward the Oval Office. "Yes,
Globe Is Warming, Even if Bush Denies It."

That same pile-on week, the world's science academies released a joint
statement: Climate change was real, with humans at the burner. "The scientific understanding of climate change is now sufficiently clear to justify
nations taking prompt action."

A week after the story broke, ExxonMobil offered Phil Cooney a job.
He's been based in Houston ever since, his title another bit of soft comedy:
corporate issues manager. This could be the label on the generic can that,
opened and upended, would pour out Phil Cooney.

Senator John McCain engaged in some *Casablanca* banter on one of
the Sunday shows. "I am shocked, shocked," the senator said. "He immediately went back to work with Exxon. . . . Maybe he should have waited a
month or two."

Four weeks after that, George Bush, whiskers singed again, finally
spoke the words every canceled press release, rejected interview request,
and edited report had been designed to insulate him from. Unescorted by
Phil Cooney's *potentially*, *maybe*, or *no*. Just a president with a fact.

"Listen, I recognize that the surface of the earth is warmer," Bush said. "And that an increase in greenhouse gases caused by humans is contributing to the problem."

Scientists persist. Deniers rotate out. Phil Cooney hasn't been heard from since. The classic story: He steps into the warming process, makes (or deletes) his mark, becomes one more figure we see through a car window at the commuting hour. Men like Phil Cooney thrive best in twilight.

EXCEPT FOR THE HEARING. The House Oversight Committee summoned Phil Cooney to Washington two years later. Captured on a live feed at last, he wore the accused's most persuasive expression: a slight buffering shield of polite incomprehension. If all these people were unhappy with him, there must be some mistake.

James Hansen was also called. He sat right next to Phil Cooney at the witness table. He seemed to enjoy himself immensely: correcting the corrections. "Mr. Hansen, you are one of the nation's leading experts on climate change," the committee chair began. "What is your view of the changes made by Mr. Cooney and his staff at the White House?"

Hansen depressed the button on his mike, and told the Representatives, and perhaps also his grandchildren, "I believe that these edits, the nature of these edits, is a good part of the reason for why there is a substantial gap between the understanding of global warming by the relevant scientific community, and the knowledge of the public and policymakers."

Hansen speaking was one of the few times Cooney broke format: his mouth hangs open, dazed to find a scientist right there beside him to edit the edits. On paper he was able to command *These changes must be made*. Moments later—a beautiful sight on that pouch of a face—Jim Hansen incrementally smiled.

ASS AND CHAIR

HERE'S HOW TO SEND AN IDEA OF YOURS ALL THE WAY UP the chain.

There once was an organization called the New York Society for International Affairs. It operated a worldwide vacation effort. Airlifting lawmakers out of the humdrum (Kentucky), releasing them into the exotic (Africa). It turned loose a team of state senators on the beaches of Costa Rica for one week of high living. Five-star accommodations; tour of the president's family fruit ranch; jungle boat ride in search of "something called the something monkey," society founder Andrew Whist recalled later to investigators, that made "a tremendous noise."

Also, for some reason, on the banquet tablecloths, crystal goblets cupping Marlboro cigarettes.

The society had deep pockets, wide reach. Around Christmas 1995, it helped send four deserving power couples—American governors, squiring four first ladies—on safari across Africa.

Lawmakers returned from such trips (as vacationers will jet home from the Magic Kingdom bearing Mickey and Princess Elsa gear) with positive vibes about tobacco. "I have a very soft spot in my heart for Philip Morris," explained one of the Costa Rican adventurers, a Colorado state senator.

Wisconsin governor Tommy Thompson sent society president Andrew Whist a really thoughtful note after Africa. (Thompson would serve in the George W. Bush cabinet. Still later, he ran for president. At safari time, he headed the National Association of Governors—so, mathematically, you also got a crack at the other forty-nine.) "I value your loyalty and friendship," Governor Thompson wrote. "And look forward to sharing many more great meals [and] stories. . . . I eagerly anticipate our next adventure together—maybe it will be Australia."

And so it came to pass. Foundations are fairy godmothers with an application process. Six months later, Governor Thompson was wet-suiting it off the coast of Australia, beside a Philip Morris lobbyist wearing flippers. Then he and Andrew Whist sat with forks in their hands and the marine breeze in their hair. "A particularly nice cruise of the Sydney Harbor," society president Andrew Whist fondly recalled. "We had lunch on board."

The scandal broke. Philip Morris was supplying the society with almost 100 percent of its operating capital. "New York Society for International Affairs" was really the politically acceptable pronunciation of "Philip Morris Travel Department."

It had been a form of sunstruck experience bribery. Andrew Whist was a Philip Morris senior vice president—a lifer. In addition to serving as New York Society president, he made up the entirety of its staff. "It's basically me," Whist admitted to government attorneys.

Scandals have their mop-up phase. Everyone stooping down to collect the last stray crumbs of fun. Reporters decided to pay the society a visit. It did not seem to exist. Andrew Whist coughed up an address. The *Wall Street Journal* discovered a luxury high-rise, a puzzled doorman. Andrew Whist finally admitted the location was entirely notional: The New York Society for International Affairs was really just "a chair in my apartment."

On the other hand, it gave the *Wall Street Journal* the chance to run this once-in-a-publication sentence: HEADQUARTERED IN A CHAIR.

So if you ever need to start a foundation, don't despair. To field the sort of organization that can bounce influential figures all around the globe, you really only need money and ambitious furniture.

WITH JUNK SCIENCE, Philip Morris took things in a more brick-and-mortar direction. Their group would still be headquartered in a chair. But that chair would be located in an actual public relations office: APCO Associates, in the Dupont Circle neighborhood of Washington, DC.

For chairman, they secured the services of another governor. A former one-termer from New Mexico named Garrey Carruthers.

Then made it look as if the whole thing had been his idea. "I am creating a coalition of scientists, academicians, former public officials, and representatives from business and industry," read the invitation APCO wrote above the governor's signature. "Its goal is to advance the principles of science."

They also needed a marble name. Something to really make you fix your hair and shoot your cuffs, the way New York Society for International Affairs did.

They batted this around for months. Leave it at "Anti-Junk Science Coalition," and you were bound to come off hectoring. (Almost no political group ever has a wholly negative name.) APCO opted for a soundalike. The American Association for the Advancement of Science is one of the world's most distinguished research organizations; among other activities, they bring out the journal *Science*. APCO called Philip Morris' new group the Advancement of Sound Science Coalition.

Immediate snag. Left to their own devices, the initials spelled out ASSC. Sound Science—which ought to walk somber and dignified—would enter public life depantsed and backward. As the ASS Coalition.

It's the way fate sometimes grades our efforts. Every move seems shrewd, calculating, ruthless. And still you end up in an Ass Coalition.

So APCO deployed some Junk grammar. Nudged *The* into the acronym. (By this logic, the United States of America would be TUSA, which sounds like a funk-rock act.) The new group marched into sunlight as the TASSC.

And now it voiced a clear, positive request: for care, judiciousness, expertise. Sound Science.

THE ASS COALITION, writes David Michaels in *Doubt Is Their Product*, became "the first entity to carry the official 'Sound Science' flag." As Michaels explains, "This cynically named movement was the creation of Big Tobacco, initially under the aegis of the Advancement of Sound Science Coalition."

It gave anti-science style. Rendering the situation comfortably bipolar, joining it to the larger national Yes-No: New or old, beach or snow, meat or vegan, dog or cat, parrot or igloo, TV versus reading, GOP versus Democrat, Android versus Apple, pop versus hip-hop, East Coast versus West Coast, America versus the World, Junk Science versus Sound Science. The result was "the vilification of any research that might threaten corporate interests as 'Junk Science,'" in Michaels' formulation. "And the sanctification of its own bought-and-paid-for research as 'Sound Science.'"

"The Advancement of Sound Science Coalition," this is the Harvard science historian Naomi Oreskes; with an ink headshake, a font-ruffling sigh. "Whose strategy was not to advance science, but to discredit it."

This is the group Ellen Merlo had had called into being to defeat the Environmental Protection Agency. To bring it down, as her strategy memo went, by gathering "all of the E.P.A.'s enemies against it at one time."

Ellen Merlo is the executive who orbited New York for the first part of her life, then came to love Philip Morris for the atmosphere: the informality, great people, the holiday drinks served every Christmas by the boss himself.

In a series of proposals, APCO promised Merlo a coalition that would secretly serve "the Philip Morris agenda." Deliver "overall credibility." Plus "a large number of potential third-party allies"—respectable faces, crowding the window—"for Philip Morris' efforts."

It's like Leonardo DiCaprio's team from the movie *Inception*. The goal was to slip an idea into the national thoughtstream, seed the dreamscape with opinions and fear. What tobacco had always wanted: "How do we get 'bad science' into leadership consciousness?" A century of research went in. What came out was Junk. But then, how could Svante Arrhenius and Roger Revelle have anticipated S. Fred Singer, Phil Cooney, Arthur Robinson, or the TASSC?

APCO ASSOCIATES was nine years old when it unveiled the ASS to the public. The firm has since hit middle-aged spread. APCO Worldwide. Thirty-one bureaus: Moscow, Mumbai, Kuala Lumpur, an empire of parking lots on which the sun never sets. In 2017, APCO received the industry SABRE Award for Best Public Relations Firm of the Year.

So it must be frustrating. What people know them for best is the ASS work. You can open up the *Times of India*, click to an online Australian magazine. "APCO," you read, is "most memorable for launching the Advancement of Sound Science Coalition."

It's a memory nobody wants to claim: left to continue its lonely circles on the baggage carousel. In 2006, the UK *Guardian* ran a George Monbiot exposé; tobacco to APCO to Sound Science to the language of climate denial.

An APCO executive wrote in to adjust this memory. The story, he said, had "misrepresented our relationship." It had really only ever been a fling: two lonely pairs of eyes meeting across a hotel bar—"an organisation we had a brief association with." Actually, the ASS chair remained in the APCO office for four years.

On the witness stand, Ellen Merlo finds a soft, chiropractic explanation for the antics her company got up to in the nineties. "I think we fell out of alignment with society," Merlo says. "We were out of step with society's expectation of us." Plus some hearing loss: "We just didn't listen." Also some emotional obstacles: "We got very defensive." Just a regrettable era; a fault of the time.

Even ASS chairman Garrey Carruthers—whose first name is the denial *spelling*. In 1994, Carruthers bragged to science-interest magazine *The Scientist* about "founding" the TASSC. In 2013, up for the university presidency at New Mexico State, Carruthers became a figure out of Philip K. Dick—he had replacement memories. "Frankly, I was not terribly engaged with them," the chairman said. (The *Albuquerque Journal* dryly observed that the governor's name had "appeared on fundraising material distributed nationwide.") Carruthers got the job. The dream of every smoker: Inhaled the product, escaped lasting harm.

And Philip Morris got the jump on everyone—disguised its involvement from the very beginning. "We thought it best to remove any possible link," went the memo. They wrote checks totaling a quarter million dollars to get the ASS ball rolling; Year Two's budget was $880,000. Before launch, Philip Morris drafted up some talking points.

Q. Isn't it true that Philip Morris created the TASSC to act as a front
group for it?
A. No, not at all.

A script from Philip Morris. For people compensated by Philip Morris.
To explain they weren't in the employ of Philip Morris.

This is the approach. Complexify that one touch past easy understanding. So that the brain—as when studying route maps in a foreign metro,
or clicking to find a very long FAQ page—grows a little cloudy pelt of
thought moss and powers down.

The sublime and beautiful solution—one that never occurred to Consulting Detective Holmes or Professor Moriarty. A mystery not too diabol-
ical, but too cluttered to solve. The best way of pulling off the heist is to pile
the jewel under explanation by the stacks and stacks.

THE FUNNY THING IS, it failed with cigarettes. It just succeeded with
climate. It was the bullet fired by the dying mobster, that strikes the policeman there on a warrant to arrest somebody else.

Formal ASS operations began in late 1993. By 1997, Governor Garrey
Carruthers could look back at four years of solid achievement. "The issue
of Junk Science," he said, "has become the topic of network news specials,
major articles in newspapers, and a key topic in Congress."

It happened fast. As the *New Republic*'s Franklin Foer wrote, it succeeded even though Sound Science was "pioneered by the tobacco lobby."
Yet despite these "cynical motives, tobacco's Sound Science coalition
acquired a broad following."

By election year 1996, it had gotten itself inscribed in the Republican
Party platform. Phrasing that might have come straight from the ASS. "We
will target resources on the most serious risks," the plank read, "rather than
on politically inspired causes ... and will [require] assessments based on
Sound Science." The GOP, like a sheriff pledging his oath to the frightened
townspeople, swore to "eliminate the use of 'Junk Science.'"

The cigarette industry had become the Republicans' number-one soft-
money donor. They sponsored the 1996 convention host committee. "The

majority G.O.P.," as *Times* commentator Frank Rich wrote, "is all but choking on tobacco loot."

Newt Gingrich had led the conservatives back to power. Philip Morris
co-hosted a rhyming Manhattan celebration—the "Salute to Newt." When
politicians found themselves without a ride, seating was made available on
the tobacco jet.

"Coincidentally or otherwise," as science journalist Chris Mooney tartly
puts it, the "pro-tobacco Congress adopted 'Sound Science' as a mantra."

It was everywhere. Knight Ridder reported that "Sound Science"
was one of the year's two "environmental buzzwords." (The news service
offered a crib sheet: " 'Sound Science' is shorthand for the notion that anti-
pollution laws have gone to extremes.") This buzz had become "a roar."
Majority Leader Newt Gingrich might have been reading from a tele-
prompter prepared by ASS hands. Environmentalists "look for the hysteria
of the year," Gingrich said. Like "global warming." What type of science did
the Speaker prefer? "Sound science," Gingrich said.

His second-in-command, Tom DeLay, swam out way past the ASS
lifeguards. Ellen Merlo's people would have been perfectly happy with
the EPA portrayed as "environmental terrorists." DeLay promoted them
to midnight visits and colorful interrogations—"the Gestapo of Government." Senate Majority Leader Bob Dole—the Republican nominee
for president—quoted ASS literature and introduced a bill mandating
Sound Science.

Overall, per OSHA head David Michaels, "a superb job in marketing
the 'Sound Science' slogan and thereby undermining the use of science in
public policy." If it profited you, Sound. If not, Junk. It was like a simple,
knee-in-groin self-defense technique, taught to a nation of policy makers
and business leaders who had been afraid to walk to their cars.

The Republican Congress conducted Sound Science hearings: ozone,
global warming. At ozone, a Republican congressman held the floor as if
promoting a Red scare. "I have found extremely misleading representa-
tions by the government and government officials," said California's John
Doolittle, "that are not founded on Sound Science." A Democrat asked
whose research he trusted. "Well," the Republican answered, "you're

going to hear from one of the scientists today. Dr. Singer." It was good old S. Fred.

Majority Whip Tom DeLay gave the Moon compañero another name-check: "My assessment is from reading people like Fred Singer."

Dr. Singer was a science advisor to the ASS Coalition. So was skeptic Patrick Michaels. (Michaels would later get on the horn with CNN and bellyache about coverage: too much consensus, not enough skepticism. So a CNN producer checked the files. The global warming expert quoted most often on the network? Skeptic Patrick Michaels, "by a factor of two.") Fred-erick Seitz took a seat on the ASS board. Governor Garrey Carruthers went barnstorming. North America, South America. At a 1996 environmental conference, he let lawmakers in on the secret: the scientists themselves did not believe in global warming. The Sierra Club picketed outside the hall. At year's end, the old attention-grabber: awards. The coalition reserved the top prize, Perennial Exaggerated Scare, for warming. "To the groups and individuals still trying to stir up fears over global climate change . . . the facts say there's no cause for fear." Years later, angling for that sweet uni-versity plum, Carruthers took a somewhat modified stance. "I don't know," the former ASS man said circumspectly. "I don't do global warming. It's a scientific judgment I can't make."

Congressional testimony—the long table, the news cameras, the whis-pering aides—underwent a change in these years. As of 1992, mainstream scientists made up almost half of all climate witnesses. Industry allies (skep-tics and lobbyists) comprised only about 10 percent. During the Sound Science decade, a shift: the corporate voice now filled more than half of tes-timony chairs; science had dwindled to just under 4 percent, a peep. Same change in news coverage—it followed the hearings. And by the late nineties, as historian Naomi Oreskes writes, "Politically, global warming was dead."

The oil giant had lumbered in. From the Union of Concerned Sci-entists: "It is not a coincidence that ExxonMobil and its surrogates have adopted the mantle of 'Sound Science.' In so doing, the company is simply emulating a proven corporate strategy when one's cause lacks credible sci-entific evidence." (And they ran with it. Exxon remains an ASS enthusiast. "ExxonMobil has always advocated for good public policy," a spokesperson would tell the *Los Angeles Times* in 2015, "that is based on Sound Science.") Cigarettes had "paved the way," the Union of Concerned Scientists reported. "The calculated call for 'Sound Science' was successfully used by tobacco firms as an integral part of a tobacco company's 'information-laundering' scheme."

The idea had entered the bloodstream. Transfusing every vein, hitching a ride to every limb. In 1999, you could attend a day-long Junk Science conference at the National Press Club. Sponsored by the conservative Independent Women's Forum. Why? Word was out: Sound Science parted tobacco wallets. The Women's Forum packaged its material, contacted Philip Morris, made a horrible *Ghostbusters* reference with an open palm. "Philip Morris has been a friend to the Independent Women's Forum in the past for good reason," went their cover letter. "After all, who 'ya gonna call when you need a sensible, intelligent woman's voice? For this important junk science project, please help. . . . You just might have a team member that can make the difference."

The big difference-making teammate was Exxon. The oil company had assumed the weight of ASS support.

It was like inheriting a ball club. You got experience, organization—an offense plus athletes who knew how to execute. As the Union of Concerned Scientists reported, Exxon took "strategy [and] tactics" from "the tobacco playbook." And even "the very same TASSC personnel." Exxon paid for groups supporting Frederick Seitz, S. Fred Singer, Patrick Michaels. And a former colleague of Jim Tozzi's, Steve Milloy, who had become director of ASS activities. All brought to the field the wiliness of the veteran.

And this gave oil "a calculated and familiar disinformation campaign," the Union reported, "to mislead the public and forestall government action on global warming."

It's been with us so long, the story went on the repeat cycle. During the Bush years, Knight Ridder ran *another* Sound Science piece.

The meme had been casting its spell "since 1992, when lobbyists for the tobacco industry argued that no 'Sound Science' showed that second-hand smoke is a health hazard." (Knight Ridder: "The Bush administration, senators, industrialists and farmers repeatedly invoke the term 'Sound Science' to delay or deep-six policies they oppose.") They sought the impressions of Stanford biologist Donald Kennedy. In his capacity as editor-in-chief of *Science*—the journal put out by the American Association for the Advancement of Science, whose name and thunder the ASS crafters had stolen for their own group. So, without knowing it, Donald Kennedy was really in conversation with his celebrity double—he was orange juice debating the merits of citrus with Sunny D.

"Sound Science is whatever somebody likes," Dr. Kennedy said. "It doesn't have any normative meaning whatsoever. My science is Sound Science, and the science of my enemies is Junk."

And when pollster Frank Luntz drafted his famous battle memo, there the old ASS idea was. Top of the marquee. Strategy tips for "Winning the Global Warming Debate." Luntz printed it—siren going, lights flashing—in bold *and* italics. (Which I think is overkill.) "*The most important principle in any discussion of global warming is your commitment to Sound Science.*" It also took first chair in Luntz's Nine Principles of Environmental Policy and Global Warming: "Sound Science must be your guide."

Frank Luntz, Elizabeth Kolbert wrote in the *New Yorker*, was "viewed by Republicans as King Arthur viewed Merlin." The *Times* called his memo the conservatives' "recipe for success." Magic, cuisine—another top-shelf metaphor would be faith. Which is just how PBS said Exxon and the White House had received the Luntz commandments. "They followed this like the Bible."

Luntz replied in terms of cost to his fundament. "It was a great memo. It was great language. I busted my ass for that memo."

It's a pleasure to encounter that word here. Because it was really tobacco's ASS. Tobacco had created the ASS to spread that language. It was a smoker's cough, communicated across the years and disciplines, from the desk of Ellen Merlo to Frank Luntz, from the TASSC to the White House.

Because the language had entered that bloodstream, too. Two ASS alums served on the George W. Bush team—secretary of the interior and head of regulation at the Office of Management and Budget. The *New Yorker*'s Elizabeth Kolbert visited Paula Dobriansky, State Department undersecretary responsible for setting our terms at international climate negotiations. Dobriansky gave the US position twice in identical language, like a political action figure fitted with a voice chip: "We predicate our policies on Sound Science."

This is what George Monbiot meant, when he said no denial group ever had as large an effect on climate. They prepared the strategies, auditioned the personnel, released the memes.

Sound Science had climbed all the way to the Oval Office. In 1987, back on Hilton Head Island, tobacco had squinted down the problem: "How do we get 'bad science' into leadership consciousness?"

They had created a magic ring, only to see it transferred to some other industry's storyline. When George Bush unveiled his global warming plan, it was with the terms the ASS planner had hoped to install in the national debate. They had met their goal: they had Advanced Sound Science.

"When we make decisions," the president announced, lovelessly, on Valentine's Day 2002, "we want to make sure we do so on Sound Science. Not what sounds good, but what is real."

His explanation for crumpling the Kyoto treaty—ratified by every other industrialized nation—would have sent a school of grins sharking around the Philip Morris table. It was "not based on Sound Science."

George Bush's was a cabinet of Sound Science apostles; everyone talked it. Karl Rove talked it. So did the secretaries of Agriculture, Commerce (oversees the National Oceanic and Atmospheric Administration: "real progress [comes] by harnessing the power of Sound Science"), Education, Energy, Health and Human Services, Transportation, and Labor talked it. (Labor? "Any course of action that we take should be based on Sound Science.") Vice President Dick Cheney talked it.

The First Lady talked it with regard to reading. A White House country-western duet. George: "Laura, my wife, the First Lady, is having a series of seminars about how to introduce the Sound Science of education into curriculum all around the country." Laura: "It's founded on Sound Science—not the latest fads."

The program the Bushes promoted turned out to be "almost comically skewed," the *Washington Post* later reported. "The billions have gone to what is effectively a pilot program for untested programs with friends in high places."

Here is the Laura Bush program's director practicing some sound hosting policy: on a friendless grant proposal—one that just showed up uninvited at the door: "Kick the shit out of them in a way that will stand up to any level [of] scrutiny. . . . They are trying to crash our party and we need to beat the shit out of them in front of all the other would-be party crashers who are standing on the front lawn."

And of course Senator James Mountain Inhofe, head of the Senate committee charged with climate, talked it. He quoted the two denial grandfathers, Drs. Singer and Seitz, in the best-known floor speech on warming. "The claim that global warming is caused by man-made emissions," the senator said, "is simply untrue and not based on Sound Science."

At the beginning, APCO had promised Philip Morris two big ASS things: allies and credibility. The great moment came when George Bush swore in his final Environmental Protection Agency head. This was in 2005, twelve years after Philip Morris' experiments with the ASS began.

EPA was the agency that had kicked off the whole business. Whose reputation the cigarette maker had sought to destroy, by bringing together all its enemies at one time.

And here the president and the new agency chief were—at EPA headquarters, deep in the belly of the old Junk beast. The two men nodded at the podium, over how the White House was meeting its responsibilities as a good steward of the Earth.

Another duet. The president noted that the new agency head would "help us continue to place Sound Scientific analysis at the heart of all major scientific decisions." The new agency chief sang back, "Experiences have taught us that Sound Science is the basis of our achievements and the genesis for our future successes."

Tobacco was already in the midst of its unbelievable losing streak. The millions of documents online, the courtroom decisions and damage awards running against them. The Department of Justice was about to find them as criminal co-conspirators; they'd lost about as big as you can lose.

But I wonder, for just that one night, if the calm of successful effort descended on their New York headquarters. The quieted brain and heavy limbs. If Philip Morris reinstituted the cocktail policy that had once so impressed Ellen Merlo. The head of the company steering his drinks cart into the conference room, high and quiet above the city. Working the martini shaker, serving the highballs. I like to think of Ellen Merlo and the chairman of Philip Morris sharing a wry little toast. It had come too late for them. But they had inserted their idea into leadership consciousness at a higher level than they could have possibly dreamed.

GLENGARRY GLEN MONCKTON

THREE SNAPSHOTS FROM THE LIFE OF HIGH ADVENTURE. As hazarded by Lord Christopher Monckton.

The crisp tan and blue of the beaches at Durban, South Africa—when out of the sky plunges the viscount. This is at the seventeenth annual United Nations climate conference; the event is called "Parachuting for Truth." Lord Monckton, just south of sixty, lands on a green banner reading THE SCIENCE IS NOT SETTLED, hair doing a crazy silvery dance around his head.

A year later—Conference Eighteen, 2012, the sands of Doha, Qatar. When out of the desert strides Monckton, leading a train of camels. Words pasted against each hump: STOP. CLIMATE. HYPE. ("My lovely wife," Monckton explains, of his camel rapport, "says animals and children are attracted to me, because I have never grown up.") Inside the conference, Monckton spots an unattended microphone. He slips behind, keys the button. He is impersonating the delegate from Myanmar, and after the traditional *salaam alaikum* delivers a one-minute denunciation of climate science. "Sorry, this last speaker is not a party," apologizes the conference chairman, Abdullah bin Hamad Al-Attiyah. "It was confuse for me, because was written 'Myanmar.' But when I look at his face, it is not the Myanmar." He is banned for life from all future United Nations climate events. By this time, the BBC has declared Lord Monckton "the high priest of climate change skepticism."

Last snapshot: the Viscount of Brenchley testifying before the United

States Congress. Already, we have witnessed the daredevil Monckton. Also the Moncktons of endurance and subterfuge.

Here, finally, is the diplomatic Monckton. It's 2010, and to this congressional hearing on climate science the Democrats have invited four researchers. These include Dr. James McCarthy, the Harvard professor who once stood on the deck of a Russian icebreaker, amid disappointed tourists, to find the North Pole gone to puddle, waves, and a sad breeze. And, as their sole witness, the Republicans have imported Christopher Monckton, who brings a degree in classics.

He looks very British. The silver coastline of hair. The well-stocked and shirt-raising belly. The upticked eyebrows that accompany you from boy's school to gentlemen's club.

He is grilled by Representative Jay Inslee—now governor of Washington. There is something about a certain type of British personality that resists entering official situations with a completely straight face. Lord Monckton is the hearing's only participant to wear a pink shirt: his tie has a sort of cheetah pattern. Inslee, in our national style, favors the kind of necktie worn by G-Men in the movies.

"Lord Monckton," Inslee demands, "when did you start serving in the House of Lords?"

A pause. At home, this is a sore point. "Sir, I have never voted or sat in the House of Lords." A moment later, Monckton leans back, with an uncomfortable smile.

"So not only have the deniers—" Inslee thunders. "They not only can't produce *one* scientist to deny this clear consensus, they can't even send us a real Lord from the House of Lords. Now I think this says a lot about the status of this debate. Which we shouldn't be having."

It's a long-standing headache. Monckton is a real lord. He's a new-growth lord, the third Viscount of Brenchley. But he is the tall man who wishes to be taller. He uses the Parliament's coat of arms as his own logo. And in the heat of the interview moment—when things get exciting and loosey-goosey—he's apt to throw caution to the winds and explain that he *is* a member of the House of Lords. Only "without the right to sit or vote."

This has become so frequent the clerk of the real-life Parliament has taken steps. Posting an official letter. "I must therefore ask again that you desist from claiming to be a Member of the House of Lords, either directly or by implication," his letter begins.

"I am publishing this letter on the parliamentary website so that any-body who wishes to check whether you are a member of the House of Lords can view this official confirmation that you are not."

It took twenty climate scientists four months and forty-four pages to disentangle the inaccuracies and falsehoods in Lord Monckton's five-minute congressional testimony. Their verdict was: "chemical nonsense."

And here is a collection of press snapshots.

Lord Monckton has become an international counter voice, a denial authority. Representative "Smokey" Joe Barton, from the well-drilled state of Oklahoma, welcomed Lord Monckton to Congress this way: "He is gen-erally acknowledged as one of the most knowledgeable—if not *the* most knowledgeable—experts from a skeptical point of view on this issue of cli-mate change." The *New York Times* calls him "a leading climate skeptic." (And lists him among the three skeptic leaders globally.) CNN has elevated him to "Al Gore's chief critic." And Australia's *Sydney Morning Herald* describes Christopher Monckton as, simply, "The most effective climate skeptic in the world."

He'd only been at it since 2006. No credentials, no training. Which meant that climate denial had entered its decadent phase.

IT IS ABOVE ALL an inventive and ingenious life. We associate titles with Jane Austen and those khaki breeches you wear tucked into riding boots. The first viscount—ennobled 1957—was Monckton's grandfather Walter. When Edward VIII renounced the throne to marry an American divorcée, it was Walter Monckton, in a castle with sheeted-over furniture, who gave his abdication speech a final polish and escorted the shaky king to the microphone. Monckton's father was an army major general who stood for office on a strict animal rights platform: "All cats to be muzzled outside to stop the agonizing torture of mice and small birds."

Monckton entered public life in 1982, aged thirty, as a member of Prime Minister Margaret Thatcher's Policy Unit. Where he was viewed as uniquely unpromising. "One must hope," sniffed the *New Statesman*, that the other "appointees will be of a rather higher calibre."

After climate made Monckton a star, he'd be written about as Mrs. Thatcher's "favourite policy advisor." Also as her "science policy advisor." Because backstories enlarge in proportion to the front story—who you were then will be retouched to match what you are now. If it's true, Mrs. Thatcher spared the lord's modesty. In her memoir *The Downing Street Years*, given 914 pages of opportunity, Monckton's name isn't mentioned once.

He lasted four years. Earning a name, the *Spectator* wrote, as "a figure of fun." Monckton became well-known for clothes. It's the way of nobles, turning everything into costume, since one is already in the deeper disguise of being an ordinary nine-to-five type person anyway. Monckton was a clothes extremist: resolutely sporting the UK power hat, the bowler, during business hours. And then, for the motorcycle ride home, full biker leathers.

By 1986, he was out. Then Monckton became an editorial writer, of the hard-truth variety.

"There is only one way to stop AIDS," Monckton wrote in 1987. "Every member of the population should be blood-tested every month. . . . And all those found to be infected with the virus, even if only as carriers, should be isolated compulsorily, immediately, and permanently."

But not unsociably. The "isolated community," Monckton said, would be open to friends and family on "carefully supervised visits." Despite this bonus, Monckton endured underappreciation. "The homosexual lobby said, 'We know you, you're a Catholic, you don't like queers,'" he explained. (It's important to note, as with many deniers, that this was not the panic of the time. British novelist Martin Amis, writing in these same years, arrived at a very different conclusion: "We are in this together now.") He complained about "dotty feminists." When the first Saddam Hussein fight rolled around, the columns echoed with Lord Monckton's "almost deranged war whoops." Politically, the satirical magazine *Punch* wrote, "he is quite simply barmy."

By 1992, he was again out. To his *Evening Standard* colleagues, he'd been a "source of continual bemusement. . . . An unusual bird, prone to bowler hats, mobile phones, wild pronouncements and motorbikes." But: new, less amused management, Monckton was sacked.

This drops a curtain on his London phase. Monckton became one of those résumé ghosts who haunt every profession—phantoms patting the curtains, chasing scripts, bothering producers, hunting for their entrance back on stage. In this period, he was reduced to an act of art criticism. "These artists are guilty of the sin of our age," Monckton wrote. "A mad thirst for publicity, a loopy craving for the sensational headline with one's own name in it." One sufferer can always spot fever symptoms in another. (Two decades later, Australia's future prime minister complained of Monckton: "He is a professional sensationalist... he says these outrageous things in order to get into the press.") His byline had the sadness of the wandering man—so lost he's forgotten his destination. "Christopher Monckton is a freelance writer, who specializes in art and politics."

Monckton and his wife had set up housekeeping in a vast lodge in the Scottish highlands. He tromped the hills and glens wearing a kilt between the leather jacket and biker boots, under a ball cap that said COCK OF THE NORTH.

A Monckton surprise. His second-most-famous editorial had denounced the people and region. (The Scots were welfare junkies—human bagpipes whining for "a fix of English taxpayers' money.") "Really I have always loved Scotland, and the Scots," he explained, "and wanted to live here."

It's the thing about deniers. Opinion and belief become site-specific. Wherever they happen to be standing turns out to be the right place. In a sense, they are the world's champion believers.

So began the era of what the *London Times* calls the lord's "many money-making wheezes." He'd floated a business consultancy, Christopher Monckton Ltd. There was his James Bond stand at the London casino tables. He and his wife started a gentlemen's clothier: "Monckton's." The lord could be found at the King's Road shop, a tape measure riding his neck, speaking not as a science advisor or AIDS commandant; as a tailor. "The boxer shorts are made of the same cotton that the shirts are," Monckton would say. "Which some people think is quite smart."

Then the shop went under. Here was a man of undeniable gifts. Committing an undeniable act of squandering.

There turned out to be a reason. Monckton had been diagnosed with Graves' disease. It gives the lord's eyes their look of perpetual emphasis, as if he is forever disputing a tennis line call, staring down an argument, or alerting a friend to an open fly.

Graves' became the floating excuse, the explanation for things left undone. Healthy, Monckton told an interviewer, and "I would have become a member of Parliament, served in a cabinet, and made a good go of it." The opinion that medicalizes the life, that delivers a personality into the hands of the doctors. "The diagnosis was a relief," Monckton said, "because people tended to say I was under stress or mentally unbalanced."

He liked to think that the illness had prevented his "rising to any number of dizzy heights," he told another interviewer. "The fact is that I can't go too many miles from this house without falling ill."

So it was Monckton who had ended up sick in an isolated community. "Possibly dying," he said. "Bedridden for the best part of four years." During this ordeal, Monckton and his wife purchased a sixty-seven-room castle—a fixer-upper with a trout lake and staff of six. And into which they poured half a million pounds and the sweat equity of a gut renovation.

During these years, you could find Lord Monckton in the press. Pulled over for speeding. He was dressed in full motorcycle leathers, and explained to police he was hurrying to court to answer a *previous* speeding ticket. It expands our definition of what "bedridden" has to mean.

But these bed years also offer an origin story. His mind, freed from the distraction of piloting a body, gorged itself on problems, ephemera, speculation, fancies. When he rose, it was with his big idea: for a toy. Monckton partnered with a game manufacturer and created a fiendish 209-piece puzzle.

Because it would take forever to finish, he gave it a time-swallower name: Eternity. And to make his toy irresistible—this Wonka touch landed him back in the newspapers—Monckton offered a prize of pounds one million to the first solver. "I think I must be mad," he said. "Solve The Puzzle And You Could Be A Millionaire," said the box. "I am challenging the world," Monckton declared.

Solution, he estimated, would chew through three years. In the meantime, he would sell millions of units and become rich. "So that's one up for Scottish inventiveness," said Monckton, who was born and raised in England.

He flew to the 1999 International Toy Fair dressed full-on Scot. Kilt, hosiery, a *skean dubh*—the traditional highlands dagger—sheathed against his calf. Airport security understandably confiscated this concealed weapon. So the lord launched Eternity with a British Airways knife down his sock. Really, I don't think any denier had more sheer fun than Christopher Monckton.

There was no predicting who'd solve his puzzle. "It could be an eight-year-old child," Lord Monckton said, embracing the full spectrum of human possibility. Or it could be "an old granny," even somebody "in hospital recovering from an operation."

But he could say who wouldn't. "It won't be a computer," he promised. "And it won't be a mathematician either." As he'd later maintain about climate, expertise offers no advantage. The puzzle had been specially designed "so a computer can't do it. . . . You can't solve this by mathematics."

Eternity was completed in seven months; by two Cambridge mathematicians using a computer. Then the lord had to pay out. "Sometimes you live in a palace," he said sportingly, "sometimes in a hut."

To raise the prize money, Monckton had to put his sixty-seven-room castle on the block. This set the newspapers a loopy headline puzzle, which they quickly and cheerfully solved. "WHO WANTS TO RUIN A MILLIONAIRE?" And "IT ONLY SEEMED LIKE ETERNITY." And "TOFF MAKES POUNDS ONE MILLION PRIZE BLUNDER."

He delivered a Monckton-style speech. Each element requiring, like a napkin, its own fact check. "Eternity was the best-selling puzzle ever." That honor belongs to Rubik's cube, four hundred million units cursed at and twisted. "A quarter of a million people bought it." The lord later upped this figure to half a million. "I'm delighted it's been solved." He sat out the prize ceremony due to illness.

Years later, Monckton would offer the puzzle as his passport-of-achievement into climate science. "I am a mathematician," he declared to an Australian interviewer. "For instance, I invented the Eternity puzzle. . . . We sold shedloads." The more famous member of the solving duo—ranked by the BBC among the world's top mathematicians—on Monckton: "Quite an interesting character. . . . Um, not a mathematician."

Seven years after Eternity, with all the pieces settled, he'd confess that castle loss and puzzle setback had been discrete phenomena. Monckton had cooked the whole thing up as a publicity stunt. "History," he reassured a journalist, "is full of stories that aren't actually true." Once climate celebrity happened, setting personal honesty at a premium, he reverted to the original, made-homeless position.

He was forty-seven now. The October age, with a phone always ringing somewhere and the hour tugging at your pant leg. He graded quietly out of the headlines. He kept up with developments in the field of mathematics. You could find a shelf of Monckton's contributions at Amazon. Puzzle books: *Sudoku X, Sudoku Xmas, Junior Sudoku X*.

It is in many ways a resourceful and creative life. In 2006, a friend with a London finance house hired Christopher Monckton Ltd. to investigate the climate issue.

"I spent a month rootling about" in the scholarship, the lord said. (He was never one for library heroics. Asked what university had taught him: "To enjoy myself, and that a lot of learning doesn't come from books.") No worries, Monckton informed his friend; there was nothing to it.

The friend passed Monckton's report along to the *Daily Telegraph*. Who published. It was the news everyone longed for. So many hits the first day, Monckton's piece crashed the server.

This is the man of broad experience and mixed fortune Republicans sat at the witness table beside Harvard's James McCarthy. A former president of the American Association for the Advancement of Science, who had taught and studied climate for thirty years.

THERE'S A LINE IN David Mamet's *Glengarry Glen Ross*. "You get these *names* come up," Ed Harris complains. "You ever get 'em? You had one you'd know. . . . They keep coming up."

Denial had hit a dry spell for applicants. Philip Morris started getting out of the business in 1997. The reason for their departure was seemly. What began their freak-out were the rules against inflight smoking. The final slide was touched off by a flight attendant named Norma Broin.

A lifelong nonsmoker, Broin pinned on stew's wings in 1976, crewing in that gray wind of ashtrays and exhalations five miles up. As one writer observed, a smoking section on the airplane was like having a peeing section in the pool. In 1989, Broin's X-ray showed cancer. She came out of the anesthetic, having lost most of a lung, and her daughter sidestepped her kiss. "Oh no, Mommy—I don't want to get what you've got."

She led a Florida class action suit—flight attendants—and became known as the Perfect Plaintiff. What you'd search for online, if you wanted to junk up tobacco. A lifelong Mormon—the church is pretty much anti-everything—Broin had never even *known* a smoker. She was also a former competitive swimmer (healthy); collegiate debater (intelligent); vegetarian (proper diet); and former model (presents well in court).

Her suit occasioned what the Miami *Herald* called "a stunning battlefield surrender." At ten a.m., on a spitty October Friday, lawyers took a good look at Norma Broin . . . and tobacco's Wall of Flesh collapsed. The award was for a third of a billion dollars. The Wall of Flesh shuffled from the bench, and dispersed. (Miami *Herald*: "the platoon of tobacco lawyers filed out of the courtroom without saying a word.") "Broin," Allan Brandt writes in *The Cigarette Century*, "marked the first time tobacco had ever settled a case."

A scholar with the Georgetown University Law Center marveled in the *Washington Post*, "They have lost their resolve in a miraculous way—and seem to have lost their will to fight." Thirteen months later, fall 1998, the industry settled with forty-six states at once. And the long, foot-dragging retreat from denial began.

Eight years later, federal judge Gladys Kessler—ruling in a Department of Justice suit brought twenty-three months after Broin—made a statement of tobacco's overall guilt. After all the denials, the clever stratagems, the unpersuadable scientists, it makes for sweet reading. "Over the course of more than fifty years, defendants lied, misrepresented, and cheated the

American public ... in order to achieve their goal—to make money with little, if any, regard for individual suffering."

In 1998's icicle months, the American Petroleum Institute hosted meetings. They were putting together a team for their own assault on climate science. The think-tankers trucked in. Exxon's main lobbyist flew the corporate flag. Also electric power's, and a specialist from the exciting world of public relations. And, on the denial side, the former tobacconists came. S. Fred Singer's Science and Environmental Policy Project sent a delegate—Dr. Singer's associate and wife, Candace Crandall. Frederick Seitz's George Marshall Institute was on hand. Also Jim Tozzi's former colleague Steve Milloy: Milloy brought to bear his new influence as director of Philip Morris' Sound Science Coalition—he came wearing his ASS hat. ("There's really only about 25 of us doing this," Milloy later told a reporter from *Popular Science*. "It's a ragtag bunch." The magazine offered Milloy's simple economic rationale: "He moved into climate denial as funding from the tobacco lobby began to dry up.") And this must have been a nostalgic thing for tobacco to read about in the papers. Like seeing snapshots of the vacation home where you passed some memorable summers, now being enjoyed by another family.

The Union of Concerned Scientists reported that, after 1998, just by itself, Exxon pumped about $32 million into a "network of ideological and advocacy organizations that manufacture uncertainty on the [climate] issue." (This excludes the ten million annual dollars a former executive said Exxon devoted to "climate 'black ops'" every year between 1998 and 2005. And the many millions pipelined into the fight by the fuel-loving Koch brothers; estimates there run even higher.)

Exxon gave generously. To established, crewcut-and-horn-rim outfits like the American Enterprise Institute. And they also funded wildcat crews with Mark Mills names: The Committee for a Constructive Tomorrow (they sent Christopher Monckton falling from the sky and driving those camels); the Free Enterprise Action Institute (it sounds like a team of accountant superheroes); Frontiers of Freedom (vigilant backpackers). All these groups evangelized doubt.

Exxon was a doubt exporter, too. Which helped scupper the program.

In 2006 an executive with England's Royal Society wrote Exxon in the British style; courtesy, with a fang.

"My analysis indicates that ExxonMobil last year provided more than $2.9 million to organizations in the United States that misinformed the public about climate change." The society, he explained, would be grateful for a European list: "So that I can work out which of these have been similarly providing inaccurate and misleading information to the public."

In January 2007, Exxon announced it was cutting denial ties. (They did not. For propriety's sake, they did slow down.) With less money around, the field became susceptible to a self-funding newcomer like Christopher Monckton.

The Petroleum Institute plan had called for fresh faces. "Individuals who do not have a long history of visibility and/or participation in the climate change debate."

Like most serious hungers, this was easier to state than satisfy. Public relations watchers Sheldon Rampton and John Stauber put the difficulty into geographic terms. "The global warming consensus," the journalists write in *Trust Us, We're Experts*, was "so strong that the oil and auto industries have been forced far afield in their search for voices willing to join in their denial."

All of which explains the reentry on stage of Lord Christopher Monckton. He became, as the *New Yorker* writes, a "professional climate-change denier." Monckton later described the work as "my calling."

DENIAL HAD BECOME the wish-granting idol. You approached on tiptoe, knelt with the offering (work, support, reputation), received your heart's desire. For Monckton: audience and consequence, readers and the loopy headline.

When he'd become famous again, Monckton told the BBC a revealing thing. Illness had kept him "out of the action for twenty-five years."

This was 2011. The inactive years were the period when he'd written editorials, suggested detentions, turned Scottish, founded a clothier, received and paid speeding violations, renovated and sold a castle, hiked in kilts, marketed a puzzle. But counting backward brings you to 1986—the year he left Margaret Thatcher's service at 10 Downing Street. It's like getting a good over-the-shoulder view of your neighbor's answer sheet, with its intimate math. The illness had been loss of influence, separation from power. Denial cured that.

Suddenly: overflow crowds. Radio duplicating his voice, newspapers reprinting his opinions, his face multiplying across screens. Success is how many places will reproduce you.

It happened fast. Because the movement needed a headliner; because the lord is a thrilling speaker. He has that gift, for making passionate agreement seem the one civilized response. He gave you the old, old stuff—the stemwinder speech, where you step into a hall to experience all of it. Moments of comedy and outrage. Tall, hissable villains; noble, outmatched heroes. And in the end, it isn't the university mathematicians on their computer who beat warming. It's the eight-year-old boy, the bedridden granny—it's Monckton who solves the puzzle. Monckton gave you climate change with the proper ending.

A speech Monckton delivered in 2009 has become the single most-clicked YouTube address on the science. His voice booms. It teases out a moment, drops to its knees, invokes statesmen and poets. "So thank you America," the lord told the hushed and packed house that night in Minneapolis. Likening a climate treaty to the threat of invasion. "You *were* the beacon of freedom to the world. . . . So I end by saying to you the words that Winston Churchill addressed to your president, at the darkest hour before the dawn of freedom in the Second World War. He quoted from your great poet Longfellow." Voice now a husky whisper. " 'Sail on, oh ship of state. . . . Humanity with all its fears, with all the hopes of future years, is hanging breathless on thy fate.' Thank you." To a stunning weather of applause.

His abilities ran the gamut—Longfellow to low. Next year, Monckton led a joyous Tea Party crowd in springtime chanting across from the White House.

"Now I know in American-speak you have a word for global warming." The lord cupped a palm by his ear. "Can someone tell me what it is?"

The crowd: ". . . Bull . . . shit . . ."

"Ba-rack Hussein Obama," the lord explained, waving a finger, "is now hiding behind the drapes of the Oval Office. He cannot hear you. Global warming is?"

The crowd: "*BULL* SHIT!"

"That's better. I think he heard that one."

This is who Monckton—onstage, irresistible—always was. When you know somebody, you generally know them as a disappointed version. It's only on rare occasions that you get to meet the original idea. Four years in, opponents were calling him denial's number-one spokesperson.

He hit notes that would charm Arthur Robinson's ear: the paranoid standards. (Climate science's eyepatch-wearing objective was to "impose a communist world government on the world.") Stuff that would bring a tear to S. Fred Singer's eye. So what if the whole thing were true? Why not greet it poolside, an umbrella in your drink? "It is better just to sit back," Lord Monckton advised, "enjoy the sunshine, and adapt."

Celebrity attracts detractors. A former Conservative Party chairman set out to correct the record. "Lord Monckton isn't taken seriously by *anyone*," he told a news program. "He was a bag carrier in Mrs. Thatcher's office." The hopeful sweaty face at the back of the photo, bobbling straps and packages. Didn't stick.

A physics professor from Minnesota's University of Saint Thomas devoted months to harvesting errors in the lord's Minneapolis speech. (Fact-checking Lord Monckton: there's a sand-counting endeavor.) The professor had gone at it from a sense of duty, outrage, and neighborliness: Monckton's big speech was in his city. "I asked myself, 'If I don't, who will?'" the professor said. "This guy is a great speaker and he is very convincing." Nearly every point in the lord's speech was wrong. The professor posted results to YouTube: Monckton's speech clocked in at ninety-five minutes. The corrections ran eighty-three, like the map in the fable that's the same size as the territory it describes.

A Brigham Young academic analyzed Lord Monckton's figures—these amounted to "a rather stunning display of complete incompetence," the professor said. "He's just making it up." And still Lord Monckton continued his endless tour. At the behest of his calling.

Lost fights exert a tug. The battlefield where your head was handed to you—something compels us to return, raise our inadequate weapons, and reawait delivery. Lord Monckton had first sent eyebrows climbing with his ideas about AIDS and housing. (These had attracted a following. After one Australian event, a wellwisher sought out Monckton by the podium. "I endorse your stand on homosexuals and AIDS," the man said thickly. "They should be locked up. They should be exterminated.")

In 2011, Lord Monckton made a surprise announcement. Though heretofore not a medical man, he had cured AIDS. Also a host of other debilities—he'd even beaten his own Graves'. "Eighteen months ago," Monckton told the BBC, "I cured myself with an invention which shows much promise." He sat with legs modestly manspread, beside a cushion whose stitching said LORD OF THE MANOR. "We're curing people with everything from HIV to malaria to multiple sclerosis."

Teammates began to depart the huddle—denial in its reputation-preserving form. "Monckton is not the only climate skeptic in town. He does not represent 'us,' whoever 'us' might be." "Christopher Monckton has had no, absolutely no contact with me . . . zero."

He couldn't leave it alone. In 2014 Lord Monckton published an editorial about gay men. Who were always advocating their lifestyle—"deathstyle," Monckton wrote, "more like." They were "cheerless," had an "average of five hundred to one thousand sexual partners." (Though the especially motivated could sample "as many as 20,000.") No mention of his cure. It was high time, Monckton wrote, "to end their eerie dominance of public life." They'd "had it their own way" for too long.

British tab the *Daily Mirror* handled response. Gay and non-gay people averaged the same humdrum number of partners: six. "Yes, this man used to be pretty important in politics. Scary." Large-scale flight. This was the old, old stuff that just did not go anymore. A former political ally wrote that there was "no place for men like Monckton" on his side of the aisle.

There was one place he'd always be welcome. That same year, Lord Monckton stopped over at the Las Vegas denial awards ceremony. The event featured a performance by Austria's Kilez More. A conspiracy rapper, who offered some provocative challenges to mainstream science.

No it ain't, I'm telling you you're off the wall
Man ain't causing climate change—yeah you!

Here was another spot zoned for irony. And our bonus final snapshots. Dr. S. Fred Singer hoisting a microphone, and in a quivery voice announcing that deniers were correct not only about climate but "the right approach to life." Arthur Robinson accepting a trophy for Voice of Reason. It is probably the only point in his life when he will have to endure that description. "The world depends on us," he told the gray-headed crowd. "This is a battle we're in. Whether we like it or not."

Not even a decade, and Christopher Monckton had secured his place among the immortals. Lord Monckton's award—the camel-driving skydiver, the computer-resistant puzzle maker, the self-curer and nonvoting member of Parliament—was for Dauntless Purveyor of Climate Truth.

THE BUSINESS CARDS
AND THE STRAIGHT NOODLE

Word of the Year

The week before Christmas—with its dream reversals: lights outside the home, trees in—and there was something extraordinary to celebrate. Action on climate had come at last.

A product detrimental to the sky and to interests on the ground would be removed from circulation. The bill rattled through Congress. It spent a few months in the Senate garage, for inspection and fine-tuning. And on December 19, 2007, everyone was wearing their best suits at the Department of Energy, rubbing their hands.

The bill, symbolically, was delivered to the White House in an oil-pinching Toyota Prius. The president, titanically, arrived at the signing ceremony with his motorcade, in the passenger cab of a heavily armored Cadillac DTS.

George W. Bush took the podium in a light, pinball grip.

He used words—this president who had skirmished and standstilled the climate scientists for seven years—that sounded a little funny on his lips: *renewable, energy efficiency, emissions, greenhouse gases.* "Today we make a major step," Bush announced, "to reducing our dependence on oil, confronting global climate change." For one thing, he'd just gone out and more or less slashed the tires of his own ride: the new thirty-five-miles-per-gallon fuel standard would garage the First Limousine by 2020.

There was the usual Last Supper craning and angling around the signing desk. The inkwork took a few seconds. America had done its "duty to
future generations."

The fireworks mood lasted all holiday season. The law's congressional sponsor called it a major advance against "the cataclysmic challenge of
our time: global warming." The Natural Resources Defense Council had
seen "the first step." The retirement of the one product would "significantly
reduce global warming pollution." General Electric's senior counsel regis-
tered his thumbs-up in *Bloomberg News*. "I think it's huge. The amount
of energy that's being saved," the executive said, "is more than has been
achieved since 1986 for all appliances combined."

Checking in with GE made sense. Because of the strange, poetical
place everyone had agreed climate action should begin. Two decades after
Jim Hansen declared the greenhouse to be in effect and changing the climate now.

WASHINGTON WEATHER ROLLS IN from surprising headlands. In
2005, Leonardo DiCaprio took the guest seat on *Oprah*. "Why are you so
passionate," the host asked him, "about global warming?"

DiCaprio had the cautious sound of actors traveling off-script: sticking
to the path worn by familiar phrases. "You know, global warming is not
only the number-one environmental problem that we're facing," the star
explained, "but one of the most important issues facing all humanity."

He'd brought along a set-warming gift. Thomas Edison's old breakthrough device—the one that consumed years, until paper lasted through
the night. Recast as symbol; an energy-saver, the compact fluorescent bulb.
"We all have to, you know," DiCaprio said, "bring environmentalism to the
American consciousness."

Oprah weighed the device in her palm. "Well," she announced, "I *love*
this light bulb." After a moment she said: "It feels heavier, though."

The actor agreed. "It *is* a little heavier."

A couple months later, under a sweaty cover line ("Global Warming:
Be Worried, Be Very Worried"), *Time* offered tips on "How to Seize the
Initiative." There was the fine example of Coldplay's second album. Pro-

duction, tour, vans delivering CDs: even music could bruise the sky. So the musicians had prepared a bandage—ten thousand mango trees, to be planted in Southern India. Eating carbon, dropping fruit. "A sweet deal all around," *Time* said. (A fifth of the saplings never arrived; nearly half the rest withered. "For a band on the road all the time," explained a source within the Coldplay organization, "it would be difficult to monitor a forest.")

Vanity Fair debuted an annual Green Issue; Julia Roberts half-smiling on the cover. "If Julia Roberts can do it, so can you. . . . Roberts is proof that it's never too late to start caring for the earth and that it can all start at home." (When the second issue dropped, Treehugger.com encouraged readers to "dog-ear your favorite exhibits of Green's staying power!")

The NFL was on a drive to bring the 2007 Super Bowl deep into carbon-neutral territory. "We have to jump in the game and take some kind of action," a league official gruffly explained. "Instead of waiting on the sidelines." (Vocabulary is homesteading—making the unfamiliar habitable. The *New Oxford American Dictionary* picked "Carbon Neutral" as 2006's Word of the Year. Its 2009 word, new fads crowding the horizon, would be "Unfriend." That same year, *Vanity Fair* folded its Green Issue.) The torture-and-rescue drama *24* declared its intention to become television's first carbon-neutral weekly series. "When appropriate," producers said, all but guaranteeing some long nights in the writers room, the show would be "incorporating the issue of global warming and the importance of carbon emission reduction in storylines." A salute to Flava Flav on Comedy Central became history's first carbon-neutral celebrity roast.

THE HOLLYWOOD DREAM is the four-quadrant movie. Reaching into the pockets of every demographic. Under twenty-five and over, female and male. Here's the concept as expressed in a trade-publication sentence: "Warner Brothers attempted to counterprogram *The Avengers*. . . . But the strategy backfired when *Avengers* turned into a four-quadrant monster."

With celebrities, TV, the NFL, and *Oprah*, climate was bridging quadrants. It picked up news in the fall of 2006.

Rupert Murdoch owned papers, publishers, studios and networks, the reliably anti-[insert mistreated noun] Fox News. Murdoch announced his change of heart from a Tokyo lectern. "I have to admit," Murdoch said. "Until recently I was somewhat wary of the [global] warming debate. I believe it is now our responsibility to take the lead on this issue."

The mogul sat for interviews. Ninety-nine percent of experts now agreed about the science; he himself now owned a hybrid car. He would no longer have any truck with the sorts of politicians who obstruct action on climate change. A journalist asked if Murdoch had any favorites for the 2008 race. It's an example of how the rich live different: they follow a private calendar. "I don't know who's sailing," Murdoch replied sunnily. He was a competitive yachting enthusiast. Then—the mental finger snap: "Ah! I thought you were talking about the America's Cup."

Part of this was the planet—which had produced some spectacularly demoralizing weather effects. Australia, Murdoch's home continent, was then squinting and baking through its worst-ever drought. In 2005, Hurricane Katrina had swamped an entire American city.

Another factor was the sort of change business leaders are equipped with special barometers to read. Shifts in the social weather. There was Jim Rogers, CEO of Duke Energy—America's third-largest power company—announcing plans to decarbonize his entire operation by mid-century. Rogers had felt it on the wind: climate legislation was sure to get written. Executives like Jim Rogers wanted some measure of influence over the pen. Rogers shared his thinking with the *New York Times*—the classic Washington dinner advisory. "If you're not at the table, you're going to be on the menu."

As they say in the horror movies, the calls were coming from inside the house. In fall 2006, Frank Luntz confessed to broadcasters, "My own beliefs have changed." The pollster now accepted global warming. In the long media buffet line, there are few dishes sweeter than a public recanting.

A couple years later, Luntz would conduct his first focus sessions on behalf of the environment. His advice: (a) Lose the frozen imagery. American paychecks meant a lot more than "glaciers or polar bears"; (b) Ditch the entire lexicon. "Sustainability" and "green jobs" were ear-closers. The absolute worst was "carbon neutral." It suggested a guest list you weren't on—"Hollywood types flying across the country and buying carbon offsets."

Better to say that polluters would be held accountable. This was the trick environmentalists never learned. You could keep giving the right answer forever; if you didn't pick up the classroom slang, you'd go on receiving the F. The Luntz Company slogan was, "It's not what you say, it's what they hear."

(An effect once noted by Friedrich Nietzsche, without benefit of focus groups. "It has caused me the greatest trouble, and still causes me the greatest trouble: to realize that what things are called is unspeakably more important than what they are.")

American business was hearing and saying some startling things. Even Walmart, the parking lot at the heart of Red State America, had gone green. Every Walmart fact is oversize: entered with a condor quill in a Superman ledger. The biggest single employer in America. The largest purchaser of manufactured goods, of electricity, in the world.

In 2006, Walmart traveled to Congress. The company would welcome regulation on greenhouse gases. "The environment," their CEO told the *New York Times*, "is begging for the Walmart business model."

The retailer appointed a vice president of sustainability. His business card, one-half normal size, was printed on 100 percent recycled paper. Walmart badgered suppliers. The budget-stretcher Hamburger Helper used to come in a box wide enough for curly noodles. Walmart prevailed upon the manufacturer—flatten those noodles out. Thinner boxes meant fewer trucks, less fuel, an 11 percent drop in Hamburger Helper-related greenhouse emissions. "That is *such* a profound story," enthused the head of footwear giant Timberland, at a sustainability conference. "Like, straighten out the freakin' *noodle*, and the world gets better?" A much-decorated Harvard political scientist would later call these months the "politically pivotal juncture in late 2006 and early 2007."

In January 2007, the heads of ten major corporations formed up and called for immediate climate legislation. They were led by General Electric's Jeff Immelt, the most influential executive in corporate America. (He'd been "picking up a change in public mood," the CEO told *Time* environmental writer Eric Pooley. "People wanted action.") These included investors and engine makers, getters and spenders, movers and shakers: General Electric, Lehman Brothers, Du Pont, British Petroleum, Caterpillar, and of course Duke Energy. The group brought their case to Washington. It was as if the companies were presenting themselves to the desk sergeant at the police station, wrists extended: "Regulate us." Menu-avoiding Jim Rogers spoke for the group. "We know enough to act now. We must act now."

And the weather—swaying trees, rattling windows—had hit the District of Columbia.

A "flurry of news conferences and proposed solutions," the *Times* reported that January. On climate, there was "evidence of the changed mood all over Washington."

Newt Gingrich's noisy crowd had been voted down. California senator Barbara Boxer now led the committee that had climate change. For more than a decade, Boxer told *USA Today*, lawmakers had orbited a single question: Is there global warming? "I'm done with that, " the senator said. "I'm over it. We need to move forward."

Nancy Pelosi now banged the gavel as Speaker of the House. "The science of global warming and its impact is overwhelming and unequivocal," Pelosi said, "Now is time to act."

There were feats of matchmaking. Pelosi starred the next spring in a climate ad. Her scene partner was Newt Gingrich—the lawmaker who had once saluted environmentalists with a dinosaur tie. "We don't always see eye-to-eye, do we Newt?" Pelosi mugged. "No," said Gingrich, smiling unhappily. "But we do agree: Our country must take action to address climate change."

It was the end of a long drive—Prius or motorcade—when you look up, and the possibilities are suddenly big in the windshield. Legislation that "once had a passionate but quixotic ring to it is now serious business," is how the *Times* put it. "The climate here has definitely changed."

IT HELPED THAT, for once, warming had a sense of timing. A new IPCC report was due. On release eve, lights were symbolically dimmed across Europe—as they'd been in America for the death of Thomas Edison. At the Eiffel Tower, the Alcalá Arch in Madrid, the Athenian Parliament, Rome's Colosseum.

The Intergovernmental Panel on Climate Change was the final reboot of George W. Bush. The program was started by the UN a few weeks after his father's election. The original George Bush welcomed members to Washington. "It is a great pleasure to be the first President to address this distinguished group," Bush said. "The United States is strongly committed to the IPCC process of international cooperation on global climate change. . . . The stakes are very high."

With the IPCC, More Research achieved its Platonic form. 130 countries, thousands of scientists. Many different cultures, languages, national sensitivities. "Conservatives and skeptics in the United States administration might have been expected to oppose the creation of a prestigious body to study climate change," writes historian Spencer Weart.

But viewed from the proper angle, the panel was secretly ideal. "A complex and lengthy study process," Weart explains, "would delay any move to take concrete steps to restrain emissions." And anything from the UN was unlikely to dazzle with organization, or set land records for speed.

It took twenty years.

The panel was understood to be the opposite of flame-eyed and wild-haired. It was sober, deliberative. A language barrier favors moderation. Thousands of passengers in the same vehicle, travel guides spread across laps, everybody trying to agree on destination and a route.

In 1990, seven hundred scientists reviewed the IPCC's First Assessment Report. "A clear scientific consensus has emerged on estimates of the range of global warming that can be expected," read what was called the Scientists' Declaration. "Countries are urged to take immediate actions."

Then five years of More Research.

The panel's Second Assessment Report, 1995, found "the balance of evidence suggests a discernable human influence on climate." Environmental writer Bill McKibben heard hoofbeats; he calls this, "Perhaps the most significant warning our species, as a whole, has yet been given."

The world's scientists had rooted through all available data points. As Charlton Heston might've put it, Global warming was made out of people.

Then six more years. As science percolated.

The IPCC is a sort of scientific Rotten Tomatoes. It's got a ratings system. The operative phrase is *words of estimative probability*. This concept comes to us from the jumpy and paranoid CIA of the fifties. If some analyst wanted to announce a "serious possibility" of unfavorable outcomes, it helped to have everybody on the same statistical page.

IPCC rankings climb the probability scale. Very Unlikely (less than a 10 percent chance) to plain old Unlikely (10 to 33 percent; no stars). More Likely than Not (above 50 percent) and Likely (66 to 90 percent; these are generally favorable reviews). To Very Likely (90–99 percent) and Virtually Certain (99 percent, plus whatever rarefied air remains above). These last two are more or less Certified Fresh.

2001's Third Assessment Report found the odds that humans were behind the previous fifty warmer years had kicked up to Likely. While deniers sulked and plotted, the IPCC had identified "new and stronger evidence that most of the observed warming [is] attributable to human activities." The UN Environment Program director told the *Washington Post*, "We should start preparing ourselves."

(*The Third Assessment Report.* 2001. Director: The IPCC. Ensemble cast. A large international team of researchers tries unsuccessfully to warn humanity of an approaching crisis. Comedy-drama. Run time: 156 months. Rating: Likely.)

You could feel it. A final answer seemed to hang low and steady on the horizon. It took six more years.

Then it was here. February 2, 2007. And the president's final rerun: The first George Bush handing off the problem to his reboot. Babies born when the IPCC began its study were posting résumés on LinkedIn. It had been half a century since Roger Revelle taped the page into his report, calling warming "a large scale geophysical experiment of a kind that could not have happened in the past nor be reproduced in the future."

The Fourth Assessment Report dropped in the early hours, Paris-time. Pale sky, damp sidewalk, clicking shoes.

Conclusions were so blunt they made the TV anchors double up, because what else was there to say? Brian Williams, NBC: "An international group of scientific experts came out today and said in no uncertain terms: global warming is real, and it's almost certainly caused by what we humans do to this planet." CNN: "The world's top climate scientists couldn't make it any more clear: Global warming is real, it's getting worse, and we're to blame." (And, for the triple, ABC: "There is no longer any question that the earth is warming . . . and that those gases are produced by us.")

Leaders from nations with dimmed monuments spoke to reporters. French president Jacques Chirac made things sound Gallic, subtitled, and existential: this was warming with a Gauloise between its fingers. "We are on the historic threshold of the irreversible," he said. Italy's environmental minister gave matters an Aesop spin: "While climate changes run like a rabbit, world politics move like a snail."

Climate Change, the Fourth Assessment reported, was "unequivocal." Our human responsibility was upgraded to Very Likely: 90 percent and above, Certified Fresh.

A celebration—an answer. A grim one—that answer was bad news. "This day marks the removal from the debate," announced the head of the Environment Program, "over whether human activity has anything to do with climate change."

The science panel co-chair was Dr. Susan Solomon. Also a rerun for her. In 1987, Solomon had helped establish that industrial gases were indeed shearing a hole in the ozone. She told reporters at the Paris press conference, "It's later than we think." Whether a fix might be found in time for her children was an open question: something could probably still be done for the grandkids. Dr. Richard Alley, who'd spent his career with the haunting blue of polar ice, was lead author. "Policy makers paid us to do good science," Alley said. "And now we have very high scientific confidence in this work. This is real, this is real, this is real. So now act. The ball's back in your court."

This effectively concluded the Bush program on global warming. Six years of smothering scientific fires under procedural blankets. With data curling around the edges anyway. Now here the answer was, from a program first welcomed by his father.

February is also Academy Awards month, with its worldwide viewership. Former Vice President Al Gore's frog-positive *An Inconvenient Truth* took Best Documentary. (Melissa Etheridge even won for Best *Song*. Working at the margins of the possible, she'd found a rhyme for the title. "Now I am throwing off the carelessness of youth / To listen to An Inconvenient

Truth.") When Gore flew to the capital for a climate speech, spectators lined up twenty-four hours in advance and the papers said he'd returned in triumph. "Gore has attained," the *Times* wrote, "what you can only call prophetic status." For the first time, warming walked the hallways with cool friends. Later that year, the vice president and the Intergovernmental Panel would share the Nobel Prize.

And in a *New York Times* poll, 90 percent of Democrats, 80 percent of Independents, and 60 percent of Republicans agreed "immediate action was required" to curb warming and "deal with the effects of global climate change."

The *New Yorker*'s Elizabeth Kolbert had taken her catastrophic field notes—jetting between melting spots—two years earlier. Now she saw non-catastrophic conditions. The IPCC report was "an important, perhaps historic event"; Washington itself was undergoing "an important political shift." And all three major presidential contenders—none weighing anchor for Rupert Murdoch's America's Cup—were on board. Sen. Hillary Clinton: "This is a problem whose time has come." Sen. John McCain: "The argument about climate change is over. Now it's time to act." Sen. Barack Obama: "Will we look back at today and say this was the moment we took a stand?"

Warming had become—culture, business, science, voting—a four-quadrant monster.

Even Exxon had packed a duffel, headed for the wharf, shipped out. A year earlier, CEO Rex Tillerson had called scientific consensus "an oxymoron." Now his public affairs guy was scheduling media calls to clarify that the oil giant had never been in the denial business. ("That is flat wrong," the publicist explained.) And two weeks after IPCC, the supertanker captain himself reflagged.

Rex Tillerson spoke at a fuel industry conference. "The risks to society and ecosystems from climate change could prove to be significant," he said. "It is prudent to develop and implement sensible strategies that address these risks."

If the story had run in a trade publication, the headline would've been, "Issue Tracking Like a Superhero: Warming Set for $140M Opening Weekend with All-Quadrant Appeal."

It's like an episode of *Gilligan's Island*, or *Lost*. You know something's going to come along and wreck it—nobody is getting off this island.

What did they do with all this momentum? They gathered it together, then applied it to the product back at the start of everything. In March 2007, Congress and environmentalists set out to ban Thomas Edison's light bulb.

2.

How to Green Your Sex Life

I wanted to see climate law in the process of creation. Because it's a rare thing, like a shy comet or the kind of reluctant flower that opens once a generation. So in the fall I headed for Washington. I also wanted to find out: All the decades, scientists, leaders, deniers, fights, warmings—and, at last, momentum. Why were we spending it on light bulbs?

It seemed an error in scale. As if an AI had won at Powerball, then blown the whole jackpot on socks.

A new outfit formed up one month after the Oscars and the IPCC: the Lighting Efficiency Coalition. High-profile members were two green groups. The Alliance to Save Energy—Duke Energy's Jim Rogers, in his ongoing quest for table space, was a co-chair. (Dedication pays. Two years later, *Newsweek* would name him one of the World's Fifty Most Powerful People.) And the National Resources Defense Council; on *Curb Your Enthusiasm*, it's the group Larry David's wife does fundraising for.

I headed to the alliance first. Staff were all incredibly nice. Lots of their garages had hybrid cars. (One couple competed to see who could tease the best performance from their Prius. Which must be a form of environmentalist loveplay.) Motion detectors were set to kill the lights if you stepped away from your desk, or just fell into a deep think. It was a tense, shipboard America they aimed to make general. The alliance even had recycled *carpeting*.

The Alliance to Save Energy offers a clear, Weight Watchers prescription: Not fitness but diet. Thin the emissions by limiting the intake. That old anxious idea from the seventies, conservation, had been given a new haircut and fresh clothes and now went around the District as *energy efficiency*.

In her nice office—I kept waiting for the lights to cut out—I sat down with longtime alliance president Kateri Callahan. Who told me I should feel bad for drinking Poland Spring.

"This, to me, is the poster child for what you *shouldn't* be doing," Callahan said. "Just the sustainability of all the plastic bottles we're creating. A lot of it is filtered municipal water anyway. It irritates the hell out of me."

Interview people at the delta of business and government, and you'll be double-teamed. They'll bring a publicist. Acting as guard dog, human thumb drive (facts and phrases), and as a local cinematographer, to catch and frame moments in which their employer especially shines.

Callahan had the brisk, broad-shouldered delivery of someone from money who has inherited a sports team. Lots of her work, as she saw it, was missionary; the pulpit was cable news. "We have to get our message out," Callahan said. There's a classic marketer's number—how many nudges it takes to bring a customer from the cushions to the sales desk. "The more you hear it—what do they say in marketing?"

"Repetition, repetition," her publicist supplied. "Seven times."

Callahan told a funny story about cycling trips with her husband. Her publicist fiercely beamed: "They *challenge* each other."

The basic message went out unassisted. "You don't create an environmental footprint with energy you're not using," Callahan said. "So the real way to have no footprint is not to use the energy in the first place." Which was why the light bulb.

Callahan, too, had felt the groundswell: action would come this year. "Polls are showing people want Congress to take action on climate," she said, "that is not a Band-Aid but a long-term fix."

The bill, as Congress hunkered and tinkered, would include a bump in vehicle fuel efficiency. But since gas mileage was an anxious fact of life and dealerships already, this wouldn't strike voters as especially climate-related.

With spray cans and the ozone layer, you got the one-to-one correspondence. The incandescent light bulb would be the first product to fall to warming. And as a message it seemed so . . . paltry.

What I really thought was that it was like a movie with a busted three-act structure. *Jaws*, if *Jaws* didn't work. Act One: Police chief identifies the shark problem. Act Two: More research, some loss of life. Until the thrilling Act Three climax—where the chief speaks out on the dangers of heat stroke and melanoma, the advantages of staying off the beach. If climate could be addressed by something this size, it meant the problem was either tiny, or too terrifyingly large for any gestures beyond the symbolic.

(A year later, journalist Michael Pollan would address a different movie in the *Times*. "I don't know about you, but for me the most upsetting moment in *An Inconvenient Truth* came long after Al Gore scared the hell out of me," Pollan explained. "No, the really dark moment came during the closing credits, when we were asked to change our light bulbs. . . . It was enough to sink your heart.")

I said I thought there might be more attention-grabbing places to start. "Oh ho ho," Callahan said. Four billion sockets on the national lampscape. Lights go frosted and tinkly after eleven months. "So making a change," she said, "could be significant."

And the incandescent bulb, an energy fatso, had always squandered 90 percent of its electricity on heat. The compact fluorescent was a sleek energy athlete—able to meet all nutritional goals with a morning sip of protein shake.

Even Walmart had jumped on the fluorescent bandwagon, Callahan said. The retailer had set a goal: put fluorescents in a hundred million shopping carts in one calendar year.

2005 had been a low-water mark for Walmart. Daily black eyes from the press: bad wages, skimpy health care, supply chains in China. Walmart's CEO gave the *Times* a frank battlefield assessment. "There is not the ability to change as much in many of those areas," he said, "as we can change in this area of environmental sustainability."

So I could understand why Walmart had gone big for the bulb. But why had the alliance, the National Resources Defense Council, Congress? Why not oil? Why not coal?

Callahan sighed. "I think it seems like an easier thing to do." Callahan had been working around people's tolerances for years; experience taught her not to set the bar very high. "If they think they can be part of the solution, they'll do it. So long as it doesn't tax them too much." She shook her head. "So give 'em easy things to do. Tell them to buy a light bulb. They can put it in once and feel good. . . . If they think it's too much of struggle, they just won't do it."

THE MOOD AT THE National Resources Defense Council was locker-room champagne. For years, environmentalists had stood by warming. It hadn't been a 95-yard pass at all. But a slog in a crooked stadium, in the rain, against roided-out opponents—with bribed officials and catcalls from the stands. And then the IPCC and Motion Picture Academy had marched across the field and handed them the ball.

With victory there'd been a sort of male enhancement. You no longer discussed experts and scientists—shy figures with lab parking spots. Everybody had been transformed into *guys*. "That was Darren, our vehicles guy." "Noah and David—the two huge light-bulb guys—are in LA." "The one you want to talk to, the guy who does the economics and really is into the questions, is Chris." "Talk to Ralph about this because he's in California and he's our regulated markets guy."

I sat at an NRDC Climate Center computer with George Peridas—a stress-inducingly attractive scientist. He had the slightly whittled European features that can make American looks feel like inefficiency, too large a facial footprint. Peridas discussed the research of another guy—"a guy who did a study about changing towels in hotels."

And here was where energy efficiency seemed destined. The end-of-the-world movies—where every drop of fuel and water is fought over and coddled.

"The various messages and cards that hotels put out," Peridas continued. "Yeah, yeah," I said.

"And the conclusion was, in short, that people will not respond very well if you tell them, 'Do your part and save the planet.'"

"Yeah," I said.

"'Please re-use towels, don't make us wash them every night. Because you can save money, energy, you can save the environment.' Blah blah blah. Didn't work. You know what *really* made the difference?" he asked. "When they told people that almost everybody else in the hotel was doing it."

Peridas let this sink in. Peer pressure could also be a force for good. Then he returned to the bulb. "You do it, then you see it grow. And that's building."

The NRDC also had speculative notions. The sorts of plans you hear once Sean Connery is trussed up and his host can relax and take the long view.

"You could have concentrated solar power," Peridas said. Concentrated? "Where you have big mirrors that channel the sun's energy." I thought of how Frederick Seitz's National Academy had once studied a military proposal: Hang a giant mirror in space, and pitch the battlefield into permanent daylight. Human thought exerts such basic, lazy gravity that eventually every idea comes swinging back around.

You could also put the ocean in harness—"tidal power, wave power." This had a great name: *intermittent energy*. George Peridas could tell I was getting skeptical.

"Have you ever swum in a tide?" he asked.

"Yes, I have swum in a tide."

"It's pretty *powerful*."

It was hard to say. The real has an unmistakable heartbeat; you know it when you hear it. This lacked the thump of things real people would ever actually do.

"The bulb is a tiny but worthy start." Peridas cleared his throat. "The message is that it's not just up to the government: you can do something as an individual," he said. "We can change the world through our choices and our market behavior."

This had been the pitch from Keep America Beautiful and their one-tear Indian. Back then, it had been the environmental groups who helped us see through that.

And I understood what seemed different about the environmentalists. They were no longer dreamy, big-hearted, and impractical. They had become disciplined, hard-headed, and impractical.

I sat in the office of Jim Presswood, an NRDC lobbyist. Why had the green community adopted energy efficiency? (It kept sounding like the classiest way of ducking the inevitable fight with oil. We can't change you. So we'll change ourselves.) How did it differ, in any non-rebranded sense, from conservation?

"Well, the iconic image is Jimmy Carter on TV wearing that sweater," Presswood explained. "That's not the message. The message is, we need to be smarter. To have technology, use technology. I mean, that's what the American ethos is: technology."

Julia Bovey, a young NRDC communications director, was shaking her head over her computer. I said how strange all this was. How, after all the years and fights, I'd expected more rousing and conclusive things.

"It *sucks*," Bovey said. "That's exactly what my whole job is trying to deal with right now."

She pushed her monitor around to show me. A blog post on best practices for the environmentally sustainable bedroom.

"It's actually an online debate I was just having. This is 'How to Green Your Sex Life.' Treehugger.com. And this is just ri*diculous*. I mean, 'Buy sex toys that are rechargeable, instead of ones that run on batteries.' I mean, we're not out there pushing Eco Jeans," Bovey snorted. "But people want information. And this is the stuff we're giving them: 'How to Green Your Wardrobe.' 'How to Green Your'—you name it."

She turned her monitor back around. "The big decisive action," she said, "it really needs to be legislative. We need to be generating the will to do that. But instead we get: 'How to Green Your Sex Life.'"

Before joining NRDC, Bovey had been a TV reporter. She understood the peaks and valleys of the news cycle—and the sudden clearings, where light falls on an opportunity. She felt this moment to be precious. Harvard's Theda Skocpol, in a widely discussed study of green failures next decade, would call these months pivotal: "The critical juncture in 2007 when Americans in general might have been persuaded of the urgency of dealing with global warming." Instead, greens would whiff it. The moment would close.

Bovey said, "I mean, global climate change is the biggest problem we've got. It's something that keeps me up at night—I feel this is our moment." She continued, "I've read histories of this movement. Never have we had the sort of market share that we have now—on people's *brains*. And what are we doing?" She shook her head again. "You feel like such a jackass. You make this compelling speech—and this happens to me all the time. About how much we have to solve this problem. And people say, 'So what we can we do?' And you're like, 'Change your light bulbs?'"

3.
Big Two

There's a Big Three for light, same as with cars. The Big Three are Sylvania; Thomas Edison's old kidnapped shop General Electric; and Philips, which is based in the Netherlands.

I drove to Danvers, Massachusetts (blurry pines, light blue road), and sat at a conference table with Charlie Jerabek, Sylvania's CEO. He had the blocky face and slab hands of the man who works his corporate way up from the engineering side. "I'm a ramblin' wreck from Georgia Tech," is how he introduced himself. I'd expected to hear industrial-strength bitching: how unfair it was to be singled out on climate change. Jerabek had an unperturbed spirit. "I think it's great," he said. "I'm kind of flattered this is where it's going to start."

We browsed his R&D lab, touring the shelf space of the future. When homes will go all Tony Stark and *Guardians of the Galaxy*: programmable light, bendable light, colors setting the mood like a playlist.

When I brought up the current bulb, Jerabek spoke without heat, as if reciting from a book. "You know, we all have to wean this country off imported energy," he said. "Oil or whatever it is."

Still, surrendering a product line to Congress—Sylvania must have some pretty strong climate convictions. "You know," he said, "we don't have an official company position." The rest treaded so lightly it was as if a tiny ghost had just wandered across the far end of the room. "But I think intuitively we all feel like anything we can do to help the environment is certainly going to help with whatever may be taking place from a global-warming standpoint."

The devices back at R&D were where the CEO's heart lay. "Let me tell you something," Jerabek said. "Products that are around that long—people don't make a ton of money off of them."

The Edison bulb belonged to a special economic weight class. "It's a commodity product," he explained.

The classic example is sugar. Gussy it up, go crazy on it, redesign whatever. You're still selling sugar.

"It's hard to add value that the customer is willing to pay more for. Newer products where you can bring that extra value," Jerabek said, "are really the more profitable products for us. I can't deny that."

And I understood: No complaint because no injury. Politically, this was a garage sale, and the bulb was in a box of outgrown stuff the industry was already setting by the curb. Jerabek tinkered with metaphors. The bulb was an old dog—but then, people love their old dogs. The bulb was vinyl in a world of MP3s—same problem. He found one that fit. "The incandescent is the old clunker," Jerabek said. "The need for energy efficiency, the visibility of global warming, has brought the new technology to the forefront. The technology was coming anyway."

I FLEW TO NELA PARK, General Electric's bulb campus in Cleveland, Ohio, to meet with their head of lighting.

Years ago, when General Electric Company hit a rough patch, they'd retained a famous ad man. Who advised the board to scrap that chilly third word, *Company*. Which did not seem especially friendly. And he told General Electric that initials were even friendlier. It's how the company became "G.E.—the Initials of a Friend." That's the thing with corporations. You're never sure how many expert hands an idea has passed through before reaching your brain. But once it takes up residence, the thought has the queer feeling of being not quite yours.

Nela Park really *is* a campus. Rolling lawns, grand buildings. It's days like this—brick and clouds, the stiff health of September grass—that make climate change feel beside the point. A scary page that can be returned safely to the middle of the pile. And on hot days, when shirt and armpit lather together, you think, We are all hell-bound fools.

Unlike Sylvania, GE had been fighting back. "The economics are better with incandescent bulbs," Michael Petras, their lighting head, complained in the *New York Times*.

Petras was a tall, deep-eyed man who seemed happiest talking sports. His manner was all playful menace—as if you'd lost a bet to him, and he'd stopped by to determine your feelings about double or nothing.

Petras showed me around GE's conference hall, where there was a run of Norman Rockwell oils. Old men bearing lamps. Family singing by lamplight. A glowing doorway. "They're originals," Petras said, standing close. "So, a little history. We commissioned Rockwell to do these, OK? If you'll notice, they all have a common theme, which is light. The one in the middle—you won't believe this story when I tell you. We used to give these out, to the former CEOs of the lighting business, OK? We *did*. And this one CEO, he had a problem: he couldn't get his to fit, in a frame he liked. So he cut off the bottom. If you'll notice, you do not see a date, nor do you see a signature." Petras laughed. This art story had a moral, which was: protecting investment. "Now, little did he know how valuable this was going to become, longer-term. We got these three here, plus four more. Let me show you these four. But they're all using light."

Little did Petras know: that advertising man, Bruce Barton (he's the genius who once rebranded God) had conceived these as an ad campaign, and Norman Rockwell had been hired to illustrate a fact. That GE was Edison, and Edison was light. It seemed fitting, stirring, and a little sad, to discuss his invention's demise under art that celebrated its arrival.

A publicist trailed us; Kim Freeman, an extremely effective corporate spokesperson. (Very funny, very awake.) Freeman did bulb outreach to green groups and reporters. GE had gone huge for climate change. It was then America's largest company, the world's third most valuable—turning out appliances, hospital equipment, jet engines, locomotives, power plants. A whole city kit you could spill out of a box.

Their CEO, Jeff Immelt, had boiled the climate situation down to a three-word poem: "Green is green." He'd launched a splashy, multi-billion-dollar campaign: Ecomagination. I said I thought this was a lame word. Petras bristled. Then he swelled with grouchy, GE patriotism. "As long as you own the stock, you'll be happy with it," he said. "Good word."

Kim Freeman pointed out, "Well, the tagline is 'Imagination at Work.' So it's Imagination and Eco."

"Kim, you know—"

"Imagination and Eco."

GE was sincerely trying. Jeff Immelt had gathered dozens of utility executives—good GE customers—told them they should pitch in with the government on climate laws. It was a brave banner to wave in that room.

The president of the Pew Center on Global Climate Change called the corporate position "incredibly gutsy." In his *New York Times* column, Thomas Friedman—because of Ecomagination—had nominated Jeff Immelt to replace Dick Cheney as vice president.

Funny thing: Immelt's background couldn't be more removed from a scientist like Jim Hansen's. Hansen was rural and telescopes—the observatory at the University of Iowa. Immelt was suburban and amenities: he'd been president of Phi Delt at Dartmouth. Both men had been invited back to speak at their alma maters. Hansen's dour Iowa topic was "A Discussion of Humanity's Faustian Climate Bargain and the Payments Coming Due." Jeff Immelt told Dartmouth's graduating class, "I am, and always will be, an optimist."

Yet they'd both ended up at the same scientific zip code. Immelt told the *Wall Street Journal* that warming was a case of open-and-shut—a "technical fact." And a month after I visited Nela Park, *Time* named Immelt one of its Heroes of the Environment. Jim Hansen ("a great scientist-statesman") was his neighbor on the list.

Now Immelt's firm was being asked to sacrifice the first product to climate change. "Who would have thought—right?—that this is where the industry would end up being?" Petras asked. The three of us had sat back down again.

Kim Freeman said, "A senior counsel told me this is the single biggest piece of legislation that will have an impact on the environment. Lighting, duh. Who would have thought?"

I made an appeal to their irony gauges. Edison had started their industry with the bulb—cut the ribbon on the parade of advances and problems that had marched everybody to the place where we were sitting now. (At his death, the *Times* had filled three pages with Edison obits. Henry Ford owned a test tube containing the inventor's last breath. The nation of Germany was so Edison-mad it urged the US government to do the sensible thing: requisition and autopsy the inventor's brain, for any additional secrets.)

It was homey. GE turned out to be the one place where Thomas Edison wasn't a sacred Wikipedia presence. He was policies you could disagree with, a boss who'd just stepped out of the room.

"At the end of the day," Petras said, "you can't sit around worrying about protecting Thomas Edison. I mean, we respect what he did for the organization. But, you know, this company's gotta evolve and change."

Kim Freeman nodded. "It's funny, when all this started happening, people were like, Oh, GE's holding on to that old technology," she said. "People forget that over the years we've done anything since Edison, quite honestly."

"Dow Jones sustainability, right?" Petras said.

I was surprised. Petras had mostly abstained from the vocabulary. (He'd told me flatly, "I am not an environmentalist.") This was an unforeseen use of the word. Or maybe, in their heart of hearts, this was the way companies like Walmart really did mean it.

"The ability to sustain on the Dow Jones," Petras elaborated. "We're the only original company still on it, the Industrial Index." He smiled. "You didn't know that, huh? We're the original, the oldest member."

Petras didn't feel flattered about going first; he felt singled out. "Just with everything that's going on with oil," the executive said, "you wouldn't have thought this would be the category—or the area of the economy—that people would gravitate to."

"I think there's a notion out there that this is easy," Kim Freeman said. "*Target*'s not the right word. But an easy solution. I think that's part of the reason this has been so popular."

It was something you kept hearing: the bulb as the simpler, winnable fight. Another Big Three executive asked me, "Would we be better off changing to a fifty-five-miles-per-gallon standard across the board? Obviously, yes. But we don't seem to be able to do that, and we *can* do this." And when I asked Sylvania's Charlie Jerabek, he'd chuckled. "You've got Oprah to thank for that."

"One of my friends," Freeman added, "I think she put it really well. She says, 'I can have two SUVs, but I can't have an incandescent bulb?'"

So why were they going along? In Washington, the bill was drawing close; with every meeting, you could make out the parted hair, the button on a cuff, the fingernails reaching for the handshake. Soon, GE would get rid of its bulb. "Just to clarify, *we're* not doing it. This is what the customers want," Petras said sourly. "You have to assume. Why would the congressmen that represent the people be pushing them, right? Right?" He waited for my nod; something else was making GE angry, it wasn't clear what. "That's our view."

4.
Well-Funded Classy Professionals

I hunted through the Senate in a shoeshine-and-neckwear pack of lobbyists.

We sat down with North Dakota senator Byron Dorgan. It was the loveliest Washington office I'd ever visited—glass and steel and sun-splashed views. And Dorgan was about the most elegant man I've seen: his suit seemed have to have descended from designer to shoulders, without ever knowing the sting of a hanger, the humiliation of a garment bag. He didn't say much. The lobbyists—Dow Chemical, Bank of America, Owings Corning, and the utility National Grid—offered things like, "We also manufacture, so how a bill deals with greenhouse-gas emissions matters for us." And, "It's hard to pass along the costs of energy efficiency." And, "The average home pollutes more than the average car—that's something people don't understand."

Senator Dorgan recrossed his legs and finally spoke. "Nothing here goes by itself. There's certainly a . . ." The senator searched himself for eloquence. ". . . good chance we'll get a bill that includes the kinds of things we're talking about." The lobbyists leaned forward hungrily, making a meal of every word. A senator, I saw, was another celebrity, one with the power to make financial wishes come true.

And I traveled in the company of National Resources Defense Council lobbyist Jim Presswood. Business gets the alpha; environment makes do with the surrogate. We didn't see a senator's office. We sat at a small wooden table in Senator John Kerry's waiting room—among the leaflets and tourists—and hashed things out with his young legislative aide. She seemed unfamiliar with the bill's particulars, and concluded our meeting by saying, "Good. All right, cool. Excellent."

The NRDC man had brought along a MacArthur Award-winning scientist, who tried to look at the bright side and not feel diminished. "Electricity is intangible," he shrugged. "People don't get it." I had the sense you sometimes have around scientists: of a stubborn, obstructed intelligence, too much to ever make it through the social pipes. He told a lab joke. "What's the difference between a crazy-ass idea and a project?" he asked. "A budget and a timeline."

I sat down with Bill Wicker, communications director for the chairman of the powerful Senate Energy Committee. One of those people it's a relief to find. You feel you've made your way to the slightly underslept, slightly overworked center of how things actually get done. "I've been up here about six years now," Wicker said. "I have watched the sausage being made."

He offered the view through the kitchen window. A lot of it was social. The Senate wasn't comfortable around one-track advocates with exciting plans. (A personality type a lobbyist described to me, with a smile, as: "True believers. In *big* things.") The NRDC and the alliance were guests the Senate could be itself with. "In the environmental food chain around here, the NRDC, they're at the top," Wicker said. "Well-funded, classy, bunch of professionals." The Alliance to Save Energy made the same becoming impression. "Well-respected and funded, plays well with the other children." The standard didn't seem to be what got done—this was Washington, everybody won some, lost some. It was about maintaining relationships, winning and losing in the proper style.

All the DC people I spoke with were frisky and kind and seemed to move through a world of soundstages and golf carts, with lunch at the commissary. Action came, when it had to, at certain strict points demanded by the narrative. It's like what they used to say about the old studio system: the only projects to get made are the ones absolutely nobody can find an excuse for saying no to.

I ate dinner with scientist Noah Horowitz—one of the National Resource Defense Council's two huge light bulb guys. We ordered burgers. (This was meat tourism. "At home, we're pretty much vegetarian," Horowitz said. "So when I'm away, I eat all the beef I can.")

Things were looking up, climate-wise. Senator Jeff Bingaman—Bill Wicker's boss, chairman of the powerful Senate Energy Committee—had decided to sponsor the lighting bill. Its name was an optimistic traffic jam: the Energy Efficient Lighting for a Brighter Tomorrow Act.

The greenhouse savings, to the senator, looked tremendous. "It is one of the best things we can do now in Congress," Bingaman said. The committee's ranking Republican was all candy and flowers. "This is a big, big event," he said. "I'm hoping the very best will happen to this bill."

And there were glad tidings from an infallible quadrant. The Vatican—breaking the tape ahead of four other nations—had just elected to become the planet's first carbon-neutral state. (The *Times* got off the best headline. "VATICAN PENANCE: FORGIVE US OUR CARBON OUTPUT.")

Horowitz patted out ketchup and discussed the future. Over the horizon, on the downhill side of the lighting bill—as we entered a world we'd improved through our individual behavior and market choices. Horowitz smiled. "So the good news here is—you can still drive your car," he said. "It's a Prius. Your TV is better, your building is better. Your air conditioner is better, your light bulb is better. Add all that up. *Plus* we don't leave things on when we're not using them, *and* we do that automatically. You add all that up," the scientist said, "and we're gonna get there."

It was hard to say. First, reality has that pretty unmistakable EKG, and this seemed arrhythmic. Second, it didn't exactly get the blood pumping. It didn't make you want to send emoji or shimmy up a lamppost. And last, it seemed to pretend the world to be other than what it is. Exxon hadn't spent millions, S. Fred Singer hadn't invested his career, because they feared your purchase of an energy-smart TV.

Horowitz was a dedicated man. He'd traveled across China, helping Walmart find just the right bulb. He'd been working with the Big Three toward this moment all year—been there in March, when the efficiency ball started rolling. "A bright idea whose time has come," is what he'd called it then.

"This is in Congress, in the Senate," I said. "How does that make you feel? Are you proud?"

Horowitz had one social inefficiency. I came to think of it as a scientist's adaptation, for spending time around the less straightforward majors. He ran statements through a subtext detector. Once he'd cleared them of cynicism and mock, he answered.

"Oh yeah," Horowitz said. "I feel great about it. It's not about getting the credit. I did the right thing for the right reasons."

5.

The Right Reasons

And I learned where successful climate legislation does come from.

A late September afternoon. My last meeting with the Big Three—a dim conference room in the upper recesses of Royal Philips Electronics. People just kept filing in. There was an indoor social blizzard of first names. Two publicists, Susan and Heather. A chemist, Dale, who used to run Philips Labs, and had the caved-in features of the man in the war movie who steps out of the car with sad news about your loved one. An executive named Paul, who'd seen the road with Sheryl Crow and climate activist Laurie David, on the Stop Global Warming! college tour. ("I listened to Laurie fourteen nights in a row," he said. Unclear whether this was brag or complaint.) Another executive called Rick. Everybody seemed to be awaiting just one more arrival. Philips' CEO had himself blown through town earlier that day, to launch a product called the Halogena. Room consensus was that the launch had gone pretty well. The overall feel was of tense positivity—negative outcomes excluded by sheer force of group will— which is the business feeling.

Dale wanted to gripe about environmentalists: their mistrustful nature.

"Where does it come from?" Dale asked. "There is a basic suspicion of industry. And many of the groups I deal with absolutely believe the companies are evil. Have you had that experience yet?"

I touched gel pen to notepaper. Dale looked a little seasick. The conversation had trucked on ahead.

"I didn't know he was taking notes, I wouldn't have said any of that stuff," Dale said. "So scratch that stuff. My comments."

One of the publicists put in, "I believe you are being *recorded* as well." There was a ripple of laughter.

And then a man in a wonderful suit brisked into the room. He had one of those big-screen, TV-communicator faces. He could project any emotion on it he wanted you to see. This was Randall Moorhead, vice president of governmental affairs. Which is the long way around to say "lobbyist."

I asked my climate question—how did it feel, going first?—and received a surprising answer.

"We're all for it," Moorhead said. "We *started* it."

The whole, year-long push. It hadn't come from the green groups at all.

"We went to them with the idea and they all said 'Hallelujah,'" Moorhead said. "And we all signed aboard."

Or it had been a green idea only in the double sense preferred by GE and Walmart. I must have looked skeptical. There was an outbreak of nods from the table. "Call the NRDC!" "Ask Kateri Callahan! Ask Noah—"

But why should the environmentalists, those doubters of industry ("I did not know you were taking notes at that time," Dale re-complained), take all the authority and momentum the year had given them, and hand it over to Philips? And then employ all their political chits and wiles to help get Philips' idea passed?

Moorhead had an overall business answer. Competition. The need to return through the gates with a law on the books—bringing a head, any size. "As we compete with GE and we compete with Sylvania, so the environmentalists compete with each other, too, OK?" He nodded. "They all want kudos for being the leader. They have a fundraising base they kind of compete for and all of that. It's not just manufacturers competing. It's environmentalists competing for the hearts and minds of their constituencies as well."

But GE's behavior—fighting to retain their bulb business—made flat commercial sense. Why should Philips suggest a law against one of its own products, then push to get that law enacted?

It was because Philips had created a more expensive product to replace it.

Moorhead smiled. He was near the end of something. He wanted the sly moves he'd made on behalf of his company to be deciphered and appreciated. There are times when a person is dangerously bursting with story.

It had to do with the Halogena—the light Philips' CEO had flown to Washington that day to introduce.

Moorhead had attended today's launch. He neatly unboxed one of the bulbs. I had an unexpected sensation. I was in the middle of an interview; the subject was handing me a bulb, I weighed it in my palm. If replicating exactly what another person has done makes you momentarily into that person, then for an instant I knew the sensation of being Oprah.

"I don't know how much you've been briefed on this or not. This Halogena is something we've had in the works for some time." Moorhead smiled. "And had planned to bring out this year. So we were ready."

He took back the bulb and looked at it. Hamlet with the skull in Act V. "So it was a business opportunity, yes," he said. A bulb you could buy, instead of the Edison—at non-commodity prices. "We all have the motivation to make money. But this was also a motivation where it's a win for consumers and a win for the environment. I mean, this is a win all around."

More first-name nods from the table.

"You have to understand that the other companies..." Moorhead waved the Halogena. "This light didn't happen yesterday. This is years of prototype development and testing. The other guys don't have that. We caught them—I can't say *napping*, in the sense that they weren't working on stuff in their laboratories. Because they would be foolish not to be doing stuff in their laboratories. But they can't bring this product out as quickly as we can," he said. "Competition is driving this." Thomas Edison had a slur for business people: the engineers only of contracts and dollars, who didn't know the feel of an honest tool or the hours at the workbench. He called them *sharps*. Thomas Edison had been bested by the sharps. And you had to admire them. They'd done a curvy thing but made it look direct; they had made a straight noodle.

THE REST OF THE STORY I got on background: from people with knowledge of the negotiations. In February—Oscars and the Intergovernmental Panel—Philips put out feelers to the green groups. Would they be interested in forming a coalition?

Philips understood they'd need the Big Two. You can win ugly or you can win clean. Ugly is more satisfying but clean costs less. "We could break them, if it was just us and the environmentalists," a Philips executive later explained. "But you have to understand the political reality: if it's just us and the environmentalists against Sylvania and GE, that's *also* a hell of a battle on Capitol Hill."

In mid-March, Randall Moorhead booked the National Press Club. With no warning to the Big Two, Philips declared the end of the Edison bulb.

This was done for two reasons. "First, to show the world that Philips was a leader—it was good for their company," I was told. "Second, to put the political pressure on their competitors to join the effort. And it worked in both cases."

Sylvania was the easier nut. Their lobbyists had the navigator's instinct for storms, coves, and shorelines. They understood which way the politics were blowing.

GE held out. "The lighting division did *not* want to make the change," I was told, and understood Michael Petras' foul mood. "They fought it tooth and nail. They've just been milking their brand—that old bulb they've been making lots of."

The Press Club announcement came at nine in the morning. By early afternoon, a startled General Electric was on the phone to Randall Moorhead. By four, a senior GE lobbyist and the company's general counsel were sitting glumly at the Philips conference table.

"They were very upset a coup had been pulled, in a public relations sense. And they were smart enough to know Philips was going to win," another Big Three executive explained. The general mood: "They knew what had happened to them that day."

The corporations fought. Lobbyist threatened lobbyist. "Randy," a GE executive said, "we are going to beat you."

But Moorhead understood this as a bluff the company would eventually fold. He'd promised the green groups from the very beginning. Because of the company position on climate change; because GE had just spent years and billions going splashily and nationally green; because of their made-up word.

"Jeff Immelt will *have* to cave," Moorhead had assured them. "The Ecomagination campaign will be dead in the water, if he insists on making those light bulbs, when we're ready to change the world."

And I understood the sour mood at GE. They'd been flipped on their back by advanced corporate judo. After a week, Moorhead received the terms of surrender in a solemn phone call: "We will have an agreement with you." Some weeks later, General Electric officially joined their coalition.

And I understood why, six months later, so many Philips people had crowded around this table. This was the winning side. "When it comes to these products, we're going to be able to have more on the market sooner than the others, because we had a head start," Moorhead said. "Philips is prepared to do all this. The other companies are not. They're not because we stole a march on them, all right? That's why."

So that was the story. How you got a law on climate change. Not because, as the movie star told the host, global warming was the most pressing problem faced by all humanity.

Not because, as the environmentalist said, of our particular American disinclination against being asked to do too much. Not because, as the corporate spokesperson maintained, bulbs were an easy target. Not, in the words of the senators, because it was one of the very best things Congress could do. And not because the green groups had pounded on doors any harder, or finally located the right ones, or because the scientists had stepped back from the blackboard, nodded at their equations, and replaced the chalk. Or not only because of these things.

But because business was ready; it was in their interest. I didn't want this to be the story—because of my own self-interest. Because what about all the situations where business *wasn't* ready, when it wasn't in their interest?

Moorhead had one last bit of strategy, near-term. "My view now is, I obviously want Philips to get credit when the legislation passes," he said. But for the present, "We're stepping back a little. So the other companies won't get jealous of us taking credit at this stage."

"But you will eventually get the credit?" I asked.

"We're going to be getting some soon."

"Well," the executive named Rick said, "the credit that we want is selling this product and making money. That will be the credit."

Or maybe Edison would've gotten a kick out of this story. Recognizing and then seizing your opportunity is what he'd done as a boy—it's what brought him to the inventor's life in the first place. With his newspapers and the Battle of Shiloh. Saw an emergency, moved his product. "Well, you know," Edison once said. "I like a hustler."

6.

Sad T-Shirt

Buyers' remorse casts its shadow across politics and Christmas. Three weeks after George Bush made things official, Al Gore told the billionaires' snow club at Davos that what people really needed to change was laws, "not just light bulbs." Gore explained, with the gravity of a fresh Nobel laureate, "It's really important, where the climate crisis is concerned, for all of us to try to move *away* from the idea that personal actions by each of us represent the solution to this crisis." At the same event, *Times* columnist Tom Friedman suggested a further step away. "My mantra is, It's much more important to change your leaders than your light bulbs." The *Times* editorial board reached the same conclusion a season later—it's right there in the headline. "IT'S ABOUT LAWS, NOT LIGHT BULBS." And at the year's back half, it's how PBS ended a two-hour climate documentary. Punchline delivery services were provided by Jerry Brown, governor of California.

Brown spoke to the camera in a tone of fond and bleak disbelief. "It's intellectually dishonest to somehow say we can get some light bulbs—or, you know, get a Prius—and we're all done," Brown said. "*No.* This is going to take massive technological innovation.... That's where we are. There's no other word except 'daunting.' I'm hopeful. I'm cautiously optimistic. But I would have to say, one has to approach this with great humility."

But the time for credit-taking was at hand. Randy Moorhead bragged to Bloomberg News the day Philips' bill passed. "We knew that twelve years of pent-up demand by Democrats," the lobbyist said, "was going to be released in a number of ways. And one of those ways was energy efficiency." Kateri Callahan and her Alliance to Save Energy awarded the organization's highest honor, the Star of Energy Efficiency, to Philips. For "spearheading the creation and success of the Lighting Efficiency Coalition." (Randy Moorhead said that together they had established "a model for all to follow.") By next decade, the halogen had become America's best-selling bulb. Money again in lighting. As the *Times* reported, the ban was an "opportunity the industry hasn't encountered since Edison first flipped a switch."

Corporate intentions weren't entirely brass. "Everyone wants to do the right thing," Randy Moorhead had told me. "And a good thing." The American ideal—follow your conscience until it leads to a reward.

When I told the Philips team that their law would address only about 1 percent of American emissions, they'd become legitimately upset. "You're *way* off." "That's bullshit. It's much *more* than that." "You should talk to more environmental groups." ("I wanted to step in," Dale interjected again. "I made that statement from our dealings in mostly the negotiations phase.") A frown clotted the smooth flow of Randall Moorhead's features. "Then you're saying we're wasting our time," he said. "We shouldn't even bother with this stuff."

But that was indeed the National Resource Defense Council's number. Passing the law required nine months; the ban wouldn't take complete effect until 2020. Thirteen years to shave 1 percent. Plans to stave off the most disruptive kind of climate change required a 20 percent cut by 2020, with a 50 to 80 percent gouge by 2050. At the bulb's pace, we would undershoot the mark by several centuries.

BECAUSE NO OTHER major legislation flourished during the great political weather of 2007; nothing would really grow for three presidencies. That NRDC scientist had called the bulb "a tiny but worthy start." Start and finish were the same line.

Four years later, GE's Jeff Immelt was regretting—like a person who'd divorced, remarried, changed jobs and coasts all in one lurch—his infatuation with a color. "If I had one thing to do over again," Immelt told an MIT conference, "I would not have talked so much about green. . . . I'm kind of over the stage of arguing for a comprehensive energy policy. I'm back to keeping my head down and working." Newt Gingrich rated his climate change ad "the stupidest thing I've done recently." As the UK *Independent* observed—with a faint, worldly smile—when *Vanity Fair* folded its Green Issue, "Three years is a long time for any trend." The moment had closed.

Because warming couldn't resist. It had resumed its usual awful timing. A kind of historical reboot. The month of light's Golden Jubilee (Edison's invention, celebrated) the Great Depression began. Eighty years later, the month the lighting bill passed (Edison's invention, axed) was the start of the Great Recession.

"Public concern plunged soon after," Harvard's Theda Skocpol writes. Nobody much worries about the condition of the sky when their anxieties are centered on the roof. "Hardly any major environmental legislation has been passed in bad economic times," sighed a vice president with the Pew Center on Global Climate Change.

The Energy Independence and Security Act was the last bill signed by

George W. Bush, his last real year as president. In 2008 there was an election on. When the current occupant becomes an afterthought, a Phantom of the White House.

But there was a bonus reboot. Bush had begun his presidency by crumpling one international climate treaty. At his final State of the Union, Bush again took the podium in his light, pinball grip and called for a new climate treaty. "Let us complete an international agreement," he announced, "to slow, stop, and eventually reverse the growth of greenhouse gases." Leaving the next president to begin the process all over again.

There were few quotes from climate scientists the week President Bush signed the bill. I like to think of it as a boycott. ("This is real, this is real, this is real," Dr. Richard Alley had said in Paris on IPCC day. "Now act. The ball's back in your court." They had bounced the ball once, then returned it to the equipment shed.) It had been more than a century since Svante Arrhenius calculated, many decades since Dave Keeling and Roger Revelle. And they were like the back-home family, shortchanged after everybody else's vacation—the ones who wear the T-shirt. WE RESEARCHED CLIMATE FOR 112 YEARS, AND ALL WE GOT WAS THOMAS EDISON.

Epilogue

THE PARROT
AND THE IGLOO

The great creators of technics, among
whom you are one of the most successful, have
put mankind in a perfectly new situation, to
which it has as yet not at all adapted itself.
—ALBERT EINSTEIN, 1929

THE IGLOO

I BECAME A VERY UNPLEASANT PERSON WRITING THIS book. There's something about reading people who are lying that makes you suspicious and argumentative company. My back was always up. I was always starting fights, demanding agreement—I would tell friends what they'd said a day or year before, contrasting it with what they'd said just now. It is proof of the decency of the world that I retained any friendships at all. I became one more voice arguing about global warming.

The quote a page back is what this book has been about. That lovely easeful position to which we haven't as yet accommodated ourselves. Einstein advanced this little trouble to Edison, via shortwave radio, on the evening of light's Golden Jubilee. Edison died twenty months later. This suave irony—a haunting compliment—may be the last words from a fellow genius the old wizard ever heard.

I was working on a different book when this one made me stop and write it. I'd bumped into that Roger Revelle story from 1956, and read with fascination. Hm: we'd known for sixty years. I got mad—book-makingly mad—when I read the 1970s stuff. (Charney Report: "If carbon dioxide continues to increase, this study group finds no reason to doubt that climate changes will result and no reason to believe that these changes will be negligible." 1979—and what could be fairer, clearer, or more direct than that?)

Everyone had known. And nothing had been done. It seemed a metaphor, the largest, for all the things we know we should fix but never get around to, or think we can't fix, or hope we'll somehow be recused from ever really having to try. I started writing this book to find out the reason.

The inventors thrilled me: they are a measure of how good we can be. (Even the ruthless Westinghouse, dying because he'd been deprived of a good thing he'd made.) Then the scientists, testing and bettering each other, and with all their shy insistent courtesy expressing that here was a thing that would be trouble. And then the ingenuity—all that dark human power—of the deniers. Used to explain that here all would be well. It is an unpleasant experience, immersion in the words of people who know they are being untruthful. I'm grateful to everyone who put up with me.

All the good and terrible people in this book: I liked and hated them, too. And I bet the same ones.

I loved Charles David Keeling, so crazy about measuring carbon dioxide he was out on a rooftop and testing the night his wife went into labor. And my apologies for having to spend so much time in the company of S. Fred Singer. Or even that refreshing country house weekend with Lord Christopher Monckton; I disliked that jaunt myself. (And I had to read lots more of it, and then pretty much everything written by an English denial journalist named James Delingpole. My own personal Year Without a Summer. Reading someone, you do throw open your internal windows to the air currents of their personality. People hated me—and I hated most people—during the months that James Delingpole blew across my thoughts.) Here's my explanation for the title: I kept it till the end.

Researching a book is all before and after. Meeting that rush of people, scrambling to sort the connections. When I started tracking the stories of people who studied and lied about climate, I came across that Union of Concerned Scientists report on Sound Science and the untruthers who became Exxon's squad, the oil cash flowing to S. Fred Singer and Frederick Seitz and everyone. All just names to me. The way, unless you really focus, the news can become just a daily wheel of names. Someone else's office party you've dropped in on; gossip overheard on vacation.

This had been done *to* me, was how it felt, since the aim was to convince people of things that were untrue. But also, because it was aimed at preserving things as they are, this had been done *for* me; done in my name. But I just saw words and no way to make pictures from them.

Years later I reread the piece. And knew pretty much everything about everyone in it.

That's the experience I've tried to give you, in this march from Thomas Edison and Nikola Tesla, to climate science, to the people who lied about that science. Specialists who established an immaculate and collaborative network of untruth, about the same kinds of things—aspirin, tobacco, ozone, smoke, climate. Mostly for money. Sometimes for less negotiable rewards. (Praise. Respect. To make you stop and spend time with them.) That's the igloo side. Cool and impacted. Icy bricks supporting each other: easy to stand up, tough to knock down. The parrot you've already seen, though perhaps forgotten. Here it is again. Igloo to follow.

IN 1956, *Time* magazine and the *New York Times* had addressed science and the possibilities. That spring, Roger Revelle gave *Time* his first climate change interview—in an issue that documented the unnerving Soviet policy of smiles and an America that no longer had enough animal-act enthusiasm to support two circuses. (Since then, the remaining outfit has also folded its tent. Trained bears and top hats we've decided belong to our entertainment past.)

"In fifty years or so," *Time* reported, summarizing Revelle's research into the carbon exhalations of our machines, "this process may have a violent effect on the earth's climate."

They had known even then about the tightness of terrestrial margins. You'd need only a one- or two-degree rise to start the ice melting, the climate changing. "When all the data have been studied," *Time* explained, scientists "may be able to predict whether man's factory chimneys and auto exhausts will cause salt water to flow in the streets of New York and London."

Five months later—cement sky, pavement leaves—the *New York Times* had speculated about "striking changes in climate." A human-caused warmth that could "convert the polar regions into tropical deserts and jungles, with tigers roaming about and gaudy parrots squawking in the trees."

A spicy thing to visualize, as light and weather thinned toward Christmas: the former frozen regions being lorded over by a mouthy bird.

And now it was here. *Time* magazine's half-century waiting period elapsed. And a great international apparatus had sprung up to capture and examine the data. The IPCC, of course. Two hundred science organizations on record; from the Academy of Athens to the Zimbabwean Academy of Science. Climate change was human-caused, and real.

With three great repositories of the world's temperature. One in Manhattan—Jim Hansen's NASA office, six stories up from the *Seinfeld* coffee shop. Another amidst the Washington gridlock and founding-father statuary at the National Oceanic and Atmospheric Agency. The third in Britain: the Jack Bauer-sounding Climatic Research Unit, at the University of East Anglia.

The Unit had amassed four million separate temperature readings, from four thousand sites on the planet, reaching back 150 years.

This was weather's week-by-week episode guide—the sun-and-rain version of *Ulysses*. No characters, no Dublin, just sky. It had become the largest story in science.

It was 2009. To scientists, with their research accepted and agreed to, the climate fight now looked "winnable." They weren't following the politics. They had their eyes on their opponents, who they believed were done.

Relief can be misleading. Especially when it is anticipatory. That just-before moment then becomes the most hazardous. When you envision the glass of water so thoroughly you wonder if you've already drunk it. When you feel yourself breaking the tape so clearly you cease to run. When actually finishing the race begins to seem like a formality.

In his memoir of these years, climate scientist Michael Mann quotes an observation by General Douglas MacArthur: "It is fatal to enter any war without the will to win it."

Mann's pasting this in tells the whole story: by 2009, the climate scientists believed they'd entered the wind-up phase, their long conflict behind them. Also, that Mann believed they had lost their killer instinct. The desire—sometimes salutary—to hear an enemy skeleton crunch beneath the boot. "By early 2009," Mann writes in *The Hockey Stick and the Climate Wars*, "a troubling complacency had emerged among the scientists."

THE DENIERS WERE HUMBLED. Philip Morris had been tamed: A once-feral cat now fit to curl in any lap. Ellen Merlo, who field-marshaled talent to discredit the EPA, had reflagged. A *BusinessWeek* cover story branded her beloved Philip Morris "America's most reviled company." (And put their corporate responsibility at two hundred thousand painful and annual deaths.) Merlo now made do-over pronouncements like, "We should have sooner embraced the public health community's position." She spearheaded the Philip Morris youth-smoking prevention drive, sounding like a lion that has turned vegan and militant on the subject of antelope security. "As long as one kid smokes," Merlo promised, "there will be a youth-smoking department."

S. Fred Singer made large statements to diminishing effect. He was the glory-days rocker who has ceased to attract new listeners, content to trot out the old chart-toppers for the fans. In 2001, his think tank was properly ID-ed by the journal *Nature* as a "pressure group." In 2008, ABC News labeled his research "fraudulent nonsense." I watched his face one night on PBS—the Oregon Petition fourteen years old, Dr. Singer again trying to retail the signatures. The host asked, "Now, are they all scientists?" As Dr. Singer inhaled to speak, it flickered across his eyes: the look of someone understanding they'd been condemned to a certain type of life, by choices they had made, perhaps ill-advisedly, as a much younger person.

Same fate befell Dr. Frederick Seitz. *New Scientist* described his George Marshall Institute as though they were reviewing a fusion restaurant—a "unique mix" of Cold War militarism and warming. They found the flavor "bizarre."

He'd been neither as lucky nor as iron-thighed at outrunning his résumé as the old denial forebearer, Clarence Cook Little. In 2006, *Vanity Fair* revealed Dr. Seitz as a former tobacco guardian. Two years later he passed away—the last great campaign of Frederick Seitz's life was explaining why a person of his caliber should stoop to cradle tobacco. (Unlike Fred

Singer, no shame on his face. Instead, the aspect of a hawk, enduring the second guesses of sparrows.)

The man Frederick Seitz drove east with, sharing that pretty moonlit slice of campus; Bill Shockley, the physicist who won a Nobel and founded Silicon Valley. Late in life, Shockley had, like Clarence Little, taken a humiliating turn at the eugenics wheel. It erased all acclaim. "Ironically," a science reporter wrote, "Shockley's old friend Seitz, now in his mid-90s, followed a similar course. He became a strident critic of global warming . . . and found himself pushed aside by the scientific community." They started the travels together and ended at the same place. This was the climate Frederick Seitz died in. One year after the IPCC's "unequivocal," and Al Gore's Nobel. He died in defeat, which makes me happy.

His passing incorporated a final salute to uncertainty. The denier internet sites were being abandoned, left to collect digital cobwebs and dust. On the faltering version of his own site, Fred Singer continued for years to list the dead Fred Seitz as the live chairman of his board.

And it must've stung. The IPCC putting the matter behind everybody like that. The chiefs of Energy, the EPA, and the National Oceanic and Atmospheric Agency gave a joint press conference. Three dark suits in the dry rattlesnake slink of the cameras.

Samuel Bodman, secretary of energry, spilled over years of Jenga-built argument with a casual sweep of his hand. The secretary went all the way: declared the IPCC "Sound Science . . . we're very pleased with it," he said. "We're embracing it, we agree with it."

The dwindling band of skeptics had fought the tide shrewdly, brilliantly, and valiantly. Now they were in it, up to their necks. It enlarges the heart—shows the impact one individual can make. That so few people could frustrate the will of science and the world for so long. (It makes you wonder if they weren't the secret heroes of this story all along: a magnet for condemnation, so the rest of us could get on with our smoky lives unrestricted.) But Exxon CEO Rex Tillerson had pulled the deniers' plug: "It is clear that something is going on," Tillerson said. "It's not useful to debate it any longer."

At a spring 2009 conference of the deniers, in a tired hotel off Times

Square, the mood was mid-wake glum. "Two years after the United Nations Intergovernmental Panel on Climate Change concluded with near certainty that most of the recent warming was a result of human influences," the *Times* reported, the skeptics were facing "internal rifts and weakening support."

One attendee sighed to the UK *Guardian*, "It's almost a lost, lost battle."

Who turned it around for them? Who's been the most effective person on their side in this story?

THREE MONTHS LATER, Britain's Climatic Research Unit was under attack: a pounding, shells in the form of sixty Freedom of Information requests over a single week. All derived from the same North American website. A review board later called this campaign "orchestrated" and "literally overwhelming." The requests were all denied. That November, a hacker snuck aboard the Unit's server.

One thousand emails, plus thousands of pages of data and code, had been forcibly liberated. All posted to a Russian host. No person is a hero to their email server. (Email is the harsh mirror in the pitiless dressing room. Reflecting us at our gray, narcissistic, self-inspecting, self-promoting, paranoid, undressed worst.) British denial journalist James Delingpole called the leak "the final nail in the coffin"—and made himself famous with one word. (I'll get there in a bit.) The whole climate "conspiracy," he wrote, had been at last "suddenly, brutally, and quite deliciously exposed." Science journalist Fred Pearce, who covered the hack for the *Guardian*, called it "a public disaster."

The final punchline, right beside the tape, the finish, and the well-deserved glass of water. The science's reputation, Pearce wrote, might "never fully recover." The *New York Times* climate writer Andrew Revkin reported, "Now the foes of limits on greenhouse gases have the ball."

LIKE SO MANY THINGS in this story, it started as an entirely unrelated issue, then got boosted by cigarettes. In 1973 Harvard was retained by the National Institute of Health for a study. What researchers discovered had to do with what are called fine particles. The big worry being not gunk you can see; it's the tiny stuff—industrial dandruff, sucked down through a windpipe, speckling the lung, powdering a trachea.

The Harvard School of Public Health undertook a massive study. Why were people living longer, healthier, in the cleaner cities? The work was crazily ambitious, one of those sprawling triple-decker novels with everything in it: eight thousand adults, fourteen thousand children, followed across one and a half decades of lives and breaths.

Harvard's six cities ranged from squalid to pristine. Steubenville, Ohio (the sootiest; it later became notorious for a high school football rape), and St. Louis, Missouri; Harriman, Tennessee, and Watertown, MA; Topeka, Kansas, and Portage, Wisconsin (the cleanest; its high school has great curling teams). Their work became famous as the Harvard Six Cities Study.

The Harvard researchers promised confidentially, then took up their positions like kindly, clipboard-bearing angels. Watched their subjects age, marry, cough, stay well, get sick. They came to focus on those fine particles—2.5 microns and under.

The micron is a unit of measure in the Ant-Man world. Human hair is 70 microns wide. The eye can see down to 40 microns—after that, it's specialized lenses or take it on faith. Ragweed pollen—hay fever—is 20 microns. The fecal matter of the household dust mite, at about 10 microns, is the source of most household allergies. (Mites dine on human skin flakes, then poop them back into the air flow. You're breathing it now.) Digital clean room workers are trained for zeal to 0.5 microns: One flake on the hard drive can erase years of data.

Which is just what the Harvard team determined particulate matter was doing to human lifespans. The particles—snowing out of tailpipes and smokestacks—could settle deeper in the lung: Inhaled over years, this could start the pulmonary disease equipment rolling, the cancer gears turning. Ohio's Steubenvillers were about 26 percent more likely to die prematurely than the curling fans of Portage, Wisconsin. Because of fine particles.

"It was much, much higher than we expected," explained Six Cities Study lead author Dennis Dockery. "The effects of air pollution were about two years' reduction in life expectancy." The *New England Journal of Medicine* published the Harvard results in December 1993.

It's since achieved landmark status: as they say, "one of the single most influential public health studies ever conducted." In November 1996, the Environmental Protection Agency announced plans to regulate fine particles at 2.5 microns. (Mite poop, however, you face without government assistance.)

There've been various estimates. Some say the ensuing fight cost industry tens of millions; others go all the way up to a hundred. The brawl brought together petroleum, automakers, the National Association of Manufacturers.

The *Times* described it as a "fierce battle"; Harvard's School of Public Health recalls it as the fight of the decade. What everyone means is that industry freaked.

The companies banded together for what the *Washington Post* called "an extraordinary multi-million dollar campaign." They fought over the airwaves—regulation, the ads warned, would disfigure America, the vital America of nostalgia and dreams. At risk was the American farmer, who might be forbidden to plow. ("In church, people were all shook up about it," one North Dakota farmer told a journalist. It wasn't true. Farm dust is big-grained.) They fought on grounds of pleasure. Regulation could mean the end of every good July thing—road trips and powerboats and chuffing lawnmowers and the can of beer sucked down beside the backyard grill. Under the EPA's particle law, it would be a whole Life Without a Summer.

The best spots piled symbolism on top of symbolism, and did all the work for you. "Imagine that," marveled the announcer. "A new regulation, that takes away our freedom to *celebrate* our freedom." At risk were fireworks. The EPA so despised America, it wanted to legislate away the Fourth of July.

And they made one charge stick. Harvard—that island of finely wrought gates and you-excluding standards—was concealing something. Industry made a sly request: to proofread Harvard's raw data.

Industry understood this to be impossible, even illegal. Harvard had promised its Six City subjects confidentiality. In a data universe, unauthorized hands grubbing across your medical file is a screen grab from nightmare. Barbecues and sparklers were the appeal to the stomach and nostalgia; secret data was the whisper to the head.

Industry recognized their *coup de théâtre*. They assembled a troupe of unemployed actors (never the hardest thing to do), costumed them in lab-coats, sent them parading outside the Harvard School of Public Health. "Give us your data!"

When Douglas Dockery, the Six Cities lead author, trotted up the Capitol steps to testify, his welcome was more protestors: "Harvard, release the data!"

"It was a painful time," Dr. Dockery recalled. The fatalism of the news subject: discovering what fresh trouble your name got into while you were asleep. "You'd wake up in the morning and look in the paper, and there you'd be again."

Industry converted this research—whether conditions might not obtain under which people could perhaps be enabled to live a bit longer—into something nefarious. "The issue is the quality of the science," charged the National Association of Manufacturers. "What are they hiding?"

Harvard's team had one especially strong motivation for not backing down: the future. A Harvard biostatistician later recalled that having industry—the very people on the hook for filtering the air—verify your figures might not result in ideal outcomes. "To have a hostile group combing through your data looking for anything to attack," the researcher explained, "was not something any of us relished."

The Six Cities leader warned against any accommodation. (Because there is that relief of agreeing with the bully, which at least puts a stop to the bullying.) They had everyone else to consider. "Biased groups," the scientist warned, would treat any surrender as a license "to undermine future research by academic institutions."

A peace was negotiated. Harvard released the data to a secure, independent nonprofit. Fact-checkers spent fifty million dollars and three years. When they were done, *Science* reported a "major victory": the Six Cities data was accurate. If anything, Harvard had slightly understated the health impacts.

By then, the 2.5-micron standard had gone into effect: it proved a victory for cold numbers and warm bodies. The Office of Management and

Budget under President Bush found that while cleanup costs ran to seven billion annual dollars, benefits—unvisited medical offices, unfiled insurance claims—lived in the neighborhood of twenty billion. "The standards have had the largest benefit, and the largest benefit-to-cost ratio," Harvard's School of Public Health reported, "of any and all regulations by the U.S. government."

Particle reduction, Six Cities lead author Douglas Dockery and his colleagues demonstrated, had increased the American lifespan by 1.6 years. Nineteen extra months under the sun. To read; to argue with and enjoy family. "That's huge," Dockery explained. "If you got rid of all cancers, the net effect on average life expectancy would be two years."

The fight also revealed a surprising osteo-educational lack: what one team member called a steel backbone. "We teach people to be statisticians, epidemiologists, lab analysts, exposure scientists," one Harvard scientist said. "We must also equip them for the big fights."

IT CONFIRMED SOMETHING different for politicians and industry. There was traction to be had in withheld data.

A year after the National Ambient Air Quality Standards for Particulate Matter rule, in 1998, Republican senator Richard Shelby snuck a new law onto the books.

He rolled it in at midnight, under a sheet. It was, as the *Times* reported, a Jim Tozzi type of operation. The law "passed quietly one evening last October, without hearings or debate." One tiny, potent thing hiding among the columns of something massive and federal: "A one-sentence amendment to the 4,000-plus-page appropriations bill."

Shelby's motivation, the lawmaker explained, was the Six Cities study. Also something as nineties as ripped-knee jeans and doomsday guitar. A desire, Senator Shelby wrote, "to minimize the possibility and opportunity for the use of 'Junk Science.' "

And, as a final motivator, there was the former Pentagon jazz nerd, nodding and prompting from the wings. "We proposed the Shelby," Jim Tozzi proudly told science journalist Chris Mooney. "We were the first ones on the Shelby Amendment."

The Shelby Amendment made important government-funded raw science data available to anyone—you just had to implement the Freedom of Information Act.

Not just that. Shelby nudged the idea into the air—where it got sucked down the windpipes of a nation of armchair skeptics. Freedom of Information Act requests would become, in these weekend hands, a powerful and versatile weapon. A decade later, a member of Jim Hansen's NASA team would survey the wreckage. "The unfortunate fact is that the 'secret science' meme is an extremely powerful rallying cry to people who have no idea about what is going on." A few days after the law passed, one farsighted biologist told *U.S. News and World Report*, "This just sounds like a nightmare."

JIM TOZZI WAS under Philip Morris contract when he crewed on the Shelby Amendment. And Philip Morris was still captive to the old dream. Liberating secondhand smoke. They'd been watching, waiting, enduring, hoping.

And then they stepped into the warm dawn and golden possibilities of the Six Cities fight. Philip Morris called the Harvard struggle "our hook."

A "unique and unprecedented opportunity," the company explained, in (chortling) strategy documents. The company would "leverage" business opposition. Tighten this to a "focus on meeting our objectives." In this way, public smoking could be "reopened." Philip Morris had become the one-track friend who grabs every conversational wheel, steers all talk to the same weedy parking lot. They would connect "access to data" with a concept everybody already had positive feelings about—"Freedom of Information."

Best case: "Federal legislation passes on data sharing" and "Environmental Tobacco Smoke is considered to be an example of junk science." Worst case (the bitterest sideline regret): "We miss this opportunity."

They understood how the whole business might look in daylight. Another worst case was, "Our attempts to influence legislation become public knowledge." In the memo, there's a lovely honest moment: Godzilla sadly contemplating fin, paunch, and claws in the mirror. "We are who we are," the executives wrote. "Need a proxy." In December 1997, they handed off responsibility for the data law to Jim Tozzi.

When the Shelby Amendment passed ten months later, the rule was lodged deep within a spending bill. The kind that passes during a thick deadline scramble—leaving, as the *Washington Post* reported, "members who voted for it in the dark about most of what it contained." Forty pounds, 3,825 pages: if it landed on your head, you'd see paramedics. "Do I know what's in this bill?" Veteran Senator Robert Byrd asked a journalist. "Are you kidding? No. Only God knows what's in this monstrosity."

God, and Philip Morris. Ten days prior to, Jim Tozzi had sent over the relevant language. With a request. "Please do not release this outside of Philip Morris until it becomes law." Day of, an exec sent around the champagne email. "Tozzi just confirmed the language is in the bill. Great news." To challenge the notion that nonsmokers were inadvertently puffing on the neighbors' Marlboro, Philip Morris had helped transform the law.

The head of the National Academy of Sciences warned the rule would have "serious, unintended consequences." (He had no idea how right he was.) The Data Access Act took effect in October 1999.

Tozzi's next assignment through Philip Morris was a Data Quality law. The Shelby Amendment only got you the data handed in; half the battle. Data Quality provided the grading curve on which to fail it.

"The Data Quality Act," trade publication *Federal Paper* wrote of Jim Tozzi, "was his baby." If not the stealthy man's "crowning achievement." It passed the following year, 2000, also lodged deep within a spending bill. President Clinton signed it into law on the first day of winter. One of his last acts as president.

Jim Tozzi bragged about timing. The end-of-administration pardons; the Clinton family moving vans approaching; the messy presidential recount in Florida, George Bush and Al Gore contesting every vote. "There was so much going on," Tozzi told the *Wall Street Journal*. "It couldn't be stopped."

This time, Jim Tozzi cut out the middleman and just plowed ahead and drafted the law himself. And what a thing, for the rule-loving jazz nerd. From reading laws to writing them—like an ornithologist taking wing after a lifelong study of birds. Jim Tozzi hid his law in the middle. "We sandwiched this in between Jerry Ford's library and something else," Tozzi later explained. "Was it something that did not have hearings? Yes. Is it something that keeps me awake at night? No." And then a career's worth of Washington in one legislative shrug. "Sometimes you get the monkey," Tozzi said. "And sometimes the monkey gets you."

"It is not clear," reported the *Post*, whether anybody beyond a handful of people "knew about the buried language."

Data Quality took effect in 2002. Immediately, a libertarian think tank used the rule to make a decade-long Adaptation study—the National Assessment on Climate Change—"vanish." ("The legacy of that," *Newsweek* explained a decade later. "State efforts are spotty, and local action is practically nonexistent.") Chris Mooney called the law "Tozzi's time bomb." Donald Hornstein, a Chapel Hill legal scholar, saw the university future as pocked with craters and trimmed in barbed wire—"a dreary world of bare-knuckled trench warfare," he wrote. Of "data wars."

The nation's official definition of quality scientific data now came from the pen of Jim Tozzi. "You have not made your case," is what he'd told the FDA, about pain relief and Reye's syndrome—before 1,500 children were buried. He once called it the best decision of his career. He'd made his aspirin story the law of the land.

WHEN JIM TOZZI GOT interviewed about the second law, he tried to dull some of the corporate shine. Fact was, he'd always been a servant of the little man. "If you were John Q. Citizen," Tozzi told Knight Ridder News Service, "and you saw components of a study that you thought were inaccurate, you couldn't do anything." Inaccurate study components being a source of well-known and widespread frustration for the average individual. "Now if you have a good case, you can do something."

John Q. Citizen would have been a good word portrait of Steve McIntyre. He was fifty-seven, lived in Toronto, and played squash. He'd been a mining executive—now semi-retired, with time on his hands, he turned to climate science in the spirit of leisure. McIntyre thought it would be "like doing a giant crossword puzzle."

The final shock news item—the kind of freak story a teenaged Thomas Edison and the other plugs might have pinned to the telegraph office wall.

The climate scientists had survived Jimmy Carter, Ronald Reagan, John Sununu, George Bush, Mark Mills, Frederick Seitz, Fred Singer, Philip Morris, Exxon, the Oregon Petition, George W. Bush, Phil Cooney, Sound Science, and Lord Christopher Monckton—all to be brought low by a squash player.

"Nothing in McIntyre's life," writes the Canadian magazine *Maclean's*, "suggested he would assume a central role in one of history's great scientific debates."

McIntyre grew up in Ontario, did great in the Toronto schools: he bested a future Nobel laureate in some kind of sudden death math competition. McIntyre studied numbers at the University of Toronto, received a scholarship to Oxford. In 1971, he was offered the next logical step in the sequence: scholarship to MIT, the academic life.

As poet John Shade writes, "Here time forked."

A Steve McIntyre who attended MIT would not have become a Steve McIntyre who attacked on climate; he would have been too busy being an MIT-trained mathematician. McIntyre's parents had divorced, in the especially ugly seventies style. McIntyre was the eldest of six children. How could he head south, dwell among the calm chill of numbers, while his family remained inside the burning building? McIntyre stayed, and helped, and entered the business world. Three decades flew. Now a grandfather, he looked back and, as he told the *Wall Street Journal*, was disturbed to find no intellectual accomplishments, "despite having a good mind." He complained to another interviewer—sounding very like the good scientist Arthur Robinson—"I am well-educated."

You couldn't have brought off what McIntyre did even five years earlier. You needed the data laws. They were the scarlet route by which McIntyre became "the great Satan of climate science," *Maclean's* noted, despite having "no credentials in applied science." Data Access and Data Quality were both in effect as of 2002. McIntyre started up that year.

As the *Guardian*'s Fred Pearce reported, the scandal "would not have happened without one man ... Steve McIntyre." The Canadian "pioneered the use of Freedom of Information legislation," Pearce writes, "to demand the raw data behind the studies."

By 2003, John Q. McIntyre was using the new law to threaten the National Science Foundation into releasing raw data from the climate scientist Michael Mann. "It is my understanding," McIntyre wrote, "that the U.S. government has established standards for disclosure.... Surely the applicable standard here is, at a minimum, that of being 'capable of being substantially reproduced,' as set out in Guidelines implementing Public Law 106-554." This was Jim Tozzi's baby, his crowning achievement: the Data Quality Act. McIntyre used it on NASA, too; hinted it to NOAA; even threatened to deploy it against an informal website maintained by climate researchers in their spare time.

Freedom of Information steroided this genial man into Dwayne The Rock Johnson—"a playground bully," wrote climate scientist Ben Santer, in one of the emails made public by the advent of Steve McIntyre.

It's an irony worthy of Greek myth. Because the Harvard Six Cities scientists had not wanted a world in which future research would be subject to raw data review by hostile eyes, they had started a chain of events resulting in just that. When a Duke University climate scientist wasn't eager to release data to Steve McIntyre through the Freedom of Information Act, the Canadian contacted his journal editors. Data Access made Steve McIntyre into the type of guy who gets you in trouble with the boss.

Dr. Thomas Crowley—Duke's Nichols Professor of Earth Systems— understood how McIntyre was playing the game. "For your information," he emailed, the project in question "was not funded by any federal grant." McIntyre shot back that "merely not receiving federal funds" was a "possible legalistic technicality." Data obligations "might well ensue."

Amazing fact: the laws allowed McIntyre to pursue this new hobby from the desk and peace of his Canadian home. A sort of real-stakes video game. You bothered some abstract name by email; then, magically, concrete data appeared on your computer. McIntyre became expert in the intimidating arts: the weapons of hint, worry, and push. Of the midday insinuation that expands to fill all thoughts at night.

Scientists had never experienced anything like him. "I'm usually happy to send people stuff," Duke's Thomas Crowley told the *Chronicle of Higher Education*. But "McIntyre comes back time and again.... It's like you have nothing better to do in your life than answer questions from Stephen McIntyre."

MCINTYRE'S SPECIAL FIXATION was the young climatologist Michael Mann. They're the soldiers depicted on a battle monument; frozen together in mid-grapple. Just as the email scandal made Steve McIntyre into climate's "great Satan," it transformed Michael Mann, too. He became "the nation's most hated climate scientist." This comes via the *Yale Alumni Monthly*. When it can be picked up by your alumni magazine, the hatred has gotten very loud indeed.

Mann, twenty years McIntyre's junior, led a smoothed version of the academic life. No parental divorce, declined scholarship, or diversion through the world of business: it's like a contrast worked up to generate maximum identification and envy.

Mann's father was a math professor at the University of Massachusetts-Amherst; McIntyre's had been a surgeon. Like McIntyre, Mann's great high school talent was numbers. "When other kids were partying on Friday nights," Mann writes in his memoir, "I was hanging out with my computer buddies writing programs." Mann graduated Berkeley with a physics and math double major, took his PhD from Yale—here time didn't fork—and returned to Amherst. (Mann received the Yale geology thesis award. McIntyre never wrote a PhD thesis.) The accelerated maturation of the sciences: you find yourself working among your inspirations. "Some of my U. Mass colleagues," Mann recalls, "were parents of kids I'd grown up with."

Asked to pick the scientist out of a lineup, you'd point at Michael Mann. Shiny-topped, with that mousepad of hair around each ear, for placement of the eyeglasses. Pot-bottomed, to help the chair hours comfortably pass. Both McIntyre and Mann are bearded and bald-headed—with Steve McIntyre, the effect is more along the lines of comic actor John C. Reilly, playing an undependable expert. (Deep blinks, then the eyes searching and anxious over whatever it was the mouth just said.)

Mann had an extraordinary rookie season. Fred Pearce calls him "a young man in a hurry."

Working with two colleagues, Mann thought to express the previous six centuries of climate visually. Combine all the data: the temperature records, which reach back only to 1861, then what are called *proxies*, the footprints climate leaves on its walk across centuries—lake-bottom sediments, tree rings, ice cores, coral. Plot a graph between then and now.

When all calculations were entered, Mann and his team had a striking image. Long, straightish, not-so-warm line, 1400 to 1900. Then, with the twentieth century, a sharp upward slash. This work became famous as the Hockey Stick Graph.

Mann published in *Nature*, got profiled by the *Times*. He was off to a galloping start.

A year later, in 1999, Mann and his colleagues plotted back to the year 1000. More time travel, same sporting-good shape. "There is a certain genius in the Hockey Stick," Fred Pearce writes in his history *The Climate Files*. The past nonthreatening handle, the sharp current blade. "If the job of scientists is to make sense of confusing reality with simple statements and images that we can all understand, then Mann and his Hockey Stick may be remembered long after his critics are forgotten."

This research was later duplicated and verified. "What counts in science is not a single study. It is whether a finding can be replicated by other groups," explains Pearce. "Here Mann is on a winning streak." A dozen studies confirmed his basic results.

Criticism began when the Hockey Stick got included in the 2001 IPCC report. It made Mann's career. Also fixed a target to his back.

Two problems. First, Mann was that figure promised by Max Planck: the next generation, for whom climate change was simply one more feature (central air, inground pool) of the property you inherit. "The goddam guy is a slick talker and super-confident," complained Wallace Broecker, one of climate's grandfather voices. "I don't trust people like that."

Second, Mann was, simply, young. It made him vulnerable. In a Fox News column, Jim Tozzi's old colleague Steve Milloy called Mann's work "the top secret 'Junk Science' behind global warming hysteria." The whole two-century edifice—all of it resting on the shoulders of a thirty-three-year-old academic.

FOR WHATEVER REASON, Steve McIntyre had partnered with a think-tank man: a professor at the University of Guelph named Ross McKitrick, an economist with great connections and TV-friendly hair. The Canadians published their first Hockey Stick attack in 2003. McIntyre explained—since his background was business—that he was "auditing" Michael Mann.

They published their second attack two years later. "McIntyre, it seems," Fred Pearce writes, "was starting to have fun."

He emailed Mann's collaborators—demanding their data. "I would NOT RESPOND," Mann wrote Dr. Phil Jones, director of England's Climatic Research Unit. "The last thing this guy cares about is honest debate—he is funded by the same people as Singer, Michaels, etc."

Mann was wrong; that wasn't McIntyre. It was the co-writer who hailed from Singer Land. Ross McKitrick "brought to the collaboration contacts in the fossil fuel industry," Fred Pearce writes, "and other organizations fanning the flames of climate skepticism."

Co-writing turned out to be a bonanza investment. When paper two appeared, "McKitrick's friends stoked up the PR campaign in earnest."

Strange turn of events. "An essentially obscure paper," Fred Pearce explains, was "splashed across the front page of Canada's *National Post*."

Also—significantly—across the front page of the *Wall Street Journal*. There were poor Michael Mann and his Hockey Stick. By now, Steve McIntyre had started his own website. Which he called Climate Audit. (Stomach-tightening name: "Have you ever *been* audited?" a science defender later asked on CNN.) And where McIntyre would post John C. Reilly fluttering-eyelid items about his own victories and growing renown.

He'd gotten the thing he wanted: intellectual achievement. "On a personal basis," McIntyre wrote, "this coverage has been very gratifying, since my family and friends are not academics and being covered in the *National Post* seems much more tangible to them."

And when he blew up in America: "The article is on the front page [of the *Wall Street Journal*], complete with picture of a certain aging Canadian," McIntyre shared with readers. "I'm gratified by the coverage, to say the least."

The *Journal* story led to hearings: Michael Mann was hauled before Congress. Charges based on the Tozzi laws—Data Access, Data Quality. The hearings were a wash, but put a stink on Mann. "The trial," as journalist Rich Cohen writes of the modern legal experience, "had become the punishment."

Congress asked the National Academy to review Mann's work. This resulted in vindication. "Science Panel Backs Study on Warming Climate," went the *Times* headline.

Exoneration only made Mann a bigger target. It was as if he were wearing a lapel pin: UNDERMINE CLIMATE SCIENCE HERE. Because denial had shifted tactics. Stop worrying the whole chain; break one link. "There are people who believe," climate modeler Ben Santer told *New Scientist* in 2006, "that if they bring down Mike Mann, they can bring down the IPCC."

STEVE McINTYRE DAWNED as something new, even unprecedented. A likable denier. In profiles, the Canadian was presented as "gentle," "mild-mannered"—"polite to, and about, climate scientists." A skeptic without invective: McIntyre "rarely makes charges personally."

That's because he was making them with another name. The classic problem. McIntyre wanted to be thought of differently than the way his brain instructed him to act. In online science forums, McIntyre posted under a pseudonym—"Nigel Persaud."

As Nigel, McIntyre would make any attack. Call Michael Mann "a flat-out" liar. A deceiver responsible for "a complete fabrication." A man of "furtive and guilty" looks.

On the other hand, Nigel could be unflaggingly—even embarrassingly—supportive of one Steve McIntyre. "The above pretty much shows that McIntyre & McKitrick were completely right."

He became subject to odd whims and impulses. Browsing the Climatic Research Unit server one day, McIntyre noticed some unprotected files. They were just sitting there; so he stole them.

Next day, McIntyre announced to his Climate Audit readers he'd been passed this data by "A Mole." Some secret and lion-hearted person, deep within big science, disgusted enough to spirit out the warm raw stuff at last.

This development thrilled his science-mistrusting readers. One even posted poetry.

McIntyre told readers the Unit had opened an investigation—then speculated on the courtesies of law enforcement. "I doubt," McIntyre said, "that I would bear up well under waterboarding." A few days later, this sad update: "More news on the Climate Research Unit molehunt." The Unit had apprehended his contact. It was now busily at work "to eradicate all traces of the mole's activities."

Readers worried: "The 'Mole' risked his career and future." Had McIntyre put a good person in jeopardy?

McIntyre later admitted he'd made the whole thing up. "I had some fun with it," he told an interviewer. "It caused quite a commotion."

He developed touchy, fancy ideas about how he ought to be treated. He complained online when climate scientists were "discourteous" to him. When researchers' behavior was "graceless." (He had happened to them. They didn't like him. They were trying to solve him.) When the conversation of one Michael Mann colleague displeased McIntyre, he filed a charge of academic misconduct.

He'd built a following of similar nonexperts. Older men calling themselves citizen scientists.

The climate fight is a story of descent. From the science journals to the opinion pages, then the talk shows, then the petitions. And with each defeat, denial regrouped on a new and lower floor. Now climate had become a matter for the blogs: What the *Guardian* called "an unanticipated outpost of the rise of 'grey power.'" Climate's new adversaries were retirees "with time on their hands."

Willis Eschenbach is a good example of the soldiery. Eschenbach's experience was some years as a construction manager on the island of Fiji, a stint with IT, some time wearing waders as an Alaskan sport-fishing guide. "I'm an amateur scientist with a lifelong interest in the weather," went Eschenbach's self-description on the Climate Audit site. "Just a guy trying to move science forwards." He additionally brought to this pursuit his skills as a massage therapist, licensed through California's Aames School of Massage.

Willis Eschenbach took his waders and massage certification to England's Climatic Research Unit, mouthing the soft threats like an enforcer dropping by for a chat about the benefits of protection money. "I am sure that we can bring this to a satisfactory resolution," he wrote, "without involving appeals or unfavorable publicity."

Steve McIntyre had schooled his readers in Freedom of Information Act requests—climate scientists came to call these "FOIA attacks." Two steps. First, "You have to ask them to do something," McIntyre wrote. Then, step two: "Publicize non-responsiveness. They don't like it, but too bad."

And online renown could skip the tracks, ascend to the analog aboveground world. The *Times* described Willis Eschenbach as "an engineer." Willis Eschenbach is not an engineer. He works as a house carpenter. (Eschenbach: "I work building houses.") The *Daily Telegraph* promoted him to "scientist Willis Eschenbach." (Eschenbach: "I have absolutely no credentials.") The restless *Telegraph* again honored him as "a very experienced computer modeler." (Eschenbach: "Just about any field of science that I look at, I know absolutely nothing about. In Buddhism they call this 'beginner's mind.'")

And by mid-decade, Steve McIntyre had become, per the *Guardian*, climate's "arch-inquisitor." The man striding into the bar with the chip already on his shoulder, retailing a science version of the standard bar story. *They withheld data from the wrong guy.*

"Needless to say, [NASA scientists] were totally unresponsive to my request for source code," he'd blog at Climate Audit. Big mistake. "They shouldn't be surprised if they get a Freedom of Information request."

A gang-leader thrilling the membership at the hideout: "Freedom of Information time is close." The pleasure of the underestimated man, demonstrating once again the size of the misestimation. "When I ask them for additional data," McIntyre sighed regally to the *Wall Street Journal*, "you can imagine how cooperative they are."

As science historian Spencer Weart writes, "Leading researchers," were now being "assaulted with countless questions and demands for information."

It wasn't just the boners in McIntyre's work; the knowing that journalists and supervisors followed the blog; the seeing their names publicly muddied. (McIntyre on a career NASA scientist: "IS GAVIN SCHMIDT HONEST?") It was the drudgery. The *hours*. The diving into closed files for old numbers, the clerical scuba. "Most saw them as a threat to their work," writes Fred Pearce, "not because they would uncover fraud, but because they took up their time."

In late 2008, McIntyre requested raw data from climate modeler Ben Santer. Santer emailed sharply back: the material was already publicly available. The rawest data—before processing. "You should have no problem in accessing exactly the same model and observational datasets that we employed," Dr. Santer explained.

There'd be some math. "You will need to do a little work in order to calculate synthetic Microwave Sounding Unit (MSU) temperatures from climate model atmospheric temperature information." But that shouldn't "pose any difficulties. . . . I see no reason why I should do your work for you."

A challenge, sneaky and direct. If McIntyre judged himself qualified to audit, he should be capable of these basic calculations. And if not, not.

McIntyre's response was to become livid. And to sound like the haughty villain in a crown-and-towers movie. Santer, McIntyre complained on his blog, had been "insolent."

So McIntyre filed two Freedom of Information requests and notified Ben Santer's employers. As additional punishment, he requested work email going back two years. Now he sounded like the big-bellied city detective on an episode of *Law & Order*.

"If Santer wants to try this sort of stunt," McIntyre told readers, "I've submitted Freedom of Information requests and we'll see what they turn up. . . . I'll also see what [his] administrators have to say. We'll see if any of Santer's buddies are obligated to produce the data. We'll see if Santer ever sent the data to any of his buddies."

Immediate effect. An administrator wrote that Santer was bringing their lab into disrepute. Ben Santer, the recipient of a MacArthur genius award; who'd served for years on the IPCC; who was among the most respected climate modelers in the world. He'd been converted by Steve McIntyre from an ornament into a blemish.

Santer wanted to resign. He wrote the director at the National Cli-

matic Data Center, "I believe that our community should no longer tolerate the behavior of Mr. McIntyre and his cronies.... He deals in the
currency of threats and intimidation." He went on, "Stephen McIntyre is
the self-appointed Joe McCarthy of climate science. I am unwilling to submit to this McCarthy-style investigation."

After a month Santer released the data. No backbone was steel enough.

McIntyre didn't find anything. This had ceased to be the point. As they
said on *Game of Thrones*, he wanted scientists to bend the knee.

"McIntyre often made it clear that he did not always have any particular scientific reason for requiring the data," Fred Pearce explains. "He just
wanted to liberate it."

Exactly what that legal scholar had predicted—data wars. It led to a
"siege mentality" among the scientists. The Climatic Research Unit dug
in. They kept the 150-year temperature record. Steve McIntyre had always
wanted it. In fact, he already *had* it. It's what he'd downloaded off their
server, while pretending to be The Mole.

"He might already have it," explains Fred Pearce, "but there was a principle at stake." That principle was compliance with Steve McIntyre. He cautioned his Climate Audit readership not to anticipate anything excitingly
damning. "Nowhere have I encouraged readers to expect any smoking guns
in this data set. Quite the opposite." It was just a hobby: something you did
instead of the crossword. Climate science had become a stage set for personal dramas of intellect, heroism, and consequence.

Ben Santer, when everything was over, and their research and reputations were chipped buildings and pouring smoke, said this was the real
story of the email scandal. Stephen McIntyre "can issue FOIA requests at
will. He is the master of his domain.... He is not a climate scientist, but
he has the power to single-handedly destroy the reputations of exceptional
men and women who have devoted their entire careers to the pursuit of
climate science." Meaning resided there, with "Mr. McIntyre's unchecked,
extraordinary power."

Fred Pearce later wrote, "It was clear that by 2009 things were coming to a head." Beginner's-minded house carpenter Willis Eschenbach filed
the first Freedom of Information request with the Climatic Research Unit

in 2007. In July 2009, McIntyre and his Climate Audit readers submitted their sixty requests. In mid-November, all requests were denied. Four days later, the Unit was hacked. The files were posted online with a mocking, defiant, argument-settling name: "FOIA.zip." By a hacker who signed himself "Mr. FOIA"—Freedom of Information Act.

Somewhere, I hope Jim Tozzi read this, recognized his fingerprint, and felt the strangeness.

What the hacked emails kicked off, according to British denial journalist James Delingpole, was "the messy unraveling of the greatest scientific scandal in the history of the world."

UNLIKELY EVENTS BRING OUT the unlikely. They are a social rain that touches odd seeds. In May 2010, James Delingpole arranged his spindly body behind a lectern. Chicago, perhaps most inward-looking of the great American cities, was hosting the Fourth International Conference on Climate Change.

"Now I want to tell you a story," the British journalist began.

That hushed the Marriott ballroom. Because for this audience James Delingpole had become the thing he'd long wanted to be: a celebrity, a hero.

"About something *extraordinary*," he said, "that happened to me late last year."

It's hard to get a fix on the exact unsatisfactory impression left by James Delingpole. He's stretched wire-thin, in a manner that suggests race horses, doping scandals, and protests from PETA. The face is surprised and petulant middle-age—as if a student had walked out of last class wearing the skin and regrets of a forty year-old—but the gestures belong to a much younger man. He swings his shoulders, splays big hands, smiles at words not yet out of his mouth. A fellow journalist put the effect right there, at oral level. "James Delingpole," he wrote in the *Guardian*, "whose teeth are better suited to newspaper journalism than to television." Then he moved on to the question, rare for a daily arts page, of whether to firebomb James Delingpole's house.

He seemed to excite a repelled, fascinated nastiness. A BBC personality introduced himself by saying he'd like to sock Delingpole. "I've been lucky enough never to meet him," a Booker-nominated novelist explained. "But Delingpole is still the best example I know of the inverse relationship between self-regard and talent. Not only would I like to see him beaten with sticks, I'd supply the sticks." Then the novelist added, "I'd do the beating." For a female journalist, the problem was social—the sticky human desire to be welcomed, and known. James Delingpole was "the intense, bespectacled student you meet on your first day of university, and spend the next three years trying hard to avoid."

Delingpole continued in Chicago, "It was an ordinary Thursday morning, and I was deciding which libtard to have a go at on my blog."

Laughter—tribal and relaxed. *Libtard* is one of the signal fires by which conservatives know each other. The international conference was sponsored by the libertarian Heartland Institute. Publishers of, among others, *Climate Change Reconsidered*: "A warmer world will be a safer and healthier world for humans and animals alike." International Conference on Climate Change was another soundalike. IPCC, ICCC; just one letter's difference.

The longtime fan could be expected to ask just what James Delingpole was doing here, among the square jaws and straight arrows. Years before, he had flipped through his own guest lists in a mood of wonder. "I have spent the past decade," wrote Delingpole in the *Evening Standard*, "ruthlessly, if subconsciously, excising from my address book anyone who isn't a serious spliffhead."

His writing, he liked to brag, contained the most harrowing acid experiences, the gentlest rolls on E. His drug journalism wedded an explorer's heart to an oenophile nose. "The trustafarian dude was very obliging and gave me several generous tokes on his fat one. I did quite like the grassy taste: it has a green intensity similar to the one you get from a Kiwi Sauvignon Blanc. Then, quite suddenly . . . Wham!" He had done time as a food writer, travel reporter, shopping correspondent, TV critic. A piece on the reality show *I'm a Celebrity . . . Get Me Out of Here* (in this dry period of aspiration, a Delingpole motto) brought together all specialties: travel, flavor, the purchasable experience. "A few years ago on a Caribbean island, I tried smoking crack. It tasted absolutely delicious, like toffee bananas."

And it wasn't love of the material. "Science is bloody boring," Delingpole complained, reviewing a Stephen Hawking biopic.

He had set out to become a literary novelist. A very specific one: he wanted to be Martin Amis. Delingpole wrote of "pathetically" Googling himself, "in the hope of finding people saying I'm the greatest English novelist since the young Martin Amis." Spotting his idol at a party, Delingpole would truck on over and ask, "People say I look a bit like you. Do you think I look like you?"

It's the pitfall of a celebrity culture. A thousand profiles a year tell us, essentially, what it is to be the Martin Amis in a variety of fields. But they aren't at all clear on what it's like *not* to be him; they go quiet on that score. Which makes most careers into voyages of lonely discovery.

He published three literary novels to discouraging effect. And Delingpole has one of those skin-and-bones constitutions that seem to register disappointment along every inch. You can imagine how it must have felt—2003, the last one he tried—reading this review: "Delingpole has a talent for something, but I don't think it's for novel writing." A review in the *Spectator* by a Delingpole friend caught another oversight. "You don't make nearly enough effort to conceal what a shameless little hustler you are."

In an interview, Delingpole got to say he was self-punishing. Which was "bloody good for the writing." *The* writing—the way a literary person like Martin Amis would say it. Every aspiration is contained in that word. What you see is a great unfairness. Passionate admiration for another's work does not make you into that person. This is among the modern disappointments. Recognize it, and you walk away clean; miss it, stay in, and you fester.

Delingpole would write that "climate change is the biggest scam in the history of the world"—a global "conspiracy" by a "cabal of activists."

But he was not an especially suspicious person. He had an open mind and trusting heart. For osteopaths, homeopathy (cured his hay fever), hypnosis, neurolinguistic programming ("my trendy new therapist"), fad diets (his wife: "James has just discovered a new health fad"), acupuncture, t'ai chi. Even for the cosmetics industry: regular facials kept him young. "Vast amounts of scientific research have gone into the various unguents and anti-ageing remedies," Delingpole wrote. "And I'm just not cynical enough to believe they're all an elaborate conspiracy."

That is, he was a man of contradictions.

Another contradiction: Delingpole believed a defeat had been administered; it just wasn't clear from where. He became the kind of grouch writer who discerns in modern life a plot to limit his freedoms. To drive fast, never recycle, speak freely, enact his gender. "In an increasingly feminized world," he wrote, you were not even allowed "to be a man." This was inconvenient because "under the skin" men remained "hunters and fighters."

On the other hand there were the facials. And given the perfect opportunity for hunting and fighting—a rat scurrying around the family baseboards—Delingpole stepped graciously aside and left his wife the honor of the wetwork. "While I cowered in the corner," he wrote, "Tiffany squashed it with a metal toolbox."

Again, a man of contradictions.

"When into my *lap* fell the story that would *change* my *life*," Delingpole continued in Chicago. "And, quite possibly, save Western Civilization from the greatest threat it has ever known."

As to changing his life: Delingpole had gone on in print about the length and impatience of this wait. In 2001, drafting his last literary novel, he suffered an attack of mid-project jitters. "It's hardly ever the really good talented people (like me, say) who get their name in lights," Delingpole wrote in *The Spectator*. "But the worthless chancers with a decent publicity campaign."

He instituted a tradition. Every January, in some British paper or other, Delingpole would publish a wish and regret list. "My resolution this year—as it is every sodding year—is finally to acquire the massive and thoroughly well-deserved fame and money which eluded me in the previous one."

Two years later, 2003, before his third novel went under: "I just so do not deserve this. I'm nice and clever and talented and funny and it's about time I had a break."

Next year, reviewing the Gordon Ramsay reality show *Kitchen Nightmares*, Delingpole was a grim thirty-eight. "With each new day, I become more and more twisted with rancor, regret and self-hatred," he wrote, "over my continuing failure to make something of my life."

At the end of 2006, Delingpole composted his wish list from the year's envy and regrets, ran it in the *Spectator*. "2007 is my last chance. I know I've said it before, but this time I mean it. If the things on my wish list don't come to pass"—fame, money, the invigorating jealousy of friends— "then there's no justice and I'm definitely going to have to kill myself. Got that, God?"

God apparently didn't take the *Spectator*. 2009, eleven months before the story that would drop obligingly into his lap, found Delingpole at another low. "Not quite suicidal but definitely getting there." More success requests—plus a resolution that might have creased foreheads at the International Conference on Climate Change. "Do heroin. I've never tried it before. Might be fun. And necessary."

As to saving Western civilization: The story Delingpole found that November made him feel enhanced. A superhero powered by links and click-throughs. Blog hits soared from 100,000 per day to 1.5 million. Delingpole couldn't decide which movie character he reminded himself of most. Later on, he said Luke Skywalker. Just after, he said he felt both like a rock star and like he'd given birth to a rock star. (There's increased feminization for you.) In a video interview, the impression he left was more sleek and black-caped. "It's sort of *made* you, hasn't it?" asked the host. "Yes," Delingpole replied. "I have become this notorious figure." He gave a media-savvy shrug, "I think you have to accept it as part of your branding." Right at the exciting moment, Delingpole told his Chicago audience, he felt he'd become the great hero of the digital world. Which made sense: he'd found the story on his computer.

"Suddenly, I felt like I was Neo in *The Matrix*." His powers were not unlimited. "I couldn't actually walk through walls or do cool martial arts moves or anything like that." And then, in a great audience misread, James Delingpole executed a few elementary chops and blocks on stage at the Chicago Marriott.

He was now a reporter and blogger with the *Daily Telegraph*. A colleague linked him to another movie character—a "terrifying" case.

"Don't let anyone fool you," wrote *Telegraph* editor Will Heaven, "into thinking James Delingpole is on the fringes of the global warming debate. He's bloody not." Delingpole was in fact "amassing a vast army of climate change skeptics behind him, who think that anthropogenic global warming is a worldwide conspiracy between climatologists and politicians."

The journalist, Heaven wrote, was "slowly turning into a rather scary Dr. Strangelove figure, whose bad science could help to usher the end of the world as we know it." Except Dr. Strangelove at least had his degree. Delingpole was simply "Mr. Strangelove (B.A.). Does this worry no one else?"

When you follow a media figure, you'll have a favorite moment—the clip you make friends watch on YouTube. My favorite John Oliver is when he hosted a scientist versus denier debate; then, to keep the proportions accurate, invited ninety-six other climate researchers on stage. My preference is for the mature Delingpole. Delingpole in ascendency: "I'm at least ten times more famous than I used to be. With readers all over the world who think I'm just great." (The *Telegraph* blog stable was a brute mix of Darwin and David Mamet: Say what you must, but Always Pull Traffic. First place was job security; last prize was, eventually you're fired. In the weekly rankings, Delingpole ran uncatchably first. "He's always number one," a colleague explained to *Private Eye*, "because he really is batshit mad.")

So here is my favorite Delingpole moment: encountering one of those marchers from his Strangelove army.

This was on the radio and internet broadcast The Vinny Eastwood Show. Eastwood, a strapping New Zealander, is built like the fan waiting outside a rugby match. Big neck, soul patch, humped arms. Delingpole had become a serial non-decliner of media invites; he didn't seem to have performed even the most minimal, self-protective research. The Eastwood Show slogan was "The Lighter Side of Genocide." Two days before, his guest had discussed the "very scary implications" of certain occult symbols flashed by Madonna at the Super Bowl halftime show. Another week, his guest brought news about "a high-level secret cloning program involving major world leaders" who engage in "torture, sex and pedophilia with cloned people."

It was the sort of artisanal, house-to-house talk show made possible by Skype. Delingpole wore a Kurt Cobain flannel in front of the Union Jack. Eastwood watched him, thick-faced and eager. About fifteen minutes in, the host used the word "exterminate." Delingpole's eyebrows popped.

It's a touching moment, even for writers with branding: when the elevator door closes, and they realize they've left the main shopping concourse—they're in the weird part of the mall, with the shadowy, unvisited stores. That here is the invitation they should have turned down.

"I do completely agree with you," Eastwood was saying. "We are on the verge of a new Pearl Harbor, World War III-type mass genocide scenario."

So here is my Delingpole moment. He could have said anything. A scholarly, *You might want to double-check your sources*. A brotherly, *Dude— no*. (It's a moment where you hope for a show of heroically bad manners.)

"Potentially," Delingpole replied. Then he added that the internet had been a blessing. "Because it means that people like us can share ideas, get information out there." He considered for a moment. "You know, I have an audience I never *had* before. . . in Australia, in Canada. In Brazil!"

The host, face Stonehenge blue from his monitor, had circled back around to his topic. "What I know is, I don't want to be exterminated," he said. "That's my base political principle." Delingpole had covered his lips with two fingers: one of those social signals the body sends, with its limited vocabulary. Best just to keep mum here.

"The great thing about being on our side," Delingpole said, "is that we have all the arguments on our side."

There was a reason the host knew James Delingpole. Because of the story he'd broken. This is what the wish-granting idol had given him. His own name in lights, finally, instead of some worthless chancer's.

In Chicago, Delingpole paused. "That story, can you guess what it was called?" He waited. Raised hands, happy shouts. "*Climategate*," he said. "Well-guessed."

WHAT MADE JAMES DELINGPOLE famous at last was a simple act of copy and paste. In those first few days—taking the word "climate," and adding the word "gate."

As Spencer Weart writes, of this tiny masterstroke: once you "dubbed the affair 'Climategate,'" it became "a scandal."

It was science emails hacked from an obscure British university: A story that did not seem, on the face of things, destined to go the distance—that didn't appear to have what it took. Delingpole gave the story its camera angle. It was juicy, inside, *gate*. The kind of career-ender that cooks somebody who isn't you, and is therefore good clean fun.

James Delingpole tapped out on his *Telegraph* blog, "Climategate: the Final Nail in the Coffin . . ."

And then the emails featured on CNN, MSNBC, Fox, NPR, *Newsweek*, and in the *Times*, as "Climategate."

Because science is difficult—it's bloody boring—but everybody knows how to cover a *gate*. (Joan Didion, on the subtleties of the approach: "Tree it, bag it, defoliate the forest for it, destroy the village for it.") And what a thing for James Delingpole, who never knew the experience of putting a book in many hands, but did get to leave a word on many tongues. "My teeny, tiny spear-carrying role in the history of language," he wrote, "was assured."

Skeptic Patrick Michaels was suddenly on Fox's "Your World with Neil Cavuto." Talking usage, and more or less babbling in his excitement. "Big big big news," Michaels said. "You know, 'Climategate' now gets twenty million hits on Google. If you Google in 'Global Warming,' you only get ten million. 'Climategate' is a new word. Twenty million hits on a new word."

"Oh that's interesting," Neil Cavuto said. "That's very interesting."

It was all very big and very interesting. (Imagine a world where James Delingpole out-polls a science.) And maybe you couldn't have hoped for calm, judiciousness, or the level head. You'd have to know what we do about the scientists, and the inventors, and the history of denial, to understand how deep the story ran.

The *Guardian*'s Fred Pearce had been reporting the Hockey Stick for three years. It was what tobacco had yearned for a decade earlier, when they called the Harvard fight an "unprecedented opportunity" and "our hook." Their big overall want for twenty years: "If the process becomes suspect, the conclusion does also."

And they'd pulled it off: only for an industry not their own. That was the big, dark, interesting joke: the climate people had stepped on the last munitions from a half-century war. "The battles fought by McIntyre and his acolytes to access the raw data," writes Fred Pearce, was "the central story of the leaked emails."

WARMING HAS THE WORST TIMING. We've come to be on good terms with email hacks in our era. (Twitter, then Representative Anthony Weiner in his underwear on Twitter, is the story of any invention.) Just a new kind of cultural mortification vehicle. The Hillary Clinton emails. Sony Pictures' massive back catalog of producer-on-movie-star cattiness, which brought down a studio president. Wikileaks and Julian Assange demonstrating the power of the private made public—the words of men and women of influence, before making themselves decent for company.

We take all this with a grain of salt now. But the climate scientists had the mosquitoes-and-rivers misfortune of being pioneers. It was 2009, and they went first.

Two more feats of horrible clockwork. One accidental, the other on purpose.

First, the weather. It was the harshest winter in a decade, what the forecasters were calling Snowmageddon. "Defined"—this is Accuweather—"by snowstorms of historic proportions and record-breaking cold.... The worst winter some people will see in their lifetimes.... Everything was extreme." Sometimes, you just have to accept the fact that God doesn't really want a solution to global warming.

Second was human. Climate negotiations were slated for Copenhagen, two weeks after the email leak, to produce a sequel to the Kyoto treaty. The scandal dealt skeptics "a powerful political card," wrote Bryan Walsh in *Time*, "at the start of perhaps the most important environmental summit in history.... At the very moment when countries around the world—including the U.S.—seem poised, finally, to begin to control greenhouse-gas emissions."

It didn't take "a conspiracy theorist," Walsh added, to spy the connection. (Eight years later, *Esquire* would report that this was the start of a "melancholy mood" among the scientists. Some gave up, left town. Because of Climategate and the busted summit.) Saudi Arabia's lead negotiator chortled to the BBC—and this was to become the central idea of the email leak James Delingpole named Climategate. "It appears from the details of the scandal," the oil delegate said, "that there is no relationship whatsoever between human activities and climate change."

TO WATCH CLIMATEGATE clank and thunder its way across the news space was to see the erasure of all the gains the scientists had made over five decades. It was a storm cloud that rained away all memory and trust.

Here's what Bill McKibben once had to say about warming in general: industry releasing all the carbon that nature had spent patient millennia accumulating. It was "as if someone had scrimped and saved his entire life and then spent every cent on one fantastic week's debauch."

Climategate was like that with faith and good will. Frank Luntz had once addressed very squarely the problem of climate research: "People are more willing to trust scientists." Luntz had advised elevating denial's own scientists. He'd been thinking small. What finally succeeded was the reverse: make the public unwilling to trust any scientist.

There it was, in *The Atlantic*: "The stink of intellectual corruption is overpowering," wrote senior editor Clive Crook. "Can I read these emails and feel that the scientists involved deserve to be trusted? No, I cannot." It was waiting for you in *Newsweek*. "Climategate has tarnished the image of climate research." Unsettling the London *Times*: "A blow against the relationship between science and the public." And in our own homegrown *Times*. Researchers faced "a crisis of public confidence.... They say the uproar threatens to undermine decades of work and has badly damaged public trust in the scientific enterprise."

Even George Monbiot—Britain's steadiest climate voice—was embarrassed, blindsided, devastated. That pinch, for the public man, of seeming to have fallen among low companions. "It's no use pretending this isn't a major blow," Monbiot wrote the first week. (*National Post*, Canada: "Monbiot has been called Britain's Al Gore.") The emails "could scarcely be more damaging.... I'm dismayed and deeply shaken."

At the other end of the thermostat, Patrick Michaels became more giddy than any denier I've ever seen. For skeptics, Climategate was a warhead with the power of a thousand Christmases. "This is not a smoking gun," Michaels told the *Times*. "This is a mushroom cloud." Later on Fox News—a different host than Cavuto; it's like they were minting new anchors to keep up—Michaels said, "My jaw dropped, dropped, *dropped*, and wound up on the floor." For at least a year, he promised, "This is going to come up and up and up."

IT BROADSIDED INTO POLITICS. Where Representative James Sensenbrenner, vice-chair of the House Science Committee, detected in climate the torches, bootsteps, and exuberant salutes of "scientific fascism." (Also "Junk Science" and "massive scientific fraud.") Senator James Mountain Inhofe, who had perhaps arrived a little in advance of events, who'd conceivably been intemperate with his assessment that "Global Warming is the greatest hoax ever perpetrated on the American public," now had the divinely in-the-loop look of a seer.

"They've been cooking science for a long time," Inhofe explained solemnly. His words scrolling across the bottom of the Fox News screen: INHOFE: THEY'VE BEEN COOKING SCIENCE FOR A LONG TIME. The senator declared himself vindicated, then launched Shelby Amendment and Data Quality investigations against the climate scientists for good measure. (Science beat the rap.)

And it sprawled and clambered across the conservative spectrum. Rush Limbaugh, of course, in his beefy, doubleheader voice. The "whole hoax has been exposed. . . . This is a giant, giant, giant scam." And the rest of that rabble-rousing America's defense force crew. Damp-eyed Glenn Beck: "The global warming hoax continues to be exposed and debunked. . . . They're just cooking the books." Boulder-jawed Sean Hannity: "So it's safe to say that global warming has been revealed and that the movement is run by hacks and frauds. . . . But the political fallout from the scandal—well, that's just beginning."

And it broke mainstream. ABC News with Charles Gibson, tweaking Al Gore: "As the world seems finally poised to do something about global warming, an inconvenient scandal." All the good feeling from 2007, gone. The steady ascent up Mount Public Trust—all the crampons, piton-hammering and enduring necessary to *reach* 2007—undone, via entanglement with Steve McIntyre. Climate change was once again reopened for interpretation.

Anderson Cooper promised CNN viewers, "We'll show you some of the emails and you'll hear from both sides of the debate. You make up your own mind." And, in the event you hadn't: "The emails contain language that indicates the figures for global warming are perhaps being twisted and manipulated, and have been for years."

CBS News went for broke. If you were a *gate* fan, we'd reached the gatiest of gates. "To anyone skeptical about the science of global warming, Climategate is the biggest scandal ever." CNN called their Climategate special "Trick or Truth?" And welcoming Steve McIntyre, introduced him as a "scientist." (McIntyre didn't correct them.) Because it improved the story; it upped the drama. Because if McIntyre wasn't a scientist, what was the point?

The bottom third of the news screen is powerful real estate. It takes care of the brain's filtering and compression work for you. People on air can gabble away—you're never quite sure what they're driving at. And then you see the point summarized by chyron in that space broadcasters call their Lower Third.

With text's greater authority; we trust things written down. *Romeo and Juliet*, assisted by the Lower Third. Juliet: "Palm to palm is holy palmers' kiss." Romeo: "O, then, dear saint, let lips do what hands do." Lower Third: "MONTAGUE TO CAPULET—LET'S KISS."

That's the message you carry away from the screen. EMAILS REVEAL COORDINATED CAMPAIGN TO HIDE DATA ON CLIMATE CHANGE (Fox News). CLIMATE CONSPIRACY? (CNN) IS THE SCIENCE BEHIND CLIMATE CHANGE LEGIT? (Et tu, MSNBC?)

Bill Nye, the Science Guy—in these wildfire months, America's last untainted scientist—appeared on broadcast after broadcast. A lone figure, upending bucket after bucket over the flames. Pointing out to CNN's Campbell Brown, "In the meantime, you guys are running the scroll across the bottom of the screen."

One poll found that 49 percent of Americans were following Climategate "very closely or somewhat closely"; 59 percent considered it "very likely or somewhat likely" that scientists had "falsified research data." *Time* magazine reported a whopping "84% of Americans believe it is at least somewhat likely that some scientists have falsified data to support their theories on global warming." Fifty years of proving and vetting, all that cautious collection and assessment by the IPCC—gone.

Deniers took sunstruck victory laps, blowing kisses to the fans. Dr. S. Fred Singer was defrosted, plugged back in, flipped back on. "We are winning the science battle," he said. "Now it turns out that global warming might have been 'man made' after all." (Eighty-five years old, and the man still had it.) Climategate rendered the deniers in heroic colors—lonely sentries who'd stood a post to defend us from an international team of cheats and swindlers.

The final director of the ASS Coalition, Steve Milloy, brought his smirk and button-down to Fox News. "These emails promote us from Skeptics to the Vindicated," Milloy said. "If we were to get *all* their emails, we would see how they've constructed global warming alarmism out of whole cloth." (Milloy is the man who later gave numbers on total denial membership. "There's really only about 25 of us doing this. . . . It's a ragtag bunch.")

Patrick Michaels had been made light of by the emailers: scientists considered him both dishonest and a kind of research Homer Simpson—a science klutz. MacArthur Award-winning climate modeler Ben Santer had written, "Next time I see Patrick Michaels at a scientific meeting, I'll be tempted to beat the crap out of him. Very tempted."

It became a routine: TV hosts would inquire about Michaels' bodily welfare. (Lower Third: HACKED EMAIL: SCIENTIST "TEMPTED TO BEAT" SKEPTIC.) "Ben Santer's not a small guy, either," Michaels deadpanned. Then he mused, "The thin-*skinned*ness. . . . Yeah, this is all gonna come out. It is not going away."

Lord Christopher Monckton, introduced on CNN by Wolf Blitzer, electrified viewers with his complete mastery of the thesaurus. "Those scientists have been fabricating, inventing, tampering with, altering, hiding, concealing, and destroying data." (Said the House of Lords member who had cured AIDS, multiple sclerosis, his own Graves' disease, and was also a mathematician.) Steve McIntyre got in some lovely, last-word payback. Commenting to the *New York Times* that he'd found the emails "quite breathtaking." Then he received a promotion from *Newsweek*—joined the carved likenesses of Frederick Seitz and S. Fred Singer on their denial Rushmore. "He has become the granddaddy," *Newsweek* reported, "of the global warming 'denial' movement."

AND THE FURY GATHERED and advanced on the scientists. They had betrayed a public trust: they had gotten us all worked up over nothing.

Andrew Brietbart—namesake of the alt-right's Brietbart.com—tweeted his recommendation: "Capital punishment for Dr. James Hansen. Climategate is high treason." This was on general principle. Hansen had not been party to the climate emails. Rush Limbaugh suggested another precautionary measure: "Making sure that every scientist at every university that's been involved in this is named and fired, drawn and quartered or whatever it is." A German climate scientist told *New Scientist* of watching a man rise from his lecture audience and brandish a noose. (In the YouTube video, the hangman is surprisingly well-dressed.) "Someday, some madman will draw a pistol and shoot you. It will happen," he warned colleagues. "I'm pretty sure about that." Over at the *Telegraph*, James Delingpole envisioned a delegation consisting of himself, Andrew B., and Steve McIntyre putting George Monbiot to death with a specialty handgun—a Luger, firing hollow-point bullets, "just to be sure of the right result."

That February, four months after Climategate, is when Andrew Revkin posted his sports update on the *Times* climate site. About the complete changeover in just two years. "Now the foes of limits on greenhouse gases have the ball."

IT WAS NEVER A SCANDAL: It was just somebody else's email—the garbage-pail thrill, the backstage *frisson* of the wall between public and private disintegrating. Email leaks are where you get to see interesting people naked.

The bulk of it was gripes. About being harried; about denial; about what it was to engage in daily professional combat with the likes of

Seed Singer, Patrick Michaels, Steve McIntyre, Willis Eschenbach, Lord Monckton.

Significantly, it was a Canadian—Steve McIntyre's neck of the woods—who stood up on their behalf. "How dare the world's media fall into the trap set by contrarians?" asked Parliament's Elizabeth May. The emails "give a picture of thoroughly decent scientists increasingly finding themselves in a nightmare."

Green groups passionately waited things out on the sideline—the sensible position. (One green executive told the *Guardian* the hack was "nothing less than catastrophic.") It was a perfect frame-up. A kind of horrible and accidental sting operation. The scandal uncovered by Steve McIntyre was the private feelings of a handful of scientists about having to interact with Steve McIntyre.

Dr. Phil Jones, director of England's Climatic Research Unit, had emailed, "Don't any of you three tell anybody that the UK has a Freedom of Information Act!" And, in the event people that *did* hear, "I think I'll delete the file rather than send to anyone."

This looked and sounded awful. And the many official committees that later investigated Climategate would chastise the scientists for "lack of openness." Especially in this digital age, when information, like creatures of the air and the human spirit, longs to be free.

But if you'd followed the story from inside, you understood: it was the shock of the career scientist confronted by an amateur—a blogger with license to barge through the door, rummage inside the desk, right up to the elbows.

And in general, it was an amazing relief, these 2009 and 2010 months, to learn that the science had been fake all along. What news could have been more welcome?

Every scandal and construction site has one tool that handles the deep business. And one line from the thousand emails took care of most of the demolition work.

The words were written by Climatic Research Unit Director Phil Jones. Who had resigned his post. Dr. Jones told the London *Times* he'd been receiving death threats: "They said they knew where I lived." And that he'd considered sparing his threateners the fancy handgun purchase and any associated hardships. "I did think about it, yes. About suicide," he said. "I thought about it several times." The sentence came from a 1999 email about Michael Mann's Hockey Stick.

"They talk of using a 'trick,'" is how CBS news reported this email. "To 'hide the decline' in world temperatures."

Bombshell. (ABC World News: "One of the most damning email exchanges credits Mann with a trick to hide the decline in temperatures.") And here's its entry in the paper of record: "In a 1999 email exchange about charts showing climate patterns over the last two millenniums," the *Times* reported, "Phil Jones, a longtime climate researcher at the East Anglia Climate Research Unit, said he used a 'trick' employed by another scientist, Michael Mann, to 'hide the decline' in temperatures."

At the end of 114 years, what could have been worse? It meant the scientists were hoaxers, and the threat a lie. Phil Jones was "finished," wrote Germany's *Der Spiegel*. "Emotionally, physically, and professionally." The scientist's "credibility has been destroyed—and so has that of his profession."

"Hide the decline" became the scandal's slogan: the words that lived to gouge the deepest marks. James Delingpole called the phrase "infamous." (Michael Mann's autobiography: "One email Phil Jones of the Climatic Research Unit sent to my coauthors and me in early 1999 has received more attention than any other.") Former GOP vice presidential candidate and Tina Fey job creator Sarah Palin could summarize for the *Washington Post*, and find herself in the same Data Quality territory as CBS, ABC, and the *New York Times*. "The emails reveal that leading climate 'experts,'" Palin wrote, "manipulated data to 'hide the decline' in global temperatures."

And it was wrong. What Phil Jones actually wrote about in 1999 was converting Michael Mann's first Hockey Stick into a chart for the World Meteorological Organization. "I've just completed Mike's Nature trick of adding in the real temps to each series for the last 20 years (i.e. from 1981 onwards)," Jones emailed, "and from 1961 for Keith's to hide the decline."

Now, for the second part. There was no cold to ignore. 1998 had been the warmest year, in the warmest decade, then on record—our thermometer number one. A colossal El Niño event had stirred weather conditions everywhere: there'd been no decline to hide.

Still, Senator James Inhofe could explain at the Copenhagen climate summit, "Of course, he means 'hide the decline in temperatures.'"

And for the first part: "Mike's Nature trick" had to do with Michael Mann's first *Nature* paper and what's called the divergence problem. The tree rings the scientist used to track temperature—among many proxy data sources—cease to yield reliable information after 1960. (There are a number of interesting theories why. The main one being pollution.) So Mann had replaced them with the lived temperature record; he'd noted doing so in *Nature*. Nothing hidden, nothing *gate*, about it.

"Trick"—Michael Mann pointed out—was a way of saying "a clever approach." Dr. Phil Jones is British. And I have reason for thinking this is a particularly British usage.

As the *Guardian*'s Fred Pearce wrote, "It is manifestly not clandestine data manipulation, nor, as claimed by Palin and Inhofe, is it a trick to hide global cooling. That charge is a lie." The journalist went on, "But it is a lie that persists."

In one of the many Climategate investigations—each ending with exonerated scientists—the English House of Commons' Science and Technology Select Committee plumbed the mysteries of slang. They determined the situation to be as Michael Mann described: "A colloquialism for a 'neat' method of handling data."

James Delingpole—who named the scandal—sputtered and fulminated. "Scientists don't *do* that," he fumed on BBC. "They don't try to hide the decline using 'Mike's Nature trick.'" That sort of thing was "now infamous."

For *Telegraph* readers enlivened by the spectacle of columns produced by the batshit mad, Delingpole wrote, "This 'trick' was a bad, naughty and very wrong thing—and not a colloquial phrase meaning 'the best way of doing something.'"

But again, I have good reason for believing James Delingpole knows that this is the way many British people use the word "trick." Because this is the way James Delingpole uses it. In the last chapter of the last literary novel he ever wrote.

It's how that last chapter begins. Delingpole being Delingpole, he used it in a sequence about getting baked. It's an aid to comprehension here to know that Rizla is the leading European brand of rolling paper.

"Norton watches me knit together the Rizlas, using this new trick I've learned," Delingpole writes. "Basically, you make it with two papers instead of three."

Delingpole continues, "I take a long drag on the joint. . . . It's good weed. Rock star quality weed almost." That *almost*. It's Delingpole's word; also the denial word, the most painful seat in the house. It's S. Fred Singer before warming, Arthur Robinson without his petition, Delingpole minus Climategate.

And James Delingpole complaining about the word "trick" is denial in its rawest data form. It is people with every reason to know better, pretending— for their wide spectrum of reward—that they know nothing of the kind.

Nine official inquiries cleared the climate scientists. Including investigations by the National Science Foundation, EPA, the Department of Commerce's inspector general. (With the National Science Foundation and the Department of Commerce additionally testing for violations of the Data Access and Data Quality Acts. All clear.) EPA determined Climategate was "simply a candid discussion of scientists working through issues that arise in compiling and presenting large complex data sets." The inspector general declared climate science "robust."

And we were back where we started in 2009—but with the essential element subtracted. With trust removed, with the doubts chained and clanking around the ankle at last. It's how things go, post-scandal. You're never again quite clean. What became fixed in the public mind was that the scientists must have done *something*.

THE CLIMATE SCIENTISTS got famous along with James Delingpole, but not in his pleasant way. (Delingpole's year-end lists ceased to feature winces or cravings. Climategate fixed him up. He didn't have anything left to mope about.) The climate scientists got famous in the unpleasant way of seeing on TV that their names and faces featured in a Neo-Nazi

death list. Of the FBI sealing an academic office because they'd received
some suspicious white powder through the mail. Of a dead thing left on the
doorstep. Of hiring private security. Of the lavish thanks from a grateful
nation. Let's open the mailbag.

> *You and your colleagues who have promoted this scandal
> ought to be shot, quartered, and fed to the pigs along with
> your families.*

> *You, sir, are a nazi. Go gargle razor blades.*

> *As a US taxpayer I want a fucking refund of all the wages
> you have fraudulently collected you asshole. Same goes for Jim
> THE FUCKING RAT Hansen.*

> *Subject: YOU FUCKING FRAUD! WATCH IT ALL
> FALL APART YOU CUNT!*

> *Nazi Bitch Whore Climatebecile.... You stupid bitch, You
> are a mass murderer and will be convicted at the Reality
> TV Grand Jury in Nuremberg, Pennsylvania.... I would
> like to see you convicted and beheaded by guillotine in the
> public square.*

That reality TV Nuremberg Grand Jury is something I'd watch, too.
(The old pre-climate James Delingpole could have reviewed it.) Down
in the Antipodes, Australians were just as virulent with their gratitude,
though they eschewed PG-13 modes of expression.

> *What a fckn load of pseudo-scaremongering turd... why
> dont ya all fck off to yr beloved europe where their economies
> are fckd because of stupid schemes like yours and leave us to
> run this great country as it should be?*

Curiously, this was the start of that "melancholy mood" *Esquire* would report on next decade. Of climate scientists "so demoralized by the accusations and investigations that they withdrew from public life." Of the climate voice, for years, leaving the stage.

It was—in the circus analogy from that ancient *Time* magazine—the end of a trail dug, laid, flagstoned, landscaped, and set out upon at that Philip Morris meeting in 1987. And, really, all the way back to 1953. Since most of us will never be scientists, public research is a trust—it's the mute hopeful look your dog gives the vet. This is what Frank Luntz knew; what industry wanted to expunge. As with Sound Science, they had succeeded beyond any reasonable dreams. "We need to challenge [the] popular notion that science is objective, value-free and without error," they had written in 1989. With this golden outcome: "If the process becomes suspect, the conclusion does also."

As *National Review* editor Jonah Goldberg explained on Fox, this was Climategate's unique and lasting contribution. The scandal "debunked" the notion, he said, "that there are these pure, gifted, wonderful scientists, riding around on unicorns just trying to say the divine truth, you know?"

James Delingpole called this perhaps "the greatest victory...won in the aftermath of Climategate." Literal trust-busting: an end, he wrote, to science's "monopoly on 'truth.'"

That Chapel Hill legal scholar who'd warned of data trench-wars had peered a little ways down the road and worked out the implications. "Whatever possible upside the Data Amendments may have in expanding the ability of nonscientists to correct the occasional and inevitable scientific mistake," he'd written in 2004, "the possible downsides of creating a world in which science is perceived as just another category of 'spin' are undoubtedly much more profound." By 2016, *Science*'s new editor-in-chief was touring the cultural wreckage: "Scientists," he said, "have been labeled as another special interest group."

The anti-vaxxers; the non-GMO foodies; the people who suspect jet trails of containing a dark toxin aimed squarely at them. The gates opened, this entered. In an *Atlantic* cover story the following year, Kurt Ander-

son explained, "the idea that much of science is a sinister scheme con-
cocted by a despotic conspiracy to oppress people" had become "popular
and respectable."

Denial never did manage to follow up on Frank Luntz's recommendation. They lacked figures to elevate. But the reverse succeeded. A friend
of mine has a theory. Somewhere deep, in the lowdown parts of us that
have escaped evolution, there remain certain hard, baked-in drives. "What's
more inspiring than a beautiful person riding through the city on a horse?"
my friend asked. "That same person pulled out of the saddle and dragged
through the mud." That was the thrill of Climategate. The scientists spat-
tered, cursing, and laundering with the rest of us.

Historian Spencer Weart wrote that the Climategate accusations had
become "permanently embedded in the public's memory." To Fred Pearce,
the science had "taken a battering from which it may never fully recover."

THAT WAS THE on-purpose part. And now, the accidental.

All that Snowmaggedon winter, it snowed and snowed. Each blizzard
a light satire on the climate scientists.

The February weekend two feet of powder fell on Washington,
the Inhofe family—the daughter and grandchildren of Senator James
Mountain—tramped across the Capitol Mall. They spent five hours rolling
together and standing up an igloo. They plonked a cardboard sign on the
roof. AL GORE'S NEW HOME! And HONK IF YOU ♥ GLOBAL WARMING.

The senator found this family construction project, he said, "really
humorous." Half a century later, and the future had turned from a parrot
to an igloo. Tourists posed with the ice house. The senator keeps a photo of
the lumpy finished product near his desk.

In a way, that's the end of the story. The scientists studied with verve,
intelligence, creativity, expertise, vigor, integrity, dedication, and courage. (I know; I have a thesaurus, too.) Voyaged to the world's forbidding
top and bottom, gloved and parka-ed, to drill and examine. They studied,
published, challenged, and corrected one another, perfecting a theory that
turned out to be true. Roger Revelle guessed carbon dioxide would not

be held by the sea. He was right; it wasn't. Dave Keeling believed the carbon dioxide would gather in the air. He measured; it was. Svante Arrhenius calculated that carbon dioxide and temperature were linked, investigators found unique wormholes into the far and measureable past, and proved this, too, from that far past to now. And the heating began. With the melt coming just where it was expected. And, in a way, the people who accepted the science overpromised. Suggested to us big changes coming soon. A parrot where there had been ice. Warm . . . where it was supposed to be cold. And day by day, men and women worked and the sky stayed blue.

Fifty years later, with the formulas and theories proved—the careers made, prizes distributed—their opponents could demonstrate the reverse. An igloo in the capital. Cold . . . where it was supposed to be warm. It was what the argument has always been. (What it always is, when there's a national argument: competing exaggerations.) The overpromise and the underpromise. Leaving the terrain in the middle, where most of us live, clear for fighting.

THE PARROT

AND THEN JIM HANSEN STARTED TURNING UP AT CLIMATE
protests and getting himself arrested. There was, in these news images, something gallant and decent. The distinguished scientist, wrists restrained, a
police officer by his side.

It made you think of a climate moment from *The Simpsons*. Nelson
Muntz, Springfield Elementary's goon, has raised his fist to Milhouse Van
Houten, the sort of eyeglass-wearing shrimp who attracts bullies as a bench
press does athletes.

"Say global warming is a myth," Nelson demands. Milhouse instantly
cowers. "It's a myth! Further study is *needed*." Fifty years of American science, business, politics, and history are contained in that flinch.

Hansen was instead taking the punch. "I'm trying to make clear," the
physicist said to a TV reporter, "what the connection is between the science and the policy." Hansen told the *New Yorker*'s Elizabeth Kolbert he'd
begun to "lose faith in the Obama administration early."

BARACK OBAMA had campaigned on warming. The "climate crisis," the
senator explained in 2008, was also a "moral challenge."

Climate and health care were the twin monsters the candidate promised, if elected, to close with and defeat. Obama indicated their positions
on the White House line. "Energy we have to deal with today," the senator
told NBC's Tom Brokaw at the second presidential debate. "Health care is
priority number two."

And maybe, in retrospect, this was the trouble's first squeak. Obama preferring, in the canny Frank Luntz style, a more soft-spoken word for "climate."

There it was again, a few hours after the election maps assumed their reassuring final color. ("I was at a lovely New York party, full of lovely people," writes Zadie Smith. "Celebrating with one happy voice as the states turned blue.") The president-elect was back on the topic. "Even as we celebrate tonight," he announced, to the weeping cheering crowd at Chicago's Grant Park, "we know the challenges that tomorrow will bring are the greatest of our lifetime."

Those challenges: a financial collapse (lava and comets, the chasm under Wall Street and ten million homes), two wars, and "a planet in peril." Jim Hansen wrote that he, too, had "moist eyes during his Election Day speech in Chicago."

Obama's second post-election appearance was at a Los Angeles climate conference. "My presidency will start a new chapter," he promised. "Now is the time to confront this challenge once and for all. Delay is no longer an option. Denial is no longer an acceptable response."

The scientists Obama named to key White House positions the newspapers called a climate Dream Team. Stanford physicist Steve Chu became the first member of any cabinet entitled to list "Nobel Prize" under "Endorsements" on his LinkedIn. Chu was appointed secretary of energy—Obama's gentle word for warming. The president described the pick as a flare; a defiant banner flown above the White House. "A signal to all that this administration will value science. We will make decisions based on facts. And we understand that those facts demand bold actions."

All this tripped the green groups into euphoria. "You see this smile on my face?" asked one green leader in *Time* magazine. "It's not going away." The National Resources Defense Council staked out the position where confidence turns surly: "Obama Means What He Says."

And then the new president rolled his chair into the desk. Where the whole thing came down to an adhesion contest.

Democrats now controlled the House, the Senate, the Oval Office. The once-in-a-generation chance to make some big change. (Which is what the job looks like from the inside, too. "Winning the election," Obama later said, "only gives you the chance to do something.") Climate or health care. Which do you pick?

Sometimes life isn't as tightly structured as we'd like. "The plan was to throw two things against the wall," a senior member of the administration later said to the *New Yorker*. "And see which looks more promising."

That was it. The tosses were made. The results were tabulated. The new president let climate slide.

"For Obama," the magazine reported, "health care had become the legislation that stuck to the wall." After that, a senior White House official explained, climate became "Obama's 'stepchild.'"

Even the president saw family planning. "It's like choosing between two of your children," he told one disappointed House leader. "But I care about health care more."

So it was health care that received the healthful, nourishing rays of presidential attention. Phone calls to the teacher, inspection of the report card, worries over every little thing. The stepchild was sent upstairs to its room, dressed in hand-me-downs, and eventually dropped out of school.

(Here's how chancy life is. In the late seventies, Jim Hansen got to know a *New York Times* science reporter at the local Chinese restaurant; it set him on a certain path. This is also true of presidencies. During primary season, health care was simply polling very strong for rival Hillary Clinton. Obama aides called it "Hillary's thing." So, at an early campaign stop, Obama promised universal coverage within his first term. "We needed something to say," an advisor later explained. "I can't tell you how little thought was given to that thought, other than it sounded good." The plan, a single line, was "scrawled hastily in January 2007. . . . Thus was born Obamacare," *Politico* reported. "A check-the-box, news-cycle expedient that would ultimately define a president." These tiny moments and strategies are the sand and grit that have gotten caught in the great turning wheels of climate.)

The first year, 2009, a climate bill managed the tricky bicameral leap; House to Senate. Where it sank.

President Obama did not extend a dry hand from the lifeboat. David Axelrod, chief Obama strategist, explained that passage of health care had

absorbed all of Congress' energy and hit points. "The horse has been ridden hard this year," he told climate supporters in 2010. "And just wants to
go back to the barn."

There was climate's chronic issue with timing: That fall, Republicans
took back the House.

And like that: The door was closed. As the *New Yorker* reported, Congress and the White House had just "missed their best chance to deal with
climate change."

On warming, the magazine continued, "Obama grew timid and
gave up." As the *Times* explained things, it was all a matter of what you
sent through, during that brief aperture when the door is open. "Mr.
Obama invested most of his political capital in his signature health care
law." Climate's failure, the paper noted, "seemed to make little impression on him."

"The Bush administration was a disaster," Jeffrey Sachs, director of
Columbia's Earth Institute, told *Newsweek*. "The Obama administration
has accomplished next to nothing either."

GLOBAL WARMING HAD proved no more convenient for this administration than any other. During the first months, Carol Browner, President Obama's climate czar, invited environmental leaders to her office and
passed around copies of the one-page White House strategy sheet. To help
pass a climate law, the best bet was, never mention "climate."

Discuss jobs. Talk about energy till you're blue in the face. Only
please—no slip-ups when it feels plain good to use what Browner called
the C-word.

It startled attendees. "This is a mistake," Bill McKibben warned the
group. "It's going to come back and haunt us."

It was only another reboot. Flip back twenty years, and it's just what
the newspapers had teased the first Bush administration for. "How can you
call a White House conference on global warming," the *Boston Globe* asked
in 1990. "And then not even use the words?"

Message discipline is a currency supported by punishment. On a South

America trip one month later, Energy Secretary Steven Chu—the Nobel-ist, the Dream Team center, the signal to all that this White House would value science—fielded a question about climate.

"I think the Caribbean nations face rising oceans," Dr. Chu said. "And they face increase in the severity of hurricanes. This is something that is very, very scary for all of us." (Eight years later, Puerto Rico would be slammed by Hurricane Maria—"one of the most violent storms ever to hit the Carib-bean," *Time* reported. Maria took three thousand lives.) Chief of Staff Rahm Emanuel telephoned down from Washington with his thoughts.

"If you don't kill [Chu]," Emanuel told Deputy Chief of Staff Jim Mes-sina, "I'm going to."

Deputies are the boss in a travel-size version. Messina caught up with Dr. Chu mid-flight, at the long Air Force One conference table. "What did you say? What were you thinking?" Messina yelled at the scientist. "And how, exactly, was this fucking on message?"

Messaging was the whole thing; it was the fight. "You had this incred-ible green cabinet of really committed people," one administration offi-cial later said. "But the only thing that really matters is what the President says—so everyone was trying to get words into his mouth. And Rahm was trying to keep the words out of his mouth."

Next to the charismatic president, Rahm Emanuel was the big story of the first Obama years. Skinny and forceful, with short gray hair that sug-gested iron filings and a magnet. The *Times* called Emanuel the most pow-erful chief of staff in twenty-five years—"a virtual prime minister."

To "appreciate the sheer reach of the Emanuel legend," the *New Repub-lic* reported, "one need look no further than the upper rungs of the Obama administration." Where, as one White House official elaborated, "Every-one in government believes they're working for Rahm."

His résumé suggested a crack-toothed brawler from the UFC: two decades of streetfights. Service as President Bill Clinton's senior advisor. (A series of confidence-building obstacles. "Well," he told a colleague with a knotty request, "I guess if I can take care of Bill Clinton's blowjobs, I can take care of that.") Then getting himself elected to Congress. Then heading the Democratic electoral committee. Rahm Emanuel looked like a lock for the next Speaker of the House.

One of those media victors: people who design their own legend. He'd sent an uncooperative pollster a dead fish, like one of Don Corleone's torpedoes. Stabbed the table at a Clinton victory party, shouting the name of a defeated campaign enemy with each thunk. *Dead. Dead. Dead.* This was politics lived as movie highlights, as the clips fans post on YouTube.

Emanuel was also from Chicago—which to fellow Chicagoans means something particular. (Rules and courtesies skipped, opponents lying face down in some gutter.) Obama came from Chicago; so did David Axelrod, his mustached and puddle-faced chief strategist. They were strangers to Washington. In 2009, perhaps no one possessed a surer feel for the city's short cuts, warm spots, and soft places than the table-stabber. Here's the *Times* on the Rahm-Obama pairing:

"A visionary outsider who is relatively inexperienced and perhaps even a tad naïve about the ways of Washington captures the White House. And, eager to get things done, hires the ultimate get-it-done insider to run his operation."

That is, it was the basic setup of the HBO comedy *Entourage.* (Coincidentally, one of the new president's favorite shows. So you got to live the situation all week, then relive it every Sunday on cable.) A movie star, newly arrived in Hollywood, hires as his agent the suavest and dirtiest streetfighter he can find. Then the agent's whole job becomes protecting the actor from his nobler instincts. Talking him out of the small, worthy pictures he longs to make; steering him into the big, dumb vehicle that will make everybody a ton of money.

"A studio picture," the agent coos. "With a lunch box." And then, because the agent is the kind of use-it-all negotiator who employs grossness as a tactic: "We're gonna get you the whole thing—an action figure with a huge cock."

Entourage was based on the Emanuel family voice. A sound that suggests a philosophy.

"At the heart of the Emanuel mystique is the family patois," the *New*

Republic explained in its Rahm cover story. "Which lurches between pronounced curtness and vivid, sometimes scatological, imagery."

The show's breakout character, superagent Ari Gold, was a close like-
ness of Rahm's brother, Ari Emanuel—CEO of the talent agency William
Morris Endeavor. One of those people you meet, then immediately crack
open the notebook. "I had never seen anybody in this business really talk
like that," explained series creator Doug Ellin. Who decided, "He has to be
in the show."

It was just how the Emanuel boys expressed themselves. Nothing
unspoken, because directness freaks out the listener. And no virtue but the
win, since beneath all the menu talk and place settings it's the one thing
everybody knows is paying for the table.

Ari on the show is a cartoon of loopy, dirty-mouthed focus. "Maybe
you've had so much come squirted in your eyes you just can't see it," Ari
tells Lloyd, his put-upon assistant. And here's Rahm Emanuel at the White
House, encouraging a tongue-tied assistant to speak his mind. "Take your
fucking tampon out and tell me what you have to say." (A staffer grumbled,
"He treats us all like we're Lloyd.") *Fuck* is Ari Gold's word. Here's the agent
discussing a colleague's poor negotiating position—one syllable repeated
so endlessly it slips below surface sense, into a kind of Gertrude Stein tone
poem. "You're fucked. Do you even know how fucked you are?" he asks. "I
mean, you are so fucking fucked. I mean, I think you are the most fucked
person I know."

It was as if *Entourage* and the chief of staff shared a writers' room. Rahm
Emanuel's pet name for the District of Columbia was "Fucknutsville." His
hot-and-cold phone sign-off was "Fuck you. I love you." The nameplate in
his office said, "Undersecretary for Go Fuck Yourself."

It was like a special Washington-themed episode of the show. And it's
how the story had run: 115 years after Svante Arrhenius, fifty years beyond
Roger Revelle, two years past the IPCC's "unequivocal"—and the brother
of the guy from *Entourage* had gotten his hands on climate change.

"The only nonnegotiable principle here is success," Emanuel told his
West Wing staff. "Everything else is negotiable."

He schooled policy makers in the mystical process whereby a single success, properly identified and cultivated, could potentially fuck its way to a whole litter of future ones: "Success breeds success," was the chief of staff's motto. Advice gets tattooed onto a subject. For a sense of how often these words must have been repeated beside climate, skip ahead six years. To 2016, and a reporter asking President Obama his hopes for a climate summit. This fired the old Rahm chip. "Success breeds success," the president said.

But it also meant the cautious, infertile reverse: one failure could put a stop to all future breeding opportunities. "We want to do this climate bill," Emanuel told green supporters. "But we need to put points on the board. We only want to do things that are successful. If the climate bill bogs down, we move on. We have health care."

Same sentiment, *Entourage* setting. "There are no asterisks in this life," Ari explains, "only scoreboards." And then, because it's how the Emanuel brothers communicate, "And ours is currently reading: 'Fucked.'"

In *The Climate Wars*, a history of these years, Eric Pooley relates a string of climate near-misses and frantic repairs that could have furnished the series with a whole season of plotlines. The agent keeping his client from inking the loser deal, signing onto the dud project.

"Emanuel and the political team saw climate as a drag on Obama's popularity and political power. . . . It would have helped if the President were doing more selling. . . . The only way for Obama to galvanize attention would have been a primetime address, but Axelrod and Emanuel didn't think that was a good idea. . . . Obama's megaphone was again missing. . . . Obama reluctantly agreed with Emanuel and Axelrod that voters didn't care about climate action enough. . . . The Energy cabinet pushing for climate action and the political and economic teams, led by Chief of Staff Rahm Emanuel, arguing against. . . . Emanuel wanted to protect his boss."

Emanuel joked privately, "Look, the dolphins will be OK for another year." Elizabeth Kolbert reported the outcome in the *New Yorker*: "President Obama was silent. He did nothing to rally public opinion on the issue." As Eric Pooley concludes, Rahm Emanuel had become "an obstacle to climate action."

It was around this time that Jim Hansen started getting himself arrested at climate protests. He had received the message: no White House rein-forcements were coming. He was out there alone, with his warnings and science. He told Elizabeth Kolbert, "This particular problem has become an emergency." He told another reporter, "The democratic process doesn't quite seem to be working."

But it was how that process had worked for three decades. Emanuel just made the situation more clear. You do not stick the president in an arthouse picture: he does summer tentpole, or he ceases to be president. The thing about the Emanuel class of advice is that it only seems braver than other schools. Beneath all the cursing, the advice is to do the safe, practical thing. ("You stupid bastard—take an umbrella.") It's also cynicism extended so far along the graph it becomes a kind of innocence. The belief that there is no course available but the familiar and ineffective one.

What voters learned was that even with Democrats, they got the same climate nothing they'd become accustomed to from Reagan, Congress, and two George Bushes. It was emotionally different: a fond, supportive, well-informed nothing. But still nothing. Jim Hansen later wrote the president had "failed miserably" on climate change.

In his first State of the Union, President Obama brought up the cli-mate crisis twice. Next year, the stepchild did not rate a single mention. 2012, one shout-out.

In that year's election, Republican challenger Mitt Romney ran expressly as a denier. "President Obama promised to begin to slow the rise of the oceans and heal the planet," Governor Romney told the GOP Convention—then had to ride out a twelve-second swell of audience laughter. "My promise is to help you and your family." The political team urged the president to keep quiet on warming. "It didn't poll well," chief strategist David Axelrod explained.

But warming, with no idea it had become unfashionable, went right on warming. 2012 was the hottest year in US history—a full 3.2 degrees above the twentieth century average. The *Times* headline had the flavor of the sports page: "NOT EVEN CLOSE."

This was the summer of the Big Heat: 2,245 counties in disaster-level drought—71 percent of the United States. "It's like farming in hell," one crop expert told Bloomberg News. An Indiana farmer shakily explained to National Public Radio how the situation could prey on the mind: "It's hard just watching the crops wither." Wildfires in Colorado, a chain of the super-storms called derechos crowding the East Coast. "This is certainly what I and many other climate scientists have been warning about," one researcher told the Associated Press. What the country was previewing, he said, "is what global warming looks like at the regional and personal level." This is the same month the *Times* reported that the issue had "all but fallen out of the national debate."

And at the actual televised debates, the stepchild was a complete no-show, for the first time in a generation. Neither candidate brought up climate; no debate moderator thought to ask.

The president had remained on message—he'd successfully avoided the C-word. "Climate change," wrote Elizabeth Kolbert in the *New Yorker*, "has become to the Obama Administration the Great Unmentionable." In the days that followed reelection, *New Yorker* editor David Remnick called for "a sustained sense of urgency." Thirty-five years after the *New York Times* first typeset the word beside climate, and the last appearance of Urgency in this book.

WITH OTHER AVENUES CLOSED, the issue had again gone circular. There was now a whole Ivy League field of study—Yale's Center for Climate Change Communication—dedicated to why Americans couldn't seem to get a handle on the subject. The program's director speculated in the *Times* that this was all a question of basic mental wiring. As of 2012, it's the thing his study had taught him. "You almost couldn't design a problem," he said, "that is a worse fit with our underlying psychology."

Never knowing—though this proved the academic's point—that he was parroting an insight from two decades earlier. Back then, the professor had been MIT, and the publication was *Time*. "If you said, 'Let's design a problem that human institutions can't deal with,' you couldn't find one better than global warming."

The Intergovernmental Panel on Climate Change released its Fifth

Assessment Report in 2013. Our human responsibility for warming was upgraded: 95 percent and above, Fresher Tomatoes. The IPCC was like a court that reconvened every six years to issue the same shadow-free verdict. (But since the culprit was always us, which way could you really root?) There was none of 2007's fanfare. These statements were reboots. "Climate change is the greatest challenge," explained the report's chairman. World leaders, the UN secretary-general told the press, must act: "Time is not on our side."

Dr. Steve Chu left the White House in 2013. He headed back to Stanford, where there were no Rahm Emanuel deputies in the vicinity to help the Nobel laureate remain on fucking message. Dr. Chu gave a speech. During Q&A, a parent asked whether there was still time to fix the situation for her children.

You can hear the frustration of government experience in Dr. Chu's response. Also the optimism that draws a scientist to that service. There will be three more mentions of Too Late in this book.

"The longer we wait," Dr. Chu said, "the bigger the hole, the deeper the hole, we dig. I can't say with certainty at this point, 'It's done. You might as well go home, whoopee it up.' . . . As a scientist, I'm not willing to do this. Because if you do *that*, you give up trying. And that's the last thing you want to do."

The scientist considered for a moment. "I will admit, though: it gets harder and harder. With each succeeding year, each succeeding decade. . . . Is there such a thing as being too late? Yes. I think, by mid-century, if we haven't gotten decidedly off this path, we're gonna cruise to three, four, five degrees. That is just *in* the cards." Beyond that, Dr. Chu added, lay somewhat less congenial terrain. "At four, five, six degrees—that's really non-adaptable *bad*."

This was the weather that made reboots sprout. Because when a condition persists for so many decades, there are only so many analogies. Without knowing it, Chu was quoting Dr. Wallace Broecker—the scientist who helped popularize the term "global warming." Broecker had once searched himself for a metaphor—and found it in the verbal GIF of a whole culture raising a handgun to its temple. "We play Russian roulette with climate," Broecker said in 1987, "hoping that the future will hold no unpleasant surprises. . . . I am less optimistic about its contents than many."

Here it was, a quarter century later. And it turns out everyone will paint the same landscape the same way, when that landscape doesn't change. "It's Russian Roulette," Dr. Chu told his Stanford audience. "Every decade you put in another bullet. And you give it to your grandchild and say 'Pull the trigger.' . . . We would never do that to our grandchildren." The physicist, a spry but graying sixty-six, spread his arms wide. "We're doing it." Dr. Chu's presentation has been viewed, on YouTube, three million fewer times than the address given by the popular climate speaker Lord Christopher Monckton.

Jim Hansen retired that year from NASA. The saddest outcome: A watchman outlasted by the night he watches. The scientist packing up and driving off, the *Times* said, left the field windswept—depriving "federally sponsored climate research of its best-known public figure."

Hansen always told interviewers he researched on behalf of his grandchildren. For the greatest of family reasons—for the proud thing it entitles you to say outside the family. Hansen never wanted them having to explain that their grandfather knew what was going to happen, but didn't make it clear.

Hansen told the Associated Press that, on reflection, he'd failed at his long task. It had been four decades. And all you had to do was look at Washington, see the shiny and exhausted faces in summer. He hadn't, Hansen now saw, been "able to make this story clear enough for the public."

The *New Yorker*'s Elizabeth Kolbert had come to sound bruised by hoping, in the manner of Jim Hansen. "As we merrily roll along, radically altering the planet . . ." she wrote And, "The United States has been officially committed to avoiding 'dangerous' climate change. One Administration after another—Bush I, Clinton, Bush II, Obama—has reaffirmed this commitment, even as they all have failed to live up to it." And, "There's no discussion of what could be done to avert the worst effects of climate change, even as the insanity of doing nothing becomes increasingly obvious."

A year after Hansen's retirement, Kolbert published *The Sixth Extinc-*

tion. Extinctions One through Five were spasms like so long to the dinosaurs—big, species-wide die-offs that cleared a generous space for us. Extinction Six was the one we were perpetrating now.

The book contained an especially vivid and disturbing image: every mouse and vine taking to its heels, the natural world in general retreat from us. "To keep pace with the present rate of temperature change, plants and animals would have to migrate poleward by thirty feet a day."

Kolbert's book received the Pulitzer Prize. Finally, somebody good, for whom warming climate had also done something good. Kolbert told an interviewer what study of the topic had taught her—the learning point overall. "We in the USA," she said, "live in a weird country."

IN THE AFTERMATH of their victory the deniers were scattered. Skeptic Pat Michaels once ran a website called World Climate Report. Originally funded by coal, it became a sort of clubhouse, a place to enjoy the latest relaxing word from S. Fred Singer, Mark Mills, Steve McIntyre. "The nation's leading publication in this realm," declared its non-modest About Us. "World Climate Report is exhaustively researched, impeccably referenced, and always timely."

The final impeccable update was October 5, 2012. Like the journal of a soldier whose position had been abruptly surrendered or bombed.

The final denying generation went out the same way; the big names on the field for the big win of Climategate. This was the Seventh Extinction— a retreat to more agreeable climes.

Lord Christopher Monckton endured a distressing episode of on-air torture. The operation took place at the banquet hall of his gentleman's club. The viscount was compelled by a BBC personality to watch video of his famous YouTube speech, then further compelled to admit what he'd gotten wrong or invented.

This was not the satisfying spectacle that bloodlust would argue. In the end, your heart goes out to the seething and diminished man. "Yes, if you like," the lord says, in a low voice. "I made a mistake." He looks small, resentful, blinking, injured. A Wonka admitting the flavoring agent was always corn syrup, and wrecked your teeth.

Same misfortune claimed James Delingpole. Sir Paul Nurse, head of Britain's Royal Academy, arrived on the journalist's doorstep with a BBC film crew. Delingpole made a horribly suave little bow: "Sir Paul."

Sir Paul had Nobeled in genetics. He was then, according to the London *Times* (the supreme *Times*, from which all other *Times*es derive) the "most influential figure in British science."

The two men sat at Delingpole's desk. "Science has *never* been about consensus," Delingpole began, with the enthusiasm of disgust. "And this is I think one of the most des*pic*able things about Al Gore's so-called consensus: consensus is not science."

The troops had advanced. From saying the consensus was bogus—the grapeshot fired by Arthur Robinson and S. Fred Singer across two skirmish decades—to saying it was real but also immaterial. Because denial, with its sharp nose, will always find the thing left to deny. And because denial is about defending, to the last feint and soldier, what territory remains.

The next part became famous. Magazines were still discussing it years later—"the magisterial tearing-down of climate skeptic James Delingpole." Sir Paul explained away his actions to the *New York Times* with the shrug of duty. "We can't sit by," he said, "without exposing bunkum."

Sir Paul asked what Delingpole would do if he fell gravely ill.

"I want to give you an analogy," the scientist gently begins. "Say you had cancer. And you went to be treated. There would be a consensual position on your treatment. And it is very likely that you would follow that consensual treatment. . . . Now, you *could* say, 'Well, I've done my research into it and I disagree with the consensual position.' But that would be a very unusual position for you to take. And I think sometimes the consensual position," Paul Nurse concludes, "is most likely to be the correct position."

Delingpole's mouth opens—and for an agony of seconds, nothing comes out. With a media figure, such silence is a prefigurement: of impotence, death, and no more TV bookings. "Yeah. Um. . . . Shall we talk about, shall we talk about Climategate? Because I—I—I—I mean, I don't, I don't accept your analogy. Uh, really. I—I—I—I think it's, uh . . ." And Delingpole concludes lamely, Sir Paul smiling from his chair, "I do slightly resent the way you're bringing in that analogy."

Delingpole had fallen to pieces on air. As he ought: this is history's most sciencey time, when miracles of technology ride in every back pocket. The journalist who named Climategate became, by his own account, the object of "a frenzied witch hunt." The irony escaped Delingpole's notice. (Maybe this has to do with the sort of talent denial attracts. Self-satisfaction erects high sea walls against any surges of self-knowledge.) These were the same experiences to which he'd introduced the climate scientists. Delingpole endured "vile online insults," read "reams" of hate email. "It was miserable," he complained. "My health suffered. I grew very depressed. . . . Five years on, the abuse hasn't stopped."

Steve McIntyre cut back on his auditing. Maybe enjoyment of the climate puzzle ceased when he was presented with the bill. McIntyre had become so engrossed, he'd missed out on the sale of an actual gold mine. His own personal loss he audited at $2.5 million. "If I had never heard of climate," McIntyre told an interviewer, "I'd have made a lot more money."

Denial's most popular surviving website is maintained by a former TV weatherman. Anthony Watts wears a Ron Burgundy mustache and never finished college; his site is called Watts Up With That? (This would be like Comedy Central reconfiguring Trevor Noah's "The Daily Show" as "You Noah What I Mean—With Trevor Noah." It's that level of joke.) In 2011, Watts staked his reputation on a final science push. Berkeley's Richard Muller is a physicist and MacArthur Award winner: an official genius. He was also a climate skeptic—Muller did not believe climate warming was caused by humans.

Dr. Muller accepted startup funds from the mischievous, fuel-loving Koch brothers, then undertook a massive reinvestigation of the temperature record. (International news: "Can a Group of Scientists In California End the War On Climate Change?") Dr. S. Fred Singer was located, plugged in, booted back up, to declare Muller's project "what needs to be done." Anthony Watts promised, "I'm prepared to accept whatever result they produce, even if it proves my premise wrong."

In 2011, Muller and his team announced results to Congress. Climate warming was caused by humans.

Anthony Watts now called the work "incomplete," "error-riddled," and "non-quality controlled." When a journalist reported this story, Watts sent a wounded letter—asking, essentially, what was up with that? "Was your intent simply to add [to] the injury done to me by Dr. Muller?"

Muller published a long essay next summer in the *New York Times*: "The Conversion of a Climate Change Skeptic."

> *Last year, following an intensive research effort involving a dozen scientists, I concluded that global warming was real and that the prior estimates of the rate of warming were correct. I'm now going a step further: Humans are almost entirely the cause.*
>
> *My total turnaround, in such a short time, is the result of careful and objective study. . . . Our results show that the average temperature of the earth's land has risen by two and a half degrees Fahrenheit over the past 250 years. . . . These findings are stronger than those of the Intergovernmental Panel on Climate Change. . . .*

Muller told *Newsweek* it took only about ninety minutes to convert the average skeptic. He offered these conversion services to Donald Trump. Factoring for extra recalcitrance, he believed he could bring in the job in under two hours.

In 2013, the *New Yorker* published a long gloomy postmortem: "What Happened to the Environmental Movement?" The piece concludes, "The science of carbon emissions is there. The politics is not."

Next year, *Times* columnist Nicholas Kristof awarded the greenhouse his trophy for Most Neglected Topic. "Here's a scary fact," he wrote, acting as a sort of editorial-page Stephen King. "We're much more likely to believe that there are signs that aliens have visited Earth (77 percent) than that humans are causing climate change (44 percent)."

2010—the last attempt at a climate law—was the warmest year on record. Then 2013 clocked in as the second-warmest. (With a kind of demoralized wonder, climate scientist Michael Mann would testify that specialized instruments were becoming unnecessary for professionals in his field. "We can see climate change, the impacts of climate change, now," Dr. Mann said. "Playing out in real time, on our TV screens, in the 24-hour news cycle.") Then three consecutive years, 2014, 2015, and 2016, each broke the record for hotness.

The *Times* ran their story on page one: "It is the first time in the modern era of global warming data that temperatures have blown past the previous record three years in a row.... Of the seventeen hottest years on record, sixteen have now occurred since 2000." (As of 2021, that record has improved—or worsened, depending on the ideological outlook—to nineteen out of twenty.)

NASA calculated that global temperatures had spiked by over half a degree, "a huge change for the surface of an entire planet to undergo in just three years." To express all that force required a nightmare math book, equations borrowed from the Manhattan Project. "The heat accumulating throughout the Earth because of human emissions," the *Times* explained, "is roughly equal to the energy that would be released by 400,000 Hiroshima atomic bombs exploding across the planet every day."

And here we were, firmly on the other side of Too Late. "The rate over time," the paper reported, "has been reasonably close to predictions that scientists first offered decades ago."

This was the outcome of what Roger Revelle once described as an unintentional, species-wide science project. "Human beings are now carrying out a large scale geophysical experiment," Revelle had written in 1957, "of a kind that could not have happened in the past nor be reproduced in the future." We'd spent thirty years putting together an unimaginably thorough answer to the discouraged question once posed by Dr. Sherwood Rowland. "What's the use of having developed a science well enough to make predictions," he'd asked, "if in the end all we're willing to do is stand around and wait for them to come true?"

Jim Hansen unintentionally gave the reply in an interview—not knowing this was a response to Dr. Rowland. "It's as if we didn't know," he said in 2014. "We might as well not know. Our fossil fuel use wouldn't be much different."

Hansen was filmed for a video history compiled by Elizabeth Kolbert.

Foresight and retirement had turned Dr. Hansen darkly glamorous—a celebrity of disappointment—and experience of Rahm Emanuel's Washington had taught him to curse. When Kolbert asked if he had a message for today's young people, a gallows humor shadow crossed the scientist's face. "Well," Dr. Hansen said, "the simple thing is, I'm sorry we're leaving such a fucking mess."

NOW IT WAS all before and after. What they told us could happen; what did. Back in 1956, the *Times* had envisioned one carbon-warmed future: the "polar regions" grown to jungles, with "tigers roaming about and gaudy parrots squawking in the trees."

On March 24, 2015, the highest-ever temperature for Antarctica was recorded: a springtime 63.5 degrees. Ideal tropical bird range is 65 to 80 degrees. That is, very nearly parrot weather—and about as meaningful as the igloo the Inhofe family once built on a Washington afternoon. Just another day's news, the fish in the pail, feeding time for the international argument.

In 2015, 195 nations negotiated the Paris Agreement; the long-awaited sequel to Kyoto. President Obama signed America in the following autumn. Three decades of the United Nations climate process, and we'd at last joined an international emissions treaty.

That November—in the midst of the record-breaking years—Donald J. Trump was elected president.

THE NEW PRESIDENT did not believe in global warming. His rawest dataset was preserved on Twitter.

A mix of American can-do, paranoia, misinformation, and decades of secondhand Fred Singer and Frank Luntz. He'd tweeted, "Is our country still spending money on the GLOBAL WARMING HOAX?" He'd tweeted, "The concept of global warming was created by and for the Chinese in order to make U.S. manufacturing non-competitive." He'd tweeted, "Give me clean, beautiful and healthy air—not the same old climate change (global warming) bullshit!"

During non-keypad hours, he had voyaged through the screen, explaining to Fox News' Bill O'Reilly, "It's a big scam for a lot of people to make a lot of money." (That *voice*. You can really hear those hollowly stressed salesman *a lot*s.) He revealed on Fox & Friends that climate change was really a form of psychological trade warfare. "This is done for the benefit of China, because China does not do anything to help climate change. They burn everything you could burn," he said. "In the meantime, they can undercut us on price. So it's very hard on our business." Climate change might or might not have benefited China. It just wasn't the reason. It was like claiming tennis existed only so Serena Williams could excel at it—the whole net-and-racket affair an elaborate conspiracy rigged up by Nike, Serena, the fans.

In June 2017—his fifth month in office—President Trump announced America's withdrawal from the Paris Agreement. And every seedling of progress was mowed under. It might as well have been 2001 again, when George W. Bush pulled us from the Kyoto treaty. Or the 1950s again, when the facts about global warming were fresh and mysterious and awaiting proof.

Syria, which had troubles of its own, signed the agreement four months later. Leaving, as the *Times* reported, "the United States as the only country" not party to the treaty on Earth.

2017 was the second-warmest year in the instrumental record. NASA's Gavin Schmidt told a reporter, "This is the new normal." Schmidt had assumed Jim Hansen's old post, steering the Goddard Institute in its offices above the *Seinfeld* coffee shop.

That fall three hurricanes smashed into America. In Miami, Irma dropped boats on roadsides. In Houston, we were treated to the sight of a highway overpass underwater: a strange, idiot image, matter out of place, like a cow in a museum gallery or a tree buckled into an airplane seat. Hurricane Maria swept the island of Puerto Rico back to the world before Thomas Edison. It took nearly a year for power to be fully restored. The *Times* reported one family's experience—and it was the reverse of the thank-you a Kansas housewife had once written Edison, about how his electricity was supporting every moment of her life. "Jazmín Méndez has lived much of the last year in the dark. No light to read by. No food cooled in the fridge. No television for her three children."

The National Oceanic and Atmospheric Agency noted that the three storms were among the five costliest in American history; all five had hit since the turn of the century. "Humanity's response must start with heading off as much unnecessary warming as possible," advised the *Washington Post* editorial board. And, a few months later, "We need to combat climate change—now. . . . This would be difficult but not impossible, if we tried."

But that same editorial board had once made a solid prediction. "Mankind's ability to protect itself," the editors had written, in 1965, about climate and the future, "is less likely to be limited by inadequate knowledge, than by its notorious inability to turn knowledge into effective public action."

The summer of 2018 was history's fourth warmest for the continental United States. A NASA spokesperson explained warming had become a daily fact: "It's now absolutely happening to millions of people around the world."

That fall, the IPCC released an interim report. We had until only about 2040 to effect "rapid and far-reaching" change. And this really was the far side of Too Late: The outcome of having stayed so long in the same water, which not even a frog will do. And of all the ingenious and high-spirited people who had found such creative and rewarding methods to help us delay.

So here is the last of the old predictions. The first we encountered. Warming has been a strange problem. Accumulating slow like a snowfall; difficult to assign blame to any individual flake. All of us suspended, turning, descending and landing together. But then, in the morning, you open the door to a world that's changed. It's that snowfall over generations.

In May 1956, the greenhouse becoming American news, *Time* magazine had interviewed Roger Revelle. The scientist explained how carbon—cars and factories—got released on the Earth's surface. How it commuted to the air, where it went about its long-range business of trapping and increasing heat. "In 50 years or so this process, says Director Roger Revelle of the Scripps Institution of Oceanography, may have a violent effect on the earth's climate." So teams of scientists would be measuring and comparing, taking a magnifying glass to the charts and figures. "When all their data have been studied, they may be able to predict whether man's factory chimneys and auto exhausts will eventually cause salt water to flow in the streets of New York and London." Half a century, from 1956. Now add another six years.

2012 was Britain's second-wettest year on record. London was spared; the city had erected a large sea gate, the Thames Barrier, for protection against storm surges. But the nation flooded.

That fall, Hurricane Sandy slammed into New York. Ocean temperatures were nine degrees above normal; warm water is hurricane food. Wind speed reached ninety miles per hour. "There were waves on Wall Street," one engineer reported. Seawater by the millions of gallons flooded the subways. (The video has a strange, unnerving look of anarchy—water pounding down staircases, jumping turnstiles.) Swells in Queens rose chest high. "Salt water," said the head of one transit line, "is a carcinogen. It's like cancer to a rail system." And eight blocks from the former headquarters of Edison Electric Light, seawater poured into a Con Edison power plant, exploded a transformer, and spilled lower Manhattan into darkness.

Because all things connect, the For Sale sign of one realtor, plucked by the storm, washed ashore half a decade later on a French beach. 1956 plus fifty. That magazine had been six years off. Because the climate doesn't care about politics, or experts, or warnings, and isn't even aware there are people. We have our days and lists and hours, our schedules and emergencies; but the climate keeps its own time.

Acknowledgments

THIS IS GOING TO BE LONG BECAUSE THE BOOK IS: GRAT-
itude toward friends and page count turn out to operate in a pretty
direct ratio.

The first person to thank—for her tremendous intelligence and equal-
size dedication to keeping this story as engaging and speedy as possible with
me at the keyboard—is my New York University colleague Lisa Gerard.
Lisa cast her wonderful editorial eye across every draft of this book; each
chapter is tighter for her taste, matchless energy, and wit.

Bill Clegg is the second. The sort of friend and agent every writer
needs: the person who makes the occasional skirmish seem not just winnable but fun, a lark not to be missed. In his company, the distressing
feels *manageable*. (A great writer and reader, Bill startled me once, hiking
through the woods, by reciting word-for-word the last paragraph of Ian
Frazier's memoir *Family*.) Bill was an occasion to rise to. It is with ample
motivation—one of those very-least-you-could-do deals—that these two
share this book's dedication.

There's the saying that a book must find friends before it can even think
about readers. This book was fed, housed, spruced up, escorted, and sometimes plunked down on the elliptical machine by many, many friends. Elizabeth Wurtzel, Sean Woods, Evelyn Shapiro, James Atlas, and Paula Berry
read early drafts, gently aiding the material in its passage from sketchy and
unpromising to medium-presentable. Leigh Feldman was a great early

reader and supporter, as was Dan Conaway. Justin Leites offered charming
been-there dissections of Gore-Clinton DC; there were also the unclouded
reading eyes of Jonathan Lipsky, Eric Pascarelli, and Dr. Jonathan Bradlow.
Emily Graff was especially sharp on the early history of the deniers, that
troubled colorful dangerous crew. And Jonathan Karp offered his structur-
ing genius—this book's final shape owes a great deal to his particular gift
for rounding out the telling of a story.

Julie Just was a wonderful reader—and a great talker *about* the material, over ashtrays, over coffee, and, after we wised up, waving away the
smoke from other people's ashtrays. So was the excellent and steadfast
Susan Golumb, also the brilliant Lili Anolik. (Being Lili, she liked the
deniers best.) Simon Toop, Cliff Benston, and the keen Ed Winstead were
excellent beta testers. Jeff Giles offered his West Coast nudges: Valerie
Van Galder was wonderful about creating just the spaces—zip code and
otherwise—that are a breeze to write in. Somebody once described the
Fitzgeralds (Zelda and Scott) as natural guests; Valerie is a natural host.
If you enjoyed this book, you have profited from the kindness and intelligence of the people it is this author's very good fortune to know.

At New York University's Creative Writing Program, Deborah
Landau has been an unflagging friend to the project: Deborah is a natural encourager and protector of all the writers—student and faculty—
entrusted to her departmental care. I'd like to also express appreciation
to the trouble-shooting Zachary Sussman and the razor-sharp Jerome
Murphy (Jerome is the ideal person to sit down and just *listen* to, at those
drizzly moments when one is apt to confuse draft exhaustion with impatience about books). Thanks also to the students who shared their semi-
nar enthusiasms and Amazon discoveries. This book could not have been
completed without the warmth, intelligence, and support of the New
York University community.

At W. W. Norton, Matt Weiland has been an overwhelming ally—
what the people in uniform call a force multiplier. Matt promised we'd
be in cahoots, and has been as good as his comradely word. (I told Julie
Just the name of my editor, and she took a breath: "Really? You're lucky—

he's been killing it.") Huneeya Siddiqui is a wonderful presence at the other end of the voice call or email: providing excellent advice, solving the problem, making clean, fun work out of the necessary. There's not much more you can ask from a colleague than that. Charlotte Kelchner proofread these pages with a magnifying glass and tweezers; and Beth Steidle provided the finished product's lovely design. At W. W. Norton marketing and publicity, Steve Colca, Kyle Radler, and Nicole Dewey have done their damnedest as sign-spinners and author-bucker-uppers.

On to the final personal round. Amy Wallace-Havens offered her tremendous gifts of intelligence and encouragement; anybody who knows her knows the size of those presents. And Sally Wallace is another of the people without whom the book simply could not exist: Sally was its first reader, and the writer came to know her edits well, so well he can recite the comments, error catches, appreciations, and advisories verbatim. The way you can with a pop song: a few seconds before I arrive at a moment she especially liked or repaired. Starting any project is like volunteering to maroon yourself on a desert island: Sally was there, during one whole season, offering her craftswoman tips on how to construct my raft and sail back home. Darin Strauss read and sharpened many drafts of this book: his suggestions are in its shoes, hairstyle, the cut of its suit. Rich Cohen offered warm reads and cool-eyed counsel. These are the friends one hopes for: people you are not afraid to be your clumsy, first-draft self with. Pat Lipsky, an artist in whatever she does, read the manuscript many times, thumbing out strong advice on her iPhone, in concession to my superstitions; paper is too easy to lose.

And Victoria Brown had to share her life for years not just with me but with piles of books and the heroes and villains of this story. (Whatever shades and traces of S. Fred Singer or Christopher Monckton still lingered over my head as I slept the computer and pinched off the desk lamp.) Her input was wonderful, and there was also just the perseverance. Of seeing someone buffeted by the internal silent weather of pushing their way through a project. This story would have been radically impoverished— weaker, in crucial ways undefended—if any the people above had wandered away from their posts. And I now understand that friends advice from the beginning. The technology lets us be more and more easily connected. So perhaps now more than ever, any book you read, the name on the jacket is really the just an approximation, a leading edge. If it's any good at all, there's an expanding fan of people standing just out of view on both sides.

A LAST SET of acknowledgments is about the material itself. To the scientists whose research this book narrates—it's been an often infuriating honor to tell these parts of their stories. I'd like to express my gratitude also to the Sterling Memorial Library at Yale and the Bobst Library at New York University. Also to the Internet Archive and its Open Library, for their ready-when-you-are holdings of difficult-to-find texts. And the Legacy Tobacco Documents Library at UC San Francisco is a great archival achievement. Also a nearly unimaginable one: like being MC of the ocean.

And finally, I would like to express my gratitude to the many writers and editors who together tell this story. This book is in a way an audience story, a genre I imagine we'll see more of, made possible by our sudden abundance of media data: how the long-term stories took shape, when they drifted across the landscape, what happened and how it felt as they reached population centers. The thing that motivated me to begin this story was the excellence of the early reporting. Then there were authors who served as models, whose names you have found throughout this book. Elizabeth Kolbert; Stuart Weart; Erik Conway and Naomi Oreskes; George Monbiot—I owe them great debts. As well as Bill McKibben, Chris Mooney, Jonathan Weiner, Jill Jonnes, James Fleming, Theda Skocpol, Allan Brandt, Siddhartha Mukhurjee, Robert Proctor, David Michaels, Eric Pooley, Fred Pearce, Jeff Goodell, Devra Davis. All wrote books to which I returned again and again; sometimes to check a fact, often for their reassuring demonstrations of excellence.

And then the hundreds of journalists and editors and publications behind the sentence-by-sentence material of this book: people who went out and interviewed, and edited, and printed, and told this story day by day, year by year, decade by decade. Working on this material was, for a journalist, reassuring. We got the story mostly right, mostly right from the start. It was a great thing to see how early these writers and publications apprehended the material, just how early they were able to pass it along. What to do with the story, of course, was always up to us.

A Note on the Sources

THIS IS A BOOK THAT TELLS LOTS OF STORIES—DRAWN from many books, newspaper and magazine articles, and broadcasts on radio and TV. As elsewhere, I would be remiss to leave unmentioned the particular debts (of inspiration and data) owed by these pages to three works in particular. Spencer Weart's wonderfully rich and concise *The Discovery of Global Warming*; Elizabeth Kolbert's calmly devastating (an effect reproduced by her title) *Field Notes from a Catastrophe*; Naomi Oreskes and Erik Conway's chiller *Merchants of Doubt*. Together these books make a kind of trilogy: the scientific discovery of climate change; the discovery of its social consequences; and the efforts by a small determined cadre to muddy up the discoveries. I encourage any reader who has enjoyed this book to consult those others: what you like here you will find there. The story they tell is discouraging. But the ability to tell that story, with such intelligence, poise, and wit, has an effect that is quite the opposite.

Each chapter of this book tends to be dominated by a few key sources, with other voices and accounts rounding out the picture. (London's smog week, S. Fred Singer's old office host the Unification Church and the tale of Mark Mills I remember having the fattest files; I was also surprised by how many speakers there were for even a short chapter like "The Frogs"—though that one is *about* sources.) In the interests of a book that would fit comfortably in the reader's hands—also a shoulder bag, finally a shelf—I have posted the endnotes online. They are to be found at www.theparrotandtheigloo.com. Often, when a moment, quote, or episode seemed too good to omit, but threatened to detour a chapter that already had much other ground to cover, I would place the side hike in the notes.

The challenge of telling the climate narrative succinctly has always benefited those with a political and commercial interest in postponing action on climate. (Which will come, as surely as ozone action did, with or without a spectacular Christopher Nolan visual effect like the ozone hole; at a certain point, the summer locker-room feel of more and more afternoons will *be* that special effect.) I used my own reading likes and dislikes as guide—the many notes and sources would have made this book seem an assignment, rather than a human story co-starring really everyone, and happening at every address at the same time. To keep things book-length, this seemed both the manageable approach and the wisest course.

For the Tesla and Edison portions, I've listed the sources relied upon by chapter. For the sections that follow—about scientists and deniers—you will find the notes organized by chapter, page, and keyword.

As above, these notes are where you'll find material that could not be comfortably squeezed into a book with even as many scenes, locations, and people as this particular climate *Decameron*. It's the home of details like the following—one of this reader's favorite observations, by the climate scientist James Hansen. A recommendation of the power of a sunny outlook. From a scientist whose subject is dark, and whose political experience has given every additional reason for grayness.

Dr. Hansen had been invited to address a White House panel that seemed bent on ignoring any recommendations. He writes:

> Being an eternal optimist (what else can be effective?) I welcomed the chance.

Optimism as pragmatism. As efficiency. And a treat to have put it in the hands-on book after all.